Active Volcanoes of the World

Series Editors

About 500 active volcanoes presently exist on the Earth's surface, of which around 50 erupt each year. Volcanoes played a crucial role in the evolution of the planet and early life, and are constantly reshaping the morphology of Planet Earth. Many active volcanoes are located in dense settlement areas, with over 500 million people living in close proximity of still active or dormant volcanoes.

On one side, volcanoes provide valuable soil and rock basis for agriculture, but often the "mountains of fire" cause disastrous societal and economical disasters caused by ash clouds, lahars, lava flows, and pyroclastic flows. Eruptions are still difficult to predict, although volcanologists around the world are constantly working on new ways to understand the character and behavior of volcanoes.

Active Volcanoes of the World is an official book series of the International Association of Volcanology and Chemistry of the Earth's Interior (IAVCEI). The series aims to be a scientific library of monographs that provide authoritative and detailed reviews of state-of-the art research on individual volcanoes or a volcanic area that has been active in the last 10.000 years, e.g. the Teide Volcano or the Chiapas Region. The books in the series cover the geology, eruptive history, petrology and geochemistry, volcano monitoring, risk assessment and mitigation, volcano and society, and specific aspects related to the nature of each described volcano.

The Active Volcanoes of the World series contains single and multi-authored books as well as edited volumes. The Series Editors, Dr. Corrado Cimarelli and Dr. Sebastian Müller are currently accepting proposals and a proposal document can be obtained from the Publisher, Dr. Annett Buettner (annett.buettner@springer.com).

Pablo J. González
Editor

El Hierro Island

Springer

Editor
Pablo J. González
Volcanology Research Group
Department of Life and Earth Sciences
Instituto de Productos Naturales y
Agrobiologia, Consejo Superior de
Investigaciones Cientificas (CSIC)
San Cristóbal de La Laguna, Spain

ISSN 2195-3589 ISSN 2195-7029 (electronic)
Active Volcanoes of the World
ISBN 978-3-031-35134-1 ISBN 978-3-031-35135-8 (eBook)
https://doi.org/10.1007/978-3-031-35135-8

This Springer imprint is published by the registered company Springer Nature Switzerland AG
The registered company address is: Gewerbestrasse 11, 6330 Cham, Switzerland

Preface

More than ten years have passed since the landmark 2011–2012 submarine eruption in El Hierro Island. This eruption represented a revolution for the Spanish volcanology. It provided with a real, sometimes unpleasant, testbed of the strengths and weaknesses in the regional volcanic eruption management systems. Almost exactly a decade later, this professional and scientific experience demonstrated to be invaluable, when another eruption, in this case on land, struck the Canary Islands. The 19th of September–13th of December 2021 Tajogaite volcano eruption in La Palma was confronted with a much more organized scientific, emergency and decision-making strategies (e.g., the PEVOLCA legal framework). Nevertheless, the La Palma volcano crisis, including pre-eruptive, eruptive and post-eruption phases, showed also new challenging aspects from which we will learn new lessons for future volcano crises. The successful response in La Palma would have not been possible without the in-depth study of the 2011–2012 El Hierro submarine eruption. The wealth of data collected and analyzed on the wake of the submarine eruption sprung into new knowledge about the behavior of the volcanic system, including the eruption impact in the environment and society. In the summer of 2022, for example, a total of 1930 results can be retrieved when searching for "El Hierro" "eruption" in Google Scholar.

The current book is a necessarily partial and biased tour of the wide knowledge of the eruption and El Hierro as volcanic system. Such information has the benefit of the hindsight and reflection over a meaningful period of time, around a decade of intense research. The scope of the current book is explained by the challenging circumstances of working during a period marked by the COVID-19 global pandemic and the, at the time, ongoing 2021 Tajogaite volcano eruption in La Palma. In fact, in October 2021, a dedicated workshop where discussions among potential book contributors had to be cancelled was planned on the same week that marked the 10th anniversary of the 2011–2012 submarine eruption in El Hierro. Despite these difficulties, the authors submitted contributing chapters, and we could carry out a rigorous peer-reviewed process resulting in the current volume—*El Hierro Island*. The topics in the book are a representative sample of what the volcano science is, knowledge always in progress. I hope that the readership should consider the covered topics as a useful starting point of discussion and opportunities of their own future research.

Last but not least, something I am very proud of editor was promoting gender equality. In fact, a glad surprise was to achieve it on our first round of list of corresponding and leading authors of the book. Such balance between male and female contributors to this book should be interpreted that the Spanish volcano science community is making positive strides. However, we should not be complacent. This is not the end of the road for a fair research environment. Less visible barriers, related with social and cultural diversity, should be considered with the aim to achieve a healthy respectful, diverse, equitable and justice community.

San Cristóbal de La Laguna, Spain Pablo J. González

Acknowledgements

Prime thanks are due to the Springer staff and the scientific book series editors, which did a superb job ensuring that every phase of the book preparation, from the months of planning to the logistical support during the editing phase, ran smoothly. Their efforts made possible a productive environment with ample time to discuss, review and re-edit the different chapters. Also, thanks to the PTI Volcanology members that embraced the project as theirs, as a community effort. We wish to thank especially the core members of the group, at PTIVolcan CSIC scientific initiative that coordinated the different teams working in each chapter, Dr. Joan Martí (CSIC-Geosciences Barcelona), Dr. Carmen López and Dr. María José Blanco (IGN), Dr. Eugenio Fraile-Nuez (IEO-CSIC) and Inés Galindo (IGME-CSIC). Without the full support of this core team, this book would have not finished. In addition, to them, I would like to thank the enthusiastic engagement of each leading and corresponding author that lifted up the large team of collaborators contributing to each chapter. The book will certainly be a collective achievement to be celebrated. Thanks to Dr. Stravros Meletlidis (IGN), Dr. Cayetano Guillén (EUTUR), Dr. Antonio M. Álvarez-Valero (University of Salamanca), Dr. Itahiza Domínguez-Cerdeña (IGN), Dr. Carmen López (IGN), Dr. Juan Tomás Vázquez (IEO-CSIC), Dr. Eugenio Fraile-Nuez (IEO-CSIC), Ms. Alba González-Vega (University of Las Palmas de Gran Canaria and IEO-CSIC), Dr. Isabel Ferrera (IEO-CSIC), Dr. Ana Sotomayor-García (ICM-CSIC), Dr. Inés Galindo (IGME-CSIC) and Dr. Juana Vegas (IGME-CSIC).

This book would have not been possible without the massive contribution of a number of reviewers. So, I would like to thank personally for their time and effort. Your comments were always positive and constructive. Your work and critical reading improved this book substantially. I am glad that all corresponding authors acknowledged your contributions and work to address and accommodate those comments. In alphabetical order Flavio Cannavo, Patricia Erfurt, Bruno Faria, Diego González, Margaret Hartley, Ulrich Keuppers, David Kirchman, Nadine Lebris, Karoly Nemeth, Adriano Pimentel, David Pyle, Rui Quartau, Ricardo Ramalho, Miriam Roemer, Anna Louise Reysenbach, Mark Thomas and Tom Winder. Thanks to all of you.

Contents

Part I
Geodynamics and Volcanology

El Hierro Island Volcanological Science: An Overview

1

Pablo J. González

Abstract

El Hierro Island, the youngest of the Canary Islands, and its latest eruption in 2011–2012 have been a playground of fruitful decade-long studies. In this book, we summarize and provide future venues of action to solve outstanding questions. The topics cover geological studies of Holocene volcanism so it informs past, present and future activity. Its active magmatic system from a petrological and geophysical lens. How monitoring of volcanic activity can be optimized and how to read the data streams in a meaningful way. The marine environmental effects of a submarine eruption are covered in detail, as well as how the society could be properly engage to reduce the risks associated to it, and appreciate and benefit from it. So, in each chapter the reader should find inspiration and future challenges waiting to be solved. Remaining puzzles pieces about how volcanism works and how it affects its environment. An effort to provide food for thoughts of future Canary Islands volcanological research, and in particular El Hierro.

Keywords

Ocean island volcano · Intraplate volcanism · Volcanology and hazards · Environmental and social effects of volcanism · Volcanic risks · Canary Islands

1.1 Introduction

El Hierro represents the emergent part of a volcanic edifice which rises ~ 5500 m from the sea bottom. Its subaerial part emerges 1501 m.a.s.l. and it has a surface area of about 269 km^2. El Hierro Island, with a surface area of approx. 270 km^2, is the smallest of the major islands of the Canary Islands archipelago. It is located around latitude 27.7 north and 18° west within the Nubian plate away from plate boundaries. El Hierro Island is the youngest of the Canary Islands, with its emerged part rising above sea level 1.2 Ma ago. Therefore, El Hierro is at an early growth stage with signs of very active eruptive activity, e.g., the largest density of preserved strombolian cones in Canary Islands, although, count-intuitively there were no credible reports of historical eruptions until 2011. The islands morphology is dominated by three rift zones, which separate massive gravitational collapses (Carracedo et al. 2001). Particularly interesting is the north collapse, El Golfo, a

P. J. González (✉)
Volcanology Research Group, Department of Life and Earth Sciences, Instituto de Productos Naturales y Agrobiología, Spanish National Research Council—Consejo Superior de Investigaciones Científicas, La Laguna, Santa Cruz de Tenerife, Spain
e-mail: pabloj.gonzalez@csic.es

P. J. González (ed.), *El Hierro Island*, Active Volcanoes of the World,
https://doi.org/10.1007/978-3-031-35135-8_1

1400 m-high scarp, thought to be the youngest massive gravitational collapse in the Canary Islands (Longpré et al. 2011).

The El Hierro Island became a scientific relevant spot for volcanology due to its recent eruption in 2011–2012. After several months of seismic unrest, a submarine eruption that lasted 5 months occurred off the southern tip of the island, near the town of La Restinga. Occurring in the 2010s and in a developed country, Spain, meant that a rapid response and monitoring effort provided with an unprecedented level of scientific data and sample collection. The datasets and samples were analyzed by local and international researchers over the next decade. Much of what we know today and eventually facilitated a successful response to the 2021 La Palma eruption (10 years later) started in 2011 and the following decade of volcanological progress.

Here, I aim to review the geodynamic context on which the following chapters should be framed, as well as an overview of covered topic, and the main recommendations for future works.

1.1.1 El Hierro: An Overview of Its Geodynamic Framework

1.1.1.1 Geodynamic Background

El Hierro Island is located in the Canary Islands. An archipelago made up of seven large islands, situated between 27°38′ and 29°25′ of north latitude and 13°20′ and 18°9′ of west longitude. The Canary archipelago forms an alignment of volcanic islands of about 500 km in length between its ends, separated by only 100 km from the African coast. Together with the archipelagos of the Azores, Cape Verde and Madeira, it forms the Macaronesia region. Except for the Azores, which are more related to the Mid-Atlantic Ridge, the rest of the archipelagos form a very active volcanic province in the last 20 Ma.

The Canary Islands are located within the African (Nubia) plate in an intraplate setting. In particular, the Canary Islands were formed over Jurassic-Cretaceous oceanic crust near the edge of Africa's continental lithosphere (Fig. 1.1). Although, the northwestern African continent margin is passive, the intense tectonic and magmatic activity of the nearby Atlas suggests an atypical origin of those volcanic archipelagos. Positioned in a transitional lithosphere, with the westernmost islands sitting on top of pure oceanic lithosphere and the eastern (Canary Ridge) islands on a transitional (thinned) continental lithosphere.

All Canary Islands are independent volcanic systems rising from 3 to 4 km depth ocean floors, except for the Lanzarote and Fuerteventura which are resting on shallower ocean floor and can be consider geologically a single "Island". It shares a common and continuous coastal abrasion platform. Considering the emerged and submerged volumes of each volcanic system, one can observe that there is an East-West trend, from larger volumes on the East towards less voluminous islands in the West. Moreover, the volume could be only an indirect indicator of the total volcanic material emitted and because the islands seem to have experience important vertical movements. In the case of Fuerteventura (in the East), uplift has resulted on outcrops of cretaceous ocean sediments ($\sim$100 Ma) now at the surface: Those rocks being the oldest outcropping rocks in the archipelago. Conversely, similar (although younger) rocks has been found at the bottom, at 2700 m depth, of a deep borehole in nearby Lanzarote. An uplift processes have also observed in some of the Western islands (La Palma). The vertical patterns are complex with uplift, and perhaps subsidence, processes reaching substantial values in spatially localized areas.

The Canary Islands magmatism is dominated by basaltic lavas, but felsic magmas are also present (trachytic to phonolitic) (Fig. 1.2). TiO_2-rich alkaline basalts predominate in mafic magmatism, which also includes hypersthene-normative tholeiitic basalts to strongly silica-undersaturated nephelinites (e.g., Schmincke 1982). On Fuerteventura, a variety of rock types, including carbonatites and ijolites, can be found in the earlier volcanic complex (e.g., Le Bas et al. 1986; Balogh et al. 1999). Lanzarote and El Hierro show the least abundance of high-silica content rocks (< 1%), followed with intermediate

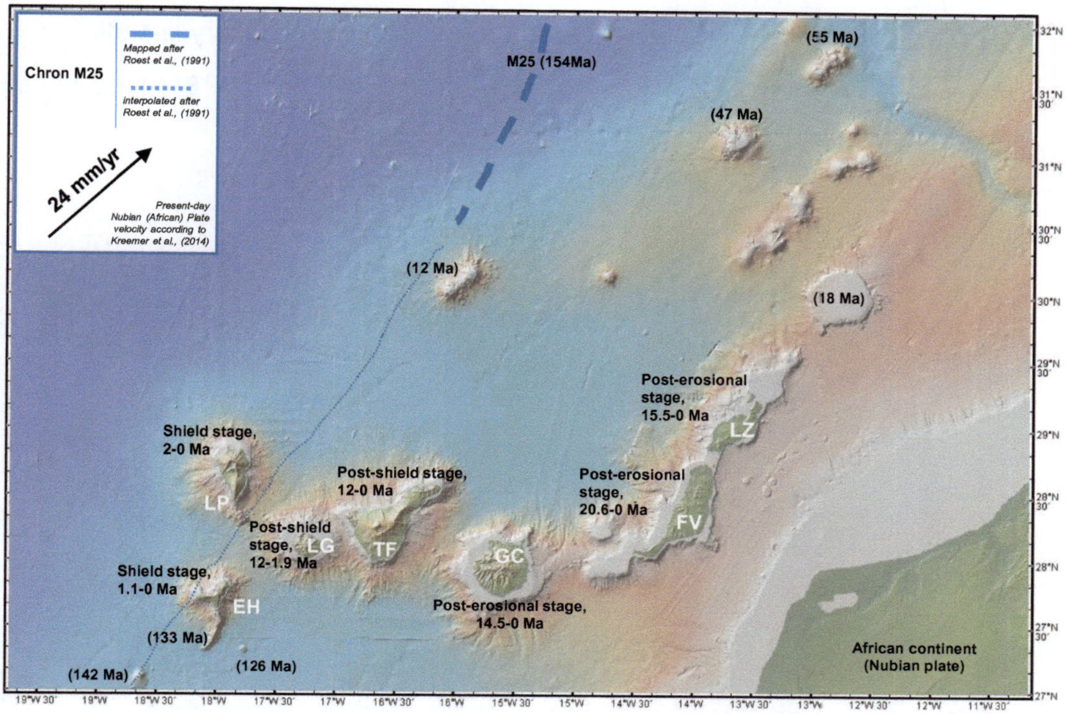

Fig. 1.1 Canary Islands archipelago geodynamic context off the passive northwestern margin of the African continent. Seamount ages as within parenthesis. Measured and interpolated segments of the last well-mapped and identified Cretaceous age geomagnetic chron of oceanic lithosphere, M25 of age 154 Ma (after Roest et al. 1992). Current present-day African plate motion towards the Northeast (Kreemer et al. 2014). LZ: Lanzarote and La Graciosa Islands; FV: Fuerteventura; GC: Gran Canaria; TF: Tenerife; LG: La Gomera; LP: La Palma; EH: El Hierro. Basemap using GeomapApp

abundance of ~ 3% in La Palma, La Gomera and Fuerteventura, and finally the most abundance in the central islands of Gran Canaria and Tenerife (>10%).

Magmatism in the Canary Islands, as other intraplate volcanic islands, can be roughly split into three main stages (Ramalho 2011): an older shield-building stage that was followed by a younger phase of activity that resulted in extensive volcanic sequences on some of the islands (such as Teide on Tenerife) as well as a large number of cinder cones (e.g., Schmincke 1982). Finally, after a pause in volcanic eruption, renewed eruption occur on the so-called post-erosional phase.

The Canary Islands volcanoes exhibits a broadly westward age-decreasing pattern. This pattern was broadly recognized since the 1970s, e.g. Abdel-Monem et al. (1971). The youngest subaerial volcanism found on the westernmost islands of La Palma and El Hierro, and the oldest in Fuerteventura, 23 Ma (Fig. 1.3). The northeastern seamounts—e.g., Dacia and Conception Bank—are generally older and show a general trend to age-decrease in the southwest (Geldmacher et al. 2005). However, there are seamounts on the southwest of the archipelago, representing a previous episode of magmatism in this region from 142 to 91 Ma and hence breaking the age progression (van den Bogaard 2013). With the western and eastern islands showing varying times of development, the Canary Islands form an E-W pattern. La Palma and El Hierro, the western islands, are currently in the juvenile shield stage (Carracedo et al. 2001). The eastern islands, Fuerteventura and Lanzarote, are undergoing a stage of erosion and are positioned parallel to the margin of NW

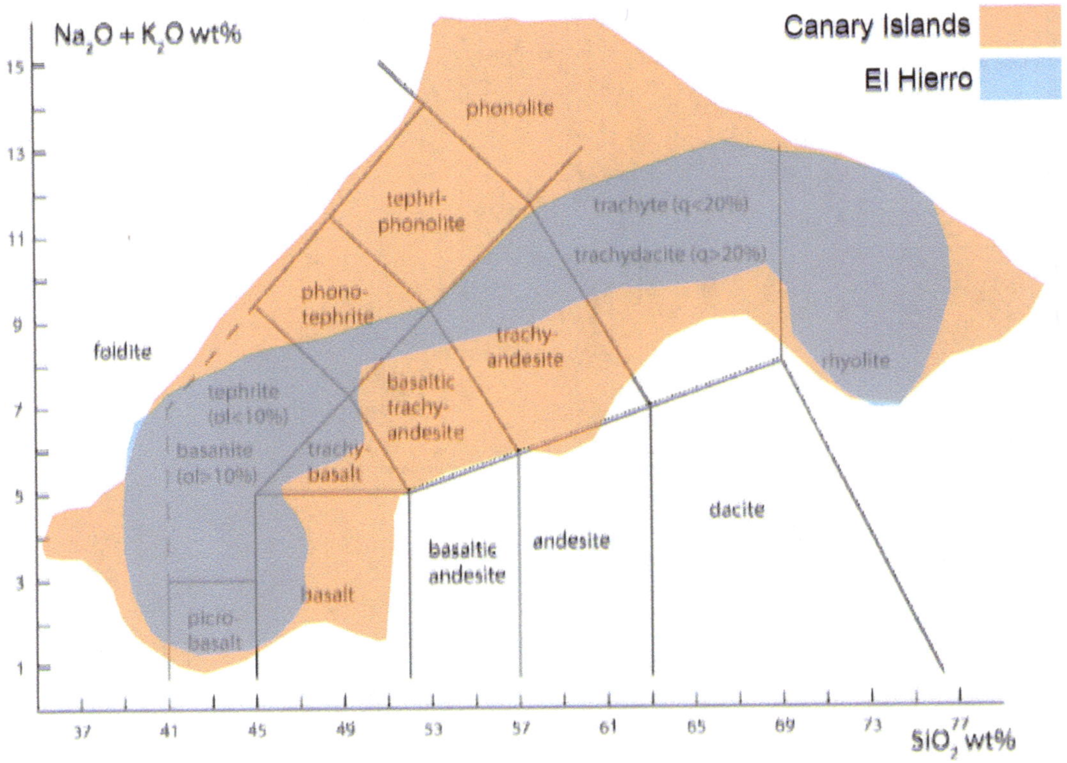

Fig. 1.2 Total alkali versus silica diagrams (TAS) with analyses of Canaries volcanic rocks, in orange colors, and on blue colors for El Hierro Island rocks (Carracedo et al. 2001)

Africa. Despite this, activity was present on all islands during the Holocene, with the exception of La Gomera (Paris et al. 2005). In particular, there has been historical eruption in Lanzarote, Tenerife, La Palma and El Hierro (Longpré and Felpeto 2021).

1.1.1.2 Lithosphere Structure

The Canary Islands are sitting over Jurassic-age oceanic lithosphere on the West, clearly evidenced by the M25 chron anomaly, 157 Ma. The central and eastern part of the archipelago lies on lithosphere without magnetic anomalies, corresponding to the Cretaceous magnetic calm. Hence, the age and nature of the lithosphere have been derived indirectly by seismic refraction lines, and it is considered that the central and, particularly the eastern islands (Canary Ridge) were built on top of thinned continental lithosphere or transitional lithosphere.

The Canary Islands' lithosphere beneath the archipelago is only relatively thinner than thicknesses expected for a normally cooling old oceanic lithosphere, 130 km (Fig. 1.4). The most recent estimates using seismic data, the surface elevation (isostasy), and the gravity field indicate a thickness of approximately 110 km (Fullea et al. 2015). The same authors also suggest a sublithospheric mantle abnormally hotter by roughly 100 K. These authors linked the widespread prevalence of melt beneath the islands to the low velocity layer. Hence, regional sublithospheric seismic models point towards a modest thermal anomaly beneath the Canary Islands, broadly agreeing with thermodynamic considerations, based on calculations of primary magma composition suggesting that Canary Islands show one of the lowest temperatures of all the ocean-island basalt lavas, ~ 1420–1480 °C (T excess: 70–130 °C, Herzberg and Asimow 2008).

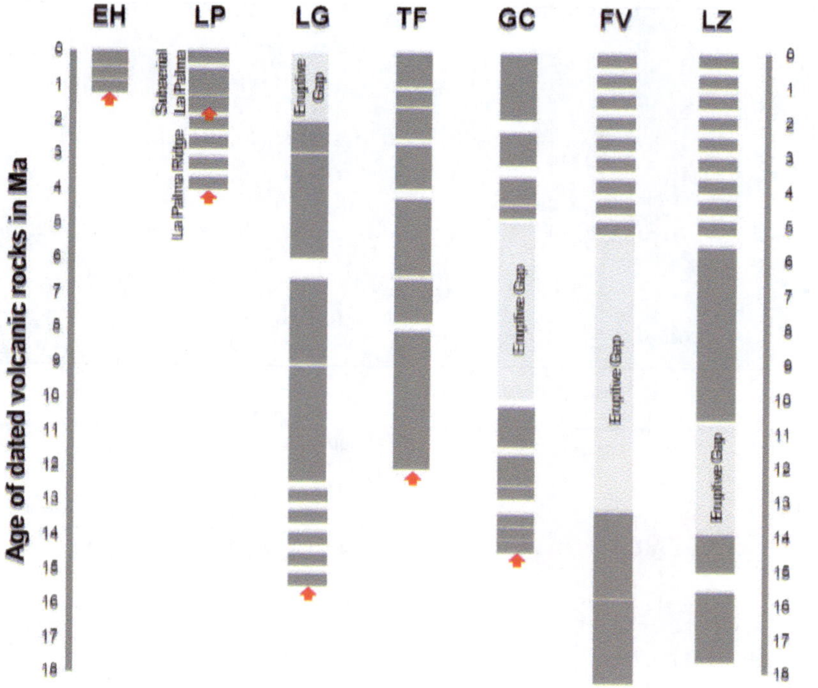

Fig. 1.3 Reconstruction of continuous and episodic volcanic eruptive phases for, mostly, subaerial volcanism in the Canary Islands. Vertical upside red arrow indicates the inception of volcanism (oldest dated rocks). In the case of La Palma ridge the dated rocks were dredged from seafloor. Figure was redraw based on results from Hunt and Jarvis (2017). LZ: Lanzarcte and La Graciosa Islands; FV: Fuerteventura; GC: Gran Canaria; TF: Tenerife; LG: La Gomera; LP: La Palma; EH: El Hierro

Seismic imaging certainly has consistently indicated that low-velocities are present beneath the Canary Islands, with very wide spread lateral extension from more localized imaging to regional extension reaching the Betics and Atlas Mountains (Duggen et al. 2009; Miller et al. 2015). Moreover, the interpretation of the seismic imaging studies discards the presence of large productivity of upper-mantle melts (Civiero et al. 2019). Tomographic models face the challenge of poor resolution due to lack of consistent local/regional seismicity and poor aperture of seismic stations. Therefore, the upper mantle structure above the 440 km discontinuity is not well mapped. Local and shallow high-rich melt may exist, and unresolved yet using seismic tomography methods. Indeed, studies of the Canary Islands, support the presence of a rich-melt zone at a depth range of 70–100 km using seismic receiver functions methods (Martinez-Arevalo et al. 2013; Lodge et al. 2012), which to a certain extent, supports geochemical models, at least in the Western Canary Islands (Day et al. 2010; Klügel et al. 2005; Neumann et al. 2002).

From a mechanical and seismo-stratigraphic point of view, the Canary Islands show a modest flexural response to the progressive volcanic loading (Collier and Watts 2001; Urgeles et al. 1998). The estimated sediment thicknesses are consistent with a roughly east-west inception of the islands, and an early barrier for continental sedimentation on the Canary Islands northern basin, that could go as far back as the Upper Cretaceous, although it is clearer from Neogene/pre-Miocene onwards (Collier and Watts 2001). The loading modelling in this study indicates an effective elastic thickness of around 35 km, in line with previous estimates using gravity and bathymetry admittance functions (Dañobeitia et al. 1994).

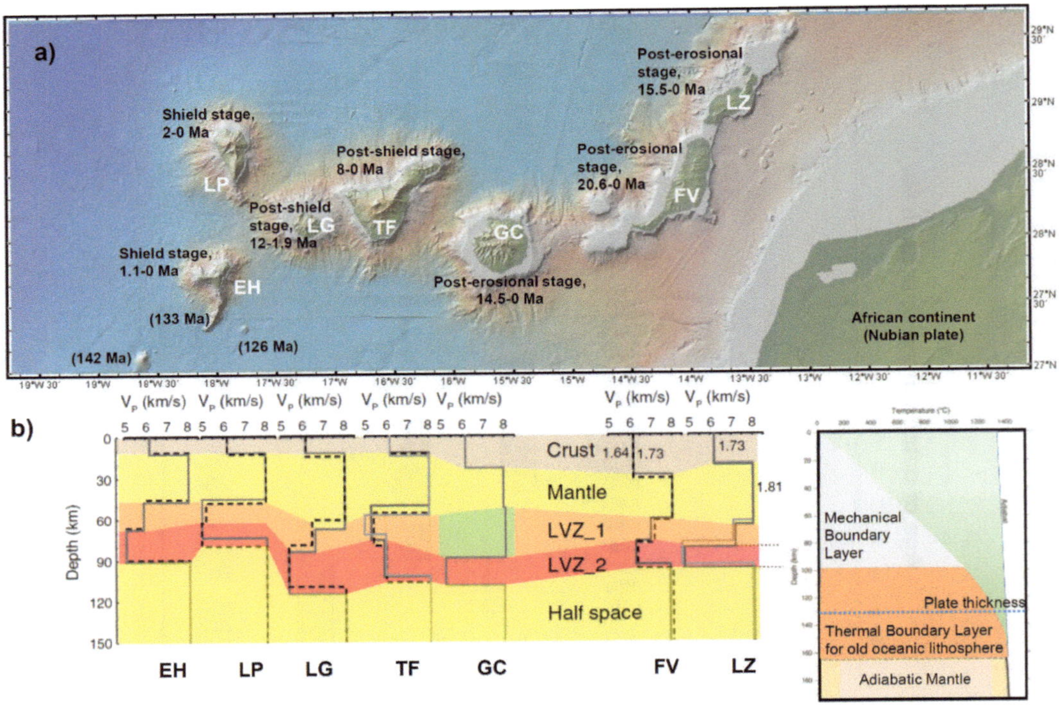

Fig. 1.4 **a** Canary Islands setting with age of oldest dated rocks. **b** (Left panel) Receiver-function seismic model of the P-wave velocities under the Canary Islands, a low velocity zone between 70 and 100 km depth is interpreted as a prevailing presence of melts (Martinez-Arevalo et al. 2013). (Right panel) Schematic of an oceanic plate model approximation to probable boundary layer structures at old ages, e.g., Jurassic age (Richards et al. 2020). LZ: Lanzarote and La Graciosa Islands; FV: Fuerteventura; GC: Gran Canaria; TF: Tenerife; LG: La Gomera; LP: La Palma; EH: El Hierro

At the crustal-scale, the structure seems to be fairly homogeneous with most of the islands showing a massive (high-density) central core, detected by gravimetric methods: Lanzarote (Camacho et al. 2001, 2018), Fuerteventura (Montesinos et al. 2005), Gran Canaria (Blanco-Montenegro et al. 2018), Tenerife (Ablay and Kearey 2000; Araña et al. 2000; Gottsmann et al. 2008; Sainz-Maza Aparicio et al. 2019), La Gomera (Montesinos et al. 2011), La Palma (Camacho et al. 2009), and El Hierro (Montesinos et al. 2006; Sainz-Maza et al. 2017). The islands' dense cores could be interpreted as network of mafic intrusions or dense cumulates. The dense cores seem to spatially disrupt the underlying structure with a flexed pre-island sediment layer. Moreover, the shallow-crustal dense cores could also possibly be correlated with evidence of underplating just beneath the crust, under Tenerife, Gran Canaria, Lanzarote and Fuerteventura using seismic methods (Dañobeitia and Canales 2000), in El Hierro using geodesy (González et al. 2013) and in El Hierro and La Palma using petrology (Klügel et al. 2005, 2015). On the other hand, most local seismic tomographic models show vertically extended bodies (Garcia-Yeguas et al. 2014; Martí et al. 2017).

1.1.1.3 Conceptual Models for the Canary Islands

The origins and genetic models of the Canary Islands remain heavily contested with multiple different variations with more or less effects of a sub-lithospheric mantle plume. However, the geological and geophysical evidence clearly violates a classical hot spot model based on Hawaii. The volcanism is long-lived (>20 Ma) for most of the islands, with a non-monotony age progression (Carracedo et al. 1998; van den Bogaard 2013), with only La Gomera island presently inactive (Paris et al. 2005). The amount of melt production is on the lower end of all ocean island volcanoes (Hoernle and Schmincke 1993; Herzberg and Asimow 2008) and there is a large temporal and spatial range of volcanic rocks derived from a wide range of magma compositions (Hoernle and Schmincke 1993). Finally, there is a lack of a broad topographic swell and of a positive gravity anomaly (Watts 1994; Collier and Watts 2001; Urgeles et al. 1998). Evidence for a low seismic velocity in the upper mantle supports sublithospheric "*thermal*" source(s); as well as recent discovery of young (Pliocene) sedimentary nanofossils beneath the western Canary Islands (Zaczek et al. 2015).

Canary Islands genetic models range from (a) those with strong to moderate influence of regional tectonics (Staudigel et al. 1986; Anguita and Hernán 1975; Araña and Ortiz 1991; Geyer et al. 2016), (b) classical mantle plume views with variations from a simple stationary source of magma (Carracedo et al. 1998) to a highly heterogeneous fusible mantle, aka the blob-model (Hoernle and Schmincke 1993) beneath a slow-moving African plate, or (c) a combination of both (Anguita and Hernán 2000; Blanco-Montenegro et al. 2018). Alternatively, and more recently, there has been renewed interest to investigate the effect of local mantle flow using two-dimensional numerical models to test edge-driven convection contributions in the context of the Canary Islands (Manjón-Cabeza Córdoba and Ballmer 2021; Negredo et al. 2022). The model proposes that upwelling can be generated due to interactions between mantle flow and lateral variations in the lithosphere. However, so far, the tests failed, in itself with only edge-driven convection component, to reproduce the location and melt production of a voluminous volcanic archipelago, such as the Canary Islands (Manjón-Cabeza Córdoba and Ballmer 2021).

So, since the inception modern geological ideas (global plate tectonics) in the 1960–1970s, a geodynamic model that fit all the peculiar aspects of the Canary Islands volcanoes remains a challenge to the community. Therefore, as Winston Churchill once said about the democracy, the *mantle plume model* is the worst geodynamic framework for the Canary Islands—except for all the others that have been tried. The formulation of an ultimate and satisfactory geodynamic model will still drive a new generation of volcanologists looking forward to solve one of the remaining puzzles of how Earth works beyond plate tectonics, continuing researching the Canary Island, and in particular El Hierro.

1.1.2 What Was Learned in the Last Decade?

This question is quite broad in its scope, and to answer it, it helps to organize such large body of knowledge. We focus on the study of its active volcanism and the impact of the volcanic activity on the ecological and social environment. Therefore, we provide the reader with a book which covers them in five broad sections. Each section covers different aspects of El Hierro Island volcanism. At the end of the book, a newcomer to the study of El Hierro Island should have a good understanding of what has been done until the end of the 2010s decade, gather the vision of the authors about what important challenges remain unsolved and those that should focus our attention in the future.

A first section of the book focuses on the past, present and future eruptive activity as evidence by land-based and near-shoreline geological studies (Chaps. 2 and 3). We continue reviewing our understanding of the magmatic architecture of the islands' volcanism and how it controls its volcanic activity (Chaps. 4 and 5); The landmark 2011–2012 El Hierro submarine eruption and its monitoring is inevitably mentioned across the book but it is the main focus in the next section (Chaps. 6, 7 and 8). The impact in the marine environment is a novel aspect that has never been studied in great detail, and this submarine eruption in El Hierro has allowed to be dissected (Chaps. 9, 10 and 11). Finally, the book ends on the effects on the society of volcanism, by describing secondary volcanic risks, such as landslides and how volcanism could be turned into a geotourism and geoheritage value (Chaps, 12 and 13).

El Hierro Island is, as presented in the previous section, the youngest island that emerged from the ocean in the Canary Islands. Hence, the study of its structure is important to reveal the architecture and magmatism, a knowledge useful to study older islands. To this end, the volcanic history during the Holocene and its volcanic structure has been revised by Meletlidis and coauthors (Chap. 2). The revision provides an overview of the volcanic activity of El Hierro Island, mainly based on the published literature, as well provide a most probable eruptive future scenario. By revising the 150 years of volcanological studies in El Hierro, they point out that the 2011–2012 eruption significance is truly a breakthrough in knowledge, not only at a local but also on a regional scale. The eruption catapulted the number and quality of results about the El Hierro volcanism, which crystalize on findings like that a uniform (radial) stress field exert a major control on the eruptive activity; a generally low magma supply rate that does not support large stalling of magmas in the crust and producing strombolian monogenetic eruptions, among other volcanic features. Those findings are informative for the design of simulations of the most expected event onshore (lava flows, ash plume dispersion and vent locations). Complementarily, the next chapter deals with an apparently minor, in volume, but probably highly overlooked and underestimate hazard, submarine and near-shoreline phreatomagmatic eruptions. This might be due to a nearly inexistent record during the written history of the Canary Islands (last 550 years), particularly because we have not witnessed phreatomagmatic/hydromagmatic eruptive activity near the shorelines. Although, that changed in 2011–2012 El Hierro submarine eruption. Moreover, Guillén et al., review and present the most up to date list of submarine eruptions in El Hierro prior to Tagoro eruption (Chap. 3). Their findings show evidence for submarine eruptions at shallow and intermediate water depths around El Hierro, even going further and suggesting that some might have occurred in historical times. A better understanding of the near shoreline and submarine eruptions should improve hazard assessments and reduce volcanic risks. The authors elegantly show that the shoreline is not a hard boundary for volcanoes, and subaerial and submarine activity is rather a continuum, that requires taking into account sea level changes in the past as well as in the future.

While the most visible effects of volcanism are the eruptive products and landforms, this activity is deeply rooted in its inner working. In the following chapters, El Hierro magma storage and transport system is discussed from a mainly geochemical (Chap. 4) and alternatively geophysical perspectives (Chap. 5). Both chapters, cross-reference the use of geophysics and geochemistry, therefore a better understanding can be achieved. In particular, Álvarez-Valero et al., make an effort to synthetize in few pages a decades long number of studies and also complement with new petrologic, geochemical and physiographic observations. This detailed analysis illuminates a complex evolution of the 2011–2012 submarine eruption. The chapter discusses the possible magma genesis processes, on the basis of classical petrology (petrography, geochemistry, and thermodynamics) on the volcanic products erupted during the submarine eruption. A multiparametric comparison complements the work reviewing how geomorphology and geophysical data (magnetometry) could be correlated with petrological changes. The conclusions are that differentiation and mixing processes occurred, the residence times at depth seems to be similar to previous eruptions. Therefore, the study of the 2011–2012 could be regarded as very informative of eruptive events in El Hierro. The companion chapter in this section, by Domínguez Cerdeña et al. reviews the geophysical data from a magma storage and migration perspective. The continuous monitoring geophysical data also allow to extend the period of observation to non-eruptive periods, 2011–2014. Seismicity and ground deformation was used to illuminate the active magmatic system during pre-eruptive (three months) unrest, eruptive (5-month), and post-eruptive stalled magmatic intrusions over a 2-year period. The geophysical analysis is consistent with independent petrological and gravimetric studies. In conclusion, the interpretation points towards the presence of a main deep magma storage system in the center of the island that is capable to feed sill-like magmatic intrusions (one of them capable to produce an eruption).

After 40 years of volcanic quiescence, the 2011–2012 Tagoro submarine eruption represented a unique opportunity and a challenge to test the Spanish volcano monitoring experts. Such challenge was exacerbated due to lack of previous crisis management experience of most involved, lack of preparedness (submarine eruptions were known but not planned for), traditionally secular lack of effective intergenerational know-how transmission of knowledge in Spain, and clear and obstructive rivalries among groups. This eruption, although a lucky near miss with little consequences for those involved, was a catalysis for the clarification of responsibilities, the proper design of a clear chain of command, and an orderly multi-institutional scientific and technical assessing board (PEVOLCA). This legacy fruitfully exploited ten years later in the 2021 La Palma eruption. A successful story steamed from the meticulous analysis of the monitoring data obtained and carefully analysed during the El Hierro submarine eruption. So, the following chapters document this effort in great detail. López et al., reviews the geophysical, geodetic and geochemical evidence for precursory Activity before the 2011–2012 Tagoro submarine eruption. An observational database that recorded the reawakening of volcanic and magmatic system beneath El Hierro. López and coworkers, in Chap. 6, show a multi-parameter monitoring of precursors with emphasis on lesson to future forecasting of eruptive activity. This experience allowed for a better interpretation of the observational data and the correct volcanic risk assessment, in La Palma, and possibly during future events. Chapter 7 provides a unique perspective of the change in morphology of a submarine eruption with a temporal sampling rate as never observed before. Vazquez et al., utilized nineteen oceanographic cruises to reconstruct the progressive rise from 400 to 88 m water depth of a submarine cone. The presented dataset documents that the construction process was non-linear with phases of construction, instabilities and destruction. The cone fed a long apron where lavas, pyroclastic and debris flows deposited in proximal, intermediate and distal zones. The geomorphology could be an analogue for similar, but long-lived volcanoes that grow in steep seafloor and are dominated by pyroclastic emissions. The monitoring of the eruption will be not completed without the intense effects that a submarine eruption has on the physical-chemical, geological and biological ecosystem (Chap. 8). A unique 10-year perspective that constitute a baseline for future and similar shallow submarine eruptions. Fraile-Nuez and co-workers' efforts document extreme changes in seawater temperatures of up to + 18.8 °C, water acidification decrease of 2.9 units in pH, deoxygenation reaching anoxic conditions and extreme metal enrichment resulting in significant alterations of the marine ecosystem. The monitoring effort is not only unique during the eruption but the long-lasting commitment to document the changes after March 2012. The marine environment has not yet recovered pre-eruptive conditions, and showing positive thermal anomalies, negative density changes, pH decrease, and high concentrations of metals and inorganic nutrients. It is worth noting that the last feature resembles conditions similar to oceanic upwelling zones. This chapter review the most complete multi-disciplinary time-series of a shallow submarine volcano, supported by the carried out of 31 oceanographic expeditions. A valuable lesson of how to conduct oceanographic studies for the better understanding future or similar submarine eruptions worldwide.

Although the physical conditions were already discussed in Chap. 8, the following three chapters dig deeper in the ecological and marine life effects. An aspect that is mostly undocumented in a mostly volcanological book series. The following Chaps. 9, 10 And 11, are testament of the dramatic although transient effects to marine life, ecosystem and environment. González-Vega et al., starts by review the wealth of data (3300 water samples) around the magnitude and timescales of the dissolved inorganic nutrients that Tagoro submarine volcano emitted during and after its eruption. The Tagoro eruption dataset is relevant not in itself but also as it is compared against other eruptions and hydrothermal systems worldwide. The importance is such that fluxes are as high as those in coastal upwelling regions. A remarkable observation that has implications for ecosystems of superficial nutrient-poor waters. Chapter 10 continues on the effect on the microbial communities. Ferrera et al., undertake the opportunity to study how microorganisms grow in underwater hydrothermal systems (hydrothermal plumes, metalliferous sediments, or near-vent zones), because as seen in the previous chapter this system is a nutrient-rich environment. The authors go further and present, discuss how a microbial community in a shallow-water environment might differ with respect to deep-sea hydrothermal vents, which have previously been studied. The 10 years span of those studies allowed to expand our knowledge of the microbiology associated with this volcano and its surrounding waters. The final chapter of the effects on the marine ecology is presented by Sotomayor-García et al. (Chap. 11). The effects of a shallow volcanic eruption in the Canary Islands' benthic habits and its associated biota has never been possible before El Hierro eruption. Hence, Sotomayor-García and coworkers go in a lot of detail to review the marine environmental perturbations. She shows that first, the eruption partly annihilated benthic and demersal pre-existing biota. However, this dramatic change was followed by a recovery phase that allow to understand the dynamics of colonization. Establishment of new marine habitats, correlates with different submarine volcanic deposits, with the most important differences on whether the substrate was harder, softer rock deposits or extreme hydrothermal vents as there were more significant variations in physical and chemical anomalies. The authors explored what the typical fauna of the region was, how the colonizers (benthic and demersal communities) dynamically recovered habitats. Those changes have implications in ichthyofauna and local fisheries. Chapter 11 concludes with future recommendations and steps to monitor trends and environmental status of the benthic and demersal communities.

The effects on the society is one of the main drivers of the study of volcanoes. Those effects are mainly thought in terms of disruption. However, this is true even for secondary risks, as such landslides (Chap. 12), less attention has been paid to the positive impact in societies living on volcanoes (Chap. 13). Galindo et al. review the citizen science strategies to identify and manage a transient and elusive secondary and indirect volcanic risk: landslides and rockfalls. She and her coworkers highlight that secondary volcanic hazards are usually overlooked, but paradoxically they may cause more damage (in terms of temporal frequency and cumulative economic damage) than primary volcanic hazards. That was the case during the El Hierro Island eruption where secondary hazards (in particular rockfalls) was the leading causes of damage. Volcanic eruption related earthquake shaking triggered multiple events of rockfall and landsliding. In other periods of volcanic eruption quiescence, such secondary risk will benefit of comprehensive inventories to characterize the triggering mechanisms and frequencies and magnitudes. However, such tasks are daugthing in terms of personal effort. In the last years, Galindo et al. has tested a new venue: citizen science. Local Civil Protection agents working along with engaged citizens provide a substantial amount of high quality data for rockfall risk analysis. This case and similar approaches could be extended to similar hazards and improve the assessment and risk reduction efforts, in other volcanic areas. Finally, the book ends with a positive note. Vegas et al., in Chap. 13, critically discuss the scientific value, geoconservation and geotourism opportunities that the Tagoro submarine volcano represents. As experts in geoheritage, an assessment clearly demonstrate the high scientific value of this eruption in the context of El Hierro. Therefore, the Tagoro submarine eruption was included as a geosite in the national inventory of geoheritage (Spanish Inventory of Geological Sites of Interest). The authors go into acknowledge the difficulties to appreciate a submarine eruption. Nevertheless, the benefits outweigh the challenges including it as a geotouristic resource. Challenges are directly addressed by the use of new technologies, e.g., virtual reality presentation at the Interpretation Center of the El Hierro UNESCO Global Geopark. In addition, a series of recommendations are provided to disseminate its value, and guaranteed its conservation. Such methodologies and strategies should benefit similar volcanoes worldwide.

1.1.3 Concluding Remarks and Future Challenges

The book chapters aim to review not only the key discoveries about each topic, but to highlight the challenges that should be tackled in the future. Chapter 1 (this chapter) should send a message that more geological, geochemical and geophysical imaging of the lithosphere and its relation with the sublithospheric mantle is necessary to formulate better geodynamic models for the El Hierro and the Canary Islands (intraplate volcanism). Chapter 2 identifies that to strength volcano monitoring and public hazard/risk awareness, an improved legal framework and a comprehensive research of monogenetic volcanism are necessary. Chapter 3 highlights the off-shore and on-shore hazards of hydromagmatism, with a stark warning that the expected sea-level rise might increase the hazard of future Surtseyan-style eruptions. Therefore, this trend combined with increase in population and coastal social and economic activities should be taken into account. Chapter 4 envisages that future exhaustive microtextural studies will allow a better understanding of the magma mixing mechanisms at depth and sediments contribution at surface. Combining this with diffusion chronometry should increase our knowledge on pre-eruptive processes and their timescales. Chapter 5 revision suggests that El Hierro magmatic system is complex with a common deep central main storage system from which radial sills emanate. Nevertheless, more need to be done to examine the collected geophysical and geodetic data for clues to understand why initial sill intrusions succeeded to progress towards the surface, while most subsequent ones stalled at depth. Chapter 6 highlights the importance of a multiparametric approach to identify precursory eruptive phenomena. It goes to suggest that previous pre-eruptive phases in the Canary Islands might underestimate the duration of preparatory volcanic processes prior to eruptions, due to the lack of the proper instrumentation. Improved monitoring working 24-7 is then a must to monitoring upcoming volcanic activity on the El Hierro and in the Canary Islands. Chapter 7 suggests that the future use of autonomous underwater vehicles (AUV) equipped with multibeam echosounders should capture future submarine eruption is much more detail than the 2011–2012 El Hierro eruption. More frequent seafloor bathymetric data and hydroacoustic sensing will capture geomorphological changes and their relationship with eruptive activity and gas emission.

Besides the recommendations about the study of physical aspects of the volcanic and magmatic system, almost half of the chapters focus on reseathe impact in the society and environment is the focus of almost half of the contributed chapters.

Chapter 8 highlights that to improve information on biological colonization, one could use autonomous underwater vehicles and similar non-invasive methods. A baseline of revisiting should, at least, be done every 2 years. This periodicity aims to detect the arrival of slow-growing organisms. Although, the 3-years after the eruption required more frequent visits to monitor small opportunistic and fast-growing taxa. Chapter 9 identifies as a pressing need that the dynamics of the inorganic nitrogen emitted from submarine eruptions, including their relative concentrations at the source and during water transport, should shed light on the impact of phytoplankton communities of inorganic emissions from volcanic and hydrothermal systems. Backing up this is the conclusion that nitrate and nitrite are not the main forms of nitrogen found, and that ammonium may be the primary form in the studied hydrothermal system. The chapter ends with a recommendation to continue with the largest datasets of inorganic nutrients from an active submarine volcano in the world. Chapter 10 follows with an interesting discrepancy in previous studies on the relative importance of the submarine emission on microorganisms. This motivates that future efforts should be focused on the design of a monitoring strategy that will test the biological effect of volcanic inputs to the water column, even under time lags, uptake and dilution processes, to distinguish between water masses affected by emissions and unaffected control waters. Chapter 11 suggests to continue with a long-term monitoring programme, including non-invasive methods. This programme is necessary to characterize benthic and demersal colonization particularly as volcanic habitats has been rarely studied in the Northeast Atlantic or even globally. Periodical monitoring of the marine resources and catches by the artisanal fleet could reveal unsolved questions about the effect of eruptive events in megafauna communities.

The final two chapters involve a substantial involvement of the wider public in their future recommendations. Chapter 12 suggests to maintain the citizen's science approach to engage and expand databases to characterize elusive secondary volcanic hazards. Expand the programme to schools in the form of challenges and games, so to maximize gain and retention of contributors to science via gamification. All those efforts to be translated into improved multi-hazards management during and after volcanic eruptions to reduce volcanic risks. Chapter 13 emphasizes different but complementary aspect of possible future public engagement. Engagement focused to protect, provide visibility, foster sustainable tourism and promote the value of be familiar with geological hazards and the geosites initiative via the El Hierro UNESCO Global Geopark.

Acknowledgements This research received funding from the Spanish Ministerio de Ciencia e Innovación research projects Volca-Motion and COMPACT and (PDI2022-139159NB-I00 and PID2019-104571RA-I00 proyectos de investigación financiados por MCIN/AEI/https://doi.org/10.13039/501100011033) and a 2020 Leonardo Fellowship Grant for Researchers and Cultural Creators, BBVA Foundation (IN[20]_CMA_CCT_0015). Additional support was received through the CSIC-PIE project with ID number PIE20223PAL008 (Real Decreto 1078/2021, de 7 de diciembre).

References

Abdel-Monem A, Watkins ND, Gast PW (1971) Potassium-argon ages, volcanic stratigraphy and geomagnetic polarity history of the Canary Islands: Lanzarote, Fuerteventura, Gran Canaria and La Gomera. Am J Sci 271:490–521

Ablay GJ, Kearey P (2000) Gravity constraints on the structure and volcanic evolution of Tenerife, Canary Islands. J Geophys Res: Solid Earth 105(B3):5783–5796

Anguita F, Hernán F (1975) A propagating fracture model versus a hot spot origin for the Canary Islands. Earth Planet Sci Lett 27(1):11–19

Anguita F, Hernán F (2000) The Canary Islands origin: a unifying model. J Volcanol Geotherm Res 103:1–26

Araña V, Camacho AG, Garcia A, Montesinos FG, Blanco I, Vieira R, Felpeto A (2000) Internal structure of Tenerife (Canary Islands) based on gravity, aeromagnetic and volcanological data. J Volcanol Geotherm Res 103:43–64

Araña V, Ortiz R (1991) The Canary Islands: tectonics, magmatism and geodynamic framework. In: Kampunzu AB, Lubala RT (eds) Magmatism in extensional structural setting: the Phanerozoic African Plate. Springer, pp 209–249

Balogh K, Ahijado A, Casillas R, Fernandez C (1999) Contributions to the chronology of the basal complex of Fuerteventura, Canary Islands. J Volcanol Geoth Res 90(1–2):81–101

Blanco-Montenegro I, Montesinos FG, Arnoso J (2018) Aeromagnetic anomalies reveal the link between magmatism and tectonics during the early formation of the Canary Islands. Sci Rep 8(1):1–14. https://doi.org/10.1038/s41598-017-18813-w

Camacho AG, Fernández J, González PJ, Rundle JB, Prieto JF, Arjona A (2009) Structural results for La Palma Island using 3-D gravity inversion. J Geophys Res 114 (B05411). https://doi.org/10.1029/2008JB005628

Camacho AG, Montesinos FG, Vieira R, Arnoso J (2001) Modelling of crustal anomalies of Lanzarote (Canary Islands) in light of gravity data. Geophys J Int 147(2):403–414. https://doi.org/10.1046/j.0956-540x.2001.01546.x

Camacho AG, Prieto Morin JF, Ancochea E, Fernández Torres J (2018) Deep volcanic morphology below Lanzarote, Canaries, from gravity inversion: new results for Timanfaya and implications. J Volcanol Geotherm Res 369:64–79. https://doi.org/10.1016/j.jvolgeores.2018.11.013

Carracedo JC, Day S, Guillou H, Badiola ER, Canas JA, Torrado FP (1998) Hotspot volcanism close to a

passive continental margin: the Canary Islands. Geol Mag 135(5):591–604
Carracedo JC, Badiola ER, Guillou H, de La Nuez J, Pérez Torrado FJ (2001) Geology and volcanology of La Palma and El Hierro, Western Canaries. Estud Geol 57:175–273
Civiero C, Custódio S, Rawlinson N, Strak V, Silveira G, Arroucau P, Corela C (2019) Thermal nature of mantle upwellings below the Ibero-western Maghreb region inferred from teleseismic tomography. J Geophys Res: Solid Earth 124(2):1781–1801
Collier JS, Watts AB (2001) Lithospheric response to volcanic loading by the Canary Islands: constraints from seismic reflection data in their flexural moat. Geophys J Int 147(3):660–676
Dañobeitia JJ, Canales JP (2000) Magmatic underplating in the Canary Archipelago. J Volcanol Geotherm Res 103:27–41
Dañobeitia JJ, Canales JP, Dehghani GA (1994) An estimation of the elastic thickness of the lithosphere in the Canary Islands using admittance function. Geophys Res Lett 21(24). https://doi.org/10.1029/94GL02552
Day JM, Pearson DG, Macpherson CG, Lowry D, Carracedo JC (2010) Evidence for distinct proportions of subducted oceanic crust and lithosphere in HIMU-type mantle beneath El Hierro and La Palma, Canary Islands. Geochim Cosmochim Acta 74(22):6565–6589
Duggen S, Hoernle KA, Hauff F, Klügel A, Bouabdellah M, Thirlwall MF (2009) Flow of Canary mantle plume material through a subcontinental lithospheric corridor beneath Africa to the Mediterranean. Geology 37(3):283–286. https://doi.org/10.1130/G25426A.1
Fullea J, Camacho AG, Negredo AM, Fernández J (2015) The Canary Islands hot spot: new insights from 3D coupled geophysical–petrological modelling of the lithosphere and uppermost mantle. Earth Planet Sci Lett 409:71–88
García-Yeguas A, Ibáñez JM, Koulakov I, Jakovlev A, Romero-Ruiz MC, Prudencio J (2014) Seismic tomography model reveals mantle magma sources of recent volcanic activity at El Hierro Island (Canary Islands, Spain). Geophys J Int 199:1739–1750. https://doi.org/10.1093/gji/ggu339
Geldmacher J, Hoernle K, Bogaard PVD, Duggen S, Werner R (2005) New 40Ar/39Ar age and geochemical data from seamounts in the Canary and Madeira volcanic provinces: support for the mantle plume hypothesis. Earth Planet Sci Lett 237(1–2):85–101
Geyer A, Martí J, Villaseñor A (2016) First-order estimate of the Canary Islands plate-scale stress field: Implications for volcanic hazard assessment. Tectonophysics 679:125–139
González PJ, Samsonov S, Pepe S, Tiampo KF, Tizzani P, Casu F, Fernández J, Camacho AG, Sansosti E (2013) Magma storage and migration associated with the 2011–2012 El Hierro eruption: implications for shallow magmatic systems at oceanic island volcanoes. J Geophys Res: Solid Earth 118:4361–4377. https://doi.org/10.1002/jgrb.50289
Gottsmann J, Camacho AG, Martí J, Wooller L, Fernández J, Garcia A, Rymer H (2008) Shallow structure beneath the central volcanic complex of Tenerife from new gravity data: implications for its evolution and recent reactivation. Phys Earth Planet Inter 168(3–4):212–230
Herzberg C, Asimow PD (2008) Petrology of some oceanic island basalts: PRIMELT2.XLS software for primary magma calculation, Geochem. Geophys Geosyst 9:Q09001. https://doi.org/10.1029/2008GC002057
Hoernle K, Schmincke HU (1993) The role of partial melting in the 15-Ma geochemical evolution of Gran Canaria: a blob model for the Canary hotspot. J Petrol 34(3):599–626
Hunt JE, Jarvis I (2017) Prodigious submarine landslides during the inception and early growth of volcanic islands. Nat Commun 8:2061. https://doi.org/10.1038/s41467-017-02100-3
Klügel A, Hansteen TH, Galipp K (2005) Magma storage and underplating beneath Cumbre Vieja volcano, La Palma (Canary Islands). Earth Planet Sci Lett 236:211–226. https://doi.org/10.1016/j.epsl.2005.04.006
Klügel A, Longpré M-A, García-Cañada L, Stix J (2015) Deep intrusions, lateral magma transport and related uplift at ocean island volcanoes. Earth Planet Sci Lett 431:140–149. https://doi.org/10.1016/j.epsl.2015.09.031
Kreemer C, Blewitt G, Klein EC (2014) A geodetic plate motion and Global Strain Rate Model. Geochem Geophys Geosyst 15:3849–3889. https://doi.org/10.1002/2014GC005407
Le Bas MJ, Rex DC, Stillman CJ (1986) The early magmatic chronology of Fuerteventura, Canary Islands. Geol Mag 123(3):287–298
Lodge A, Nippress SEJ, Rietbrock A, Garcia-Yeguas A, Ibáñez J (2012) Evidence for magmatic underplating and partial melt beneath the Canary Islands derived using teleseismic receiver functions. Phys Earth Planet Inter. https://doi.org/10.1016/j.pepi.2012.09.004
Longpré MA, Felpeto A (2021) Historical volcanism in the Canary Islands; part 1: a review of precursory and eruptive activity, eruption parameter estimates, and implications for hazard assessment. J Volcanol Geoth Res 419:107363
Longpré MA, Chadwick JP, Wijbrans J, Iping R (2011) Age of the El Golfo debris avalanche, El Hierro (Canary Islands): new constraints from laser and furnace 40Ar/39Ar dating. J Volcanol Geotherm Res 203:76–80
Manjón-Cabeza Córdoba A, Ballmer MD (2021) The role of edge-driven convection in the generation of volcanism–Part 1: a 2D systematic study. Solid Earth 12(3):613–632
Martí J, Villaseñor A, Geyer A, López C, Tryggvason A (2017) Stress barriers controlling lateral migration of

magma revealed by seismic tomography. Sci Rep 7:40757. https://doi.org/10.1038/srep40757
Martinez-Arevalo C, Mancillade FL, Helffrich G, Garcia A (2013) Seismic evidence of a regional sublithospheric low velocity layer beneath the Canary Islands. Tectonophysics 608:586–599. https://doi.org/10.1016/j.tecto.2013.08.021
Miller MS, Driscoll LJO, Butcher AJ, Thomas C (2015) Imaging Canary Island hotspot material beneath the lithosphere of Morocco and southern Spain. Earth Planet Sci Lett 431:186–194. https://doi.org/10.1016/j.epsl.2015.09.026
Montesinos FG, Arnoso J, Vieira R (2005) Using a genetic algorithm for 3-D inversion of gravity data in Fuerteventura. J Earth Sci 94(2):301–316. https://doi.org/10.1007/s00531-005-0471-6
Montesinos FG, Arnoso J, Benavent M, Vieira R (2006) The crustal structure of El Hierro (Canary Islands) from 3-D gravity inversion. J Volcanol Geotherm Res 150(1–3):283–299. https://doi.org/10.1016/j.jvolgeores.2005.07.018
Montesinos FG, Arnoso J, Vieira R, Benavent M (2011) Subsurface geometry and structural evolution of La Gomera island based on gravity data. J Volcanol Geoth Res 199(1–2):105–117
Negredo AM, van Hunen J, Rodríguez-González J, Fullea J (2022) On the origin of the Canary Islands: insights from mantle convection modelling. Earth Planet Sci Lett 584:117506
Neumann ER, Wulff-Pedersen E, Pearson NJ, Spencer EA (2002) Mantle xenoliths from Tenerife (Canary Islands): evidence for reactions between mantle peridotites and silicic carbonatite melts inducing Ca metasomatism. J Petrol 43(5):825–857
Paris R, Guillou H, Carracedo JC, Torrado FP (2005) Volcanic and morphological evolution of La Gomera (Canary Islands), based on new K-Ar ages and magnetic stratigraphy: implications for oceanic island evolution. J Geol Soc 162(3):501–512
Ramalho R (2011) Vertical movements of ocean island volcanoes: insights from a stationary plate. In: Building the Cape Verde islands. Springer Theses. Springer, Berlin, Heidelberg
Richards F, Hoggard M, Crosby A, Ghelichkhan S, White N (2020) Structure and dynamics of the oceanic lithosphere-asthenosphere system. Phys Earth Planet Inter 309:106559
Roest WR, Dañobeitia JJ, Verhoef J, Collette BJ (1992) Magnetic anomalies in the Canary Basin and the Mesozoic evolution of the central North Atlantic. Mar Geophys Res 14(1):1–24
Sainz-Maza S, Montesinos FG, Martí J, Arnoso J, Calvo M, Borreguero A (2017) Structural interpretation of El Hierro (Canary Islands) rifts system from gravity inversion modelling. Tectonophysics 712:72–81
Sainz-Maza Aparicio S, Martí J, Montesinos FG, Gómez AB, de Pablo JP, Fernández PV, García-Maroto MC (2019) Gravimetric study of the shallow basaltic plumbing system of Tenerife, Canary Islands. Phys Earth Planet Inter 297:106319
Schmincke HU (1982) Volcanic and chemical evolution of the Canary Islands. In: Geology of the Northwest African continental margin. Springer, Berlin, Heidelberg, pp 273–306
Staudigel H, Feraud G, Giannerini G (1986) The history of intrusive activity on the island of La Palma (Canary Islands). J Volcanol Geotherm Res 27(3–4):299–322
Urgeles R, Canals M, Baraza J, Alonso B (1998) Seismostratigraphy of the western flanks of El Hierro and La Palma (Canary Islands): a record of Canary Islands volcanism. Mar Geol 146(1–4):225–241
van den Bogaard P (2013) The origin of the Canary Island Seamount Province-New ages of old seamounts. Sci Rep 3(1):2107. https://doi.org/10.1038/srep02107
Watts AB (1994) Crustal structure, gravity anomalies and flexure of the lithosphere in the vicinity of the Canary Islands. Geophys J Int 119(2):648–666
Zaczek K, Troll VR, Cachao M, Ferreira F, Deegan FM, Carracedo JC, Meade FC, Burchardt S (2015) Nannofossils in 2011 El Hierro eruptive products reinstate plume model for Canary Islands. Sci Rep 5:7945. https://doi.org/10.1038/srep07945

Past, Present and Future Volcanic Activity on El Hierro

2

Stavros Meletlidis, Laura Becerril, and Alicia Felpeto

Abstract

El Hierro, the youngest island of the Canary Islands, emerged from the ocean floor only 1.2 Ma ago and it is still in its youngest stage (shield volcanism). During its growth, the island passed from the submarine stage to the subaerial activity, constructing an island that had been modified by at least six large destructive flank collapses or giant landslides, a common feature in the development of the basaltic oceanic islands. Throughout the period of its evolution, it showed different eruptive styles, developing two main edifices. The last 160 kyrs the activity has been characterized by the rift volcanism, resulting in a three-arms rift construction. This chapter provides with a comprehensive picture of the evolution of the island and generate a baseline of knowledge that will encourage a further study on volcanic processes related to the hot spot volcanism. In this work, we describe, not only the geological evolution of the island, but also review the existed petrological, geochemical and geochronological data. Furthermore, we use this data to generate the probable future eruptive scenarios. We conclude with an updated and synthesized review of the recent 2011–2012 submarine eruption.

Keywords

Volcanic hazards · Future scenarios · Strombolian activity · Volcano-tectonics · El Hierro

2.1 Introduction

One of the most studied volcanic zones in the world is the Canary archipelago. Located off northwest Africa, it has attracted many kinds of scientists during the last five hundred years. The volcanic activity in the region, a good example of oceanic intraplate magmatism, started about 60 Ma ago (Carracedo et al. 1998), and is still controversial as far as the genesis and evolution of the islands is concerned. With the exception of La Gomera, the rest of the islands have experienced activity during the Holocene, and some of them (Lanzarote, Tenerife, La Palma and El Hierro, inset in Fig. 2.1) were also the site of historical eruptions (1405 AD to today). The last onshore eruption recently occurred in La Palma

S. Meletlidis (✉)
Centro Geofísico de Canarias, Instituto Geográfico Nacional, C/La Marina 20, 2°, 38001 Santa Cruz de Tenerife, Spain
e-mail: smeletlidis@mitma.es

L. Becerril
Instituto de Ciencias de La Ingeniería, Universidad de O'Higgins, Rancagua, Chile

A. Felpeto
Observatorio Geofísico Central, Instituto Geográfico Nacional, Alfonso XII, 3, 28014 Madrid, Spain

P. J. González (ed.), *El Hierro Island*, Active Volcanoes of the World,
https://doi.org/10.1007/978-3-031-35135-8_2

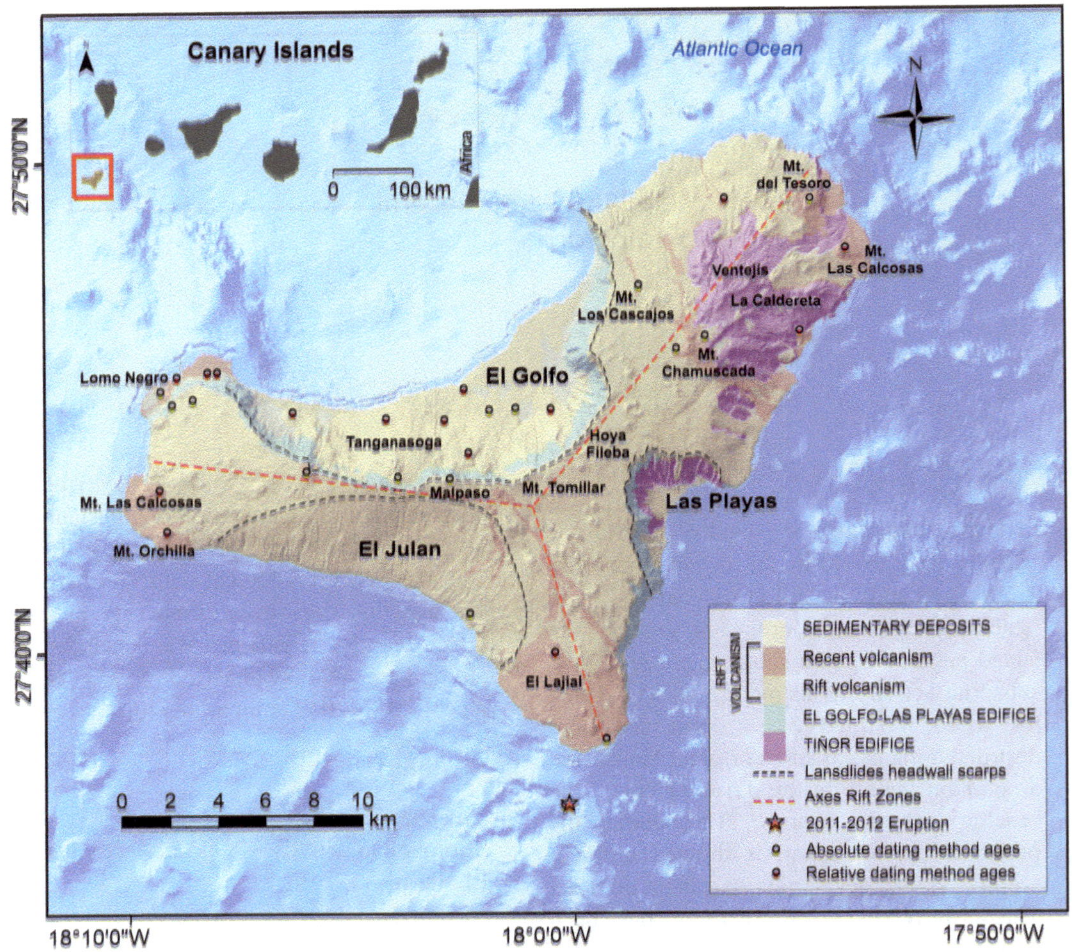

Fig. 2.1 Simplified geological map of El Hierro Island including eruptions less than 20 kyrs (see more details in Becerril et al. 2016a, b); the recent eruption offshore the island, the rift axis (red dashed lines), and the main landslide headwall scars (black dashed lines)

Island (Cumbre Vieja volcanic area, 2021), with the last offshore event of the archipelago taking place in El Hierro (submarine Tagoro volcano 2011–2012).

The Canaries, despite the fact that they emerged and evolved independently from the Atlantic seafloor, share some similar features in the magma's geochemistry, eruption types and products, but mainly in their growth stages that can be identified as a common occurrence in their evolution. In this journey, since the magma first arose from the mantle to the last erosion stage, the islands experienced multiple intervals of volcanic activity and quiescence periods, involving an extreme range of magma compositions and eruptive styles. However, during their entire evolution we can distinguish three main stages (Carracedo et al 2001) that represent a summary of the stages proposed by Stearns (1940) in Hawaii and revised and expanded afterwards by Ramalho (2011): (1) a submarine stage that constructed a shield volcano, elevated from the ocean floor and emerged on the sea surface, with high productivity of basaltic eruptions; (2) a pause of the activity which resulted in the erosion of the volcanic edifice and, finally (3) a post erosive rejuvenation stage of the island, where eruptions commonly take place involving lower volumes of commonly less evolved magmas that means more basaltic compositions. The youngest and westernmost islands of the Canaries—El Hierro and La Palma—are still in the first evolutionary stage of this cycle, where subaerial basaltic eruptions are taking place quite frequently (in a geological time frame) still growing and developing the insular edifice. El Hierro represents the emergent part of a volcanic edifice which rises ~ 5500 m from the sea bottom. Its subaerial part emerges 1501 m asl and it has a surface area of about 269 km^2.

This chapter presents an overview of the volcanic activity of El Hierro Island, mainly based on the published literature review. We describe the evolution of the island, the variety of forms, eruptive styles, products and deposits of the eruptions, its volcanic structure, and analyse the most probable eruptive future scenario on the island with their associated hazards through the simulation of the most expected forthcoming event onshore. Petrological, geochemical and geochronological data is also reviewed, with the aim of understanding its eruptive dynamics. Finally, we review and synthesise the 2011 El Hierro eruption, based on the available petrological data and the proposed models of the plumbing system beneath the island, that will help the reader to comprehend the processes that took place before and during the eruption, which are, quite likely, the same ones that have driven the volcanic activity of the island in this particular phase of its youth.

2.2 Volcanic Activity and Evolution of the Island

To rapidly understand the geological and tectonic setting of El Hierro, which has controlled its evolution and its volcanic activity, it is important to mention that its origin is related to oceanic intraplate hot spot magmatism with strong influences of the regional tectonic environment, built in the presence of regional stress fields (Staudigel et al. 1986); Anguita and Hernán 2000; Blanco-Montenegro et al. 2018). In this particular frame, El Hierro is allegedly located directly on the top of a stationary source of magma (mantle plume) on a slow-moving African plate, thus with a 1.2-million-year-old emerged edifice still in its juvenile stage (Carracedo et al. 2001).

The island rests as an independent edifice on the Jurassic crust, rising 3700 m from the ocean floor (Masson et al. 2002), as the summit of a largely submarine volcanic shield edifice. The most characteristic feature of the island is its triple-armed edifice, moulded by several landslides that gave origin to wide, horseshoe-shaped embayments, possibly as a result of a fast volcanic growth (Masson 1996; Urgeles et al. 1997, 1999; Carracedo et al. 1999, 2001; Masson et al. 2002; Longpré et al. 2011).

During the short subaerial volcanic history of the island (1.12 Ma; Guillou et al. 1996), three main constructive cycles have been proposed, separated by erosion, short pauses of volcanism and partial sector collapses. At least 6 large destructive events have been identified, being the El Golfo landslide one of the most recent (~ 87–40 kyrs) volcano flank collapses in the Atlantic region (Longpré et al. 2011 and references therein).

The oldest part of the island corresponds to the Tiñor Edifice (Fig. 2.1) which has an age of about 1.12–0.88 Ma (Guillou et al. 1996). It can be mainly identified in the NE flank of the island and the interior of the Las Playas embayment (Fig. 2.1). The volcanic succession shows subaerial outcrops with no compositional variation with time, although three units can be recognized, mainly associated with the morphological evolution of the volcanic edifice. These can be identified as: a basal unit of steep-dip lava flows, an intermediate unit with lava flows, during the shield construction, and an upper unit, with dispersed emission vents and their associated lava flows (Carracedo et al. 2001). The construction of this first edifice or volcano ended with a collapse (Carracedo et al. 2001; IGME 2010), in whose scar the second edifice, El Golfo-Las Playas volcano, began to develop, widening the island towards the west. The growth of this second volcano lasted approximately 380 kyrs (from 545 to 176 ka) (Carracedo et al. 2001;

Guillou et al. 1996; IGME 2010). In that case, two units can be identified, a basal one comprised of strombolian and surtseyan pyroclastic products and cut by numerous dykes, and an upper one composed mainly by lava flows with differentiated lava flows of trachybasalts and trachytes found at the top of the edifice and interpreted as part of its later stage. (Carracedo et al. 2001).

The island's recent volcanic subaerial activity (last 160 kyrs, Guillou et al. 1996) is characterised by monogenetic volcanism within the three rifts simultaneously active (Carracedo et al. 2001). The recent rift series products can be found broadly on the top of the last major activity (El Golfo-Las Playas volcano) and have filled the embayments and covered most of the current surface of the island (Fig. 2.1).

A radial strike distribution of dykes and eruptive fissures is observed in the on-land and submerged volcano-structures, which has been related to uniform stress fields during constructive episodes maintained during the growth of the whole volcano edifice (Becerril et al. 2015). The offshore continuation of the southern volcanic rift is interpreted as an old edifice (Gee et al. 2001; Schmincke and Graf 2000; van den Bogaard 2013).

According to the models proposed for the plumbing system and magma storage beneath the younger islands of the Atlantic region (Klügel et al. 2005; Galipp et al. 2006; Stroncik et al. 2009; Jefferey and Gertisser 2018), small intermittent magma pockets are generated in the upper crust, from the mantle-derived mafic magmas. New incoming magma batches probably mix with the previously emplaced magmas and end up triggering further ascents and result in an eruption. Eruptions tend to be strombolian in style, with an explosive initial phase and minor phreatomagmatic episodes mainly due to interference with underground water reservoirs or during the onset of the eruptions in shallow sea waters (Pedrazzi et al. 2014). Throughout its subaerial history, the island showed a low magma supply, with an estimated growth rate of $\sim$ 0.2 km^3/ka for the last 1.12 Ma (Carracedo 1999). This condition is expected to inhibit the development of large and/or long-lasting shallow magma reservoirs, explaining the absence of significant volumes of evolved volcanic products in the island (Zanon et al. 2020).

As a typical expression of monogenetic volcanism, the recent volcanic activity on the island resulted in small volcanic features such as cinder cones, lava flows and not very extended pyroclastic deposits (e.g., Mt. Tesoro, Mt. Orchilla, El Lajial or Mt. Chamuscada; Fig. 2.1). In this regard, more than 220 cones onshore belonging to the Rift Series were identified on the island, making El Hierro the island in the archipelago with the highest number of monogenetic volcanoes per km^2 (Becerril et al. 2015). Most vents are located along the rift zones. Nevertheless, several emissions centres are located inside the El Golfo landslide depression and on its scar, with their products partially refilling the collapse embayment. This volcanic activity could have been developed as a response of the plumbing system to the flank collapse (Maccaferi et al. 2017). The reduction of the load above the magma pockets, dykes and sills, possibly led to disturbances in the state of stress of the island magmatic system and generated the conditions of magma degassing, remobilization and ascent, favouring the formation of new eruptive centres (Manconi et al. 2009; Longpré et al. 2009). The most voluminous recent cone in the island is the Tanganasoga volcano (Fig. 2.1), located almost on the conjunction of the three rifts in the centre of the island, inside the landslide depression of El Golfo. This volcano exhibits multiple vents that emitted large volumes of lava and ejected large blocks due to phreatomagmatic episodes (Clarke et al. 2009, Troll and Carracedo 2016).

The recent emission of lava flows on the island, frequently resulted in the construction of coastal platforms or cascaded over pre-existing eroded cliffs, therefore younger than the last glacial maximum (approximately 20,000 years ago; Carracedo et al. 2001). The application of such geomorphological criteria, in conjunction with the available geochronological information, allowed the identification of more than 20 recent volcanic events (Becerril et al. 2016a, b), which helped to determine more than 30 areas with recent activity, some of which experienced multiple eruptive events (Fig. 2.1).

Regarding pyroclastic deposits, mafic cinder and spatter cones are spread out along the rifts, and lapilli and ash fall deposits are found around them. The only explosive episode of felsic volcanism can be found on the top of the island and is known as Malpaso Member (Pedrazzi et al. 2014).

Before the last offshore 2011–2012 eruption, only one other eruption was assumed as historical (last 600 yrs.): the Lomo Negro eruption (Hernández-Pachecho 1982), located on the western rift (Fig. 2.1). Villasante-Marcos and Pavón-Carrasco (2014) did a thorough paleomagnetic study of the lavas of Lomo Negro obtaining three possible age ranges, and re-computed the ^{14}C age obtained by Hernández Pacheco (1982) with an up-to-date calibration curve. Taking into account both the palaeomagnetic and recomputed ^{14}C results, the most probable date for the eruption is the interval 1499–1602 CE.

2.3 The Compositional Signature of the Rocks

First petrological studies on the island date from the XIX century. In 1863 Karl von Fritsch published the first map of El Hierro, introducing the primary geomorphological and geological features of the island and collecting samples for their posterior study. Some years later came the first attempt to study the geology of El Hierro (Walter 1894) using the samples that were taken 30 years earlier. This was the first comprehensive geological and petrological study of the island. Since then, a small number of scientists have focused on the geology of the island, either as a part of an extensive study of the Canary archipelago, where only some lines were written in relation to the island, or as an intent to explain the origin of the El Golfo embayment. Among them were: Simony (1890); von Knebel (1906); Fernandez Navarro (1908, 1924, 1926); Gagel (1910); von Wolff (1931); Jérémine (1935); Benitez Padilla (1945). The 1908 report of Navarro was likely the first one dealing with the different rocks of the island and Jérémine (1935) included the first analysis of a volcanic rock from El Hierro, a sample of ankaratrite.

The interest in the volcanic history of the island increased after the 1960s, mainly in the frame of a global model regarding the origin of the Canary Islands. The global research progress in areas such as geochronology, oceanography, seismology and petrographic techniques gave a tremendous boost in the studies and investigation on the evolution of the island. The first island lavas dating was provided by Abdel-Monem and co-workers in 1972, followed by Fúster et al. (1993) that were revised afterward by Guillou et al. (1996). It took more than a century, since the first bibliographic reference, to see the publication of a more complete petrological and geochemical study of El Hierro rocks that helped to better understand the magmatic system beneath the island. This study published by Pellicer (1977, 1979), providing a comprehensive geological and volcano-structural map along with a petrological study, was delivered in the late 1970s. Pellicer, through 91 lava flows and dykes samples determined three rock series in the evolution of the island that were called Ancient, Intermediate and Recent.

Petrological studies of the earlier stages on the island evolution, published during the last two decades and those on the emitted products during the 2011–2012 eruption, propose the presence of two separate magma storage regions, lying in the lower oceanic crust and in the uppermost mantle (Carracedo et al. 2001; Stroncik et al. 2009; Meletlidis et al. 2012; Becerril et al. 2013b; González et al. 2013; Martí et al. 2013a, b; Longpré et al. 2014; Klügel et al. 2015; Carracedo et al. 2015; Zaczek et al. 2015).

As mentioned above, the island is still in the shield volcano stage of development, according to the intraplate oceanic islands model (Walker 1990) characterised by monogenetic basaltic activity. In this subaerial shield building stage, where mafic to intermediate magmas are involved, their products comprise picrite basalts to mafic trachytes, with abundant intermediate members, erupted during its three volcanic cycles. Studies of major and trace elements, together with Sr and Nd isotopes, show the OIB origin of the magmas (Day et al. 2010; Aulinas et al. 2019). Furthermore, the observed mineral phases and textures support magma mixing/mingling together with fractional crystallization and partial assimilation processes (Stroncik et al. 2009).

The first constructive cycle, the Tiñor Edifice, that represents the oldest subaerial lavas of El Hierro, is characterised by typical mafic basalts, with porphyric to subaphyric texture with microcrystals of plagioclase, clinopyroxene, Fe-Ti oxides and phenocrystals of olivine (Carracedo et al. 2001) (Fig. 2.2). The same composition continues in the lower sequence of the El Golfo-Las Playas volcano products, containing augite crystals, with more evolved products such as trachytes (Fig. 2.2) with abundant phenocrystals of plagioclase feldspar during the latest stage of this cycle of activity (Carracedo et al. 2001). The Rift Series or Recent volcanism shows a varied composition, comprising alkali picrites and basanites to tephrites (Fig. 2.2), with lack of trachytes. The increase of alkalinity with respect to the two older cycles has been highlighted by many authors, attributing it to a gradual decrease of partial melting degree in the mantle (Pellicer 1977), and also to the effect of the El Golfo gravitational collapse to the magmatic plumbing system (Manconi et al. 2009).

The composition of the Holocene products ranges from picrobasalts to phonotephrites (Fig. 2.2), being silica under saturated rocks (40–50 wt.%). Pumice deposits of Malpaso Member in the central part of the island, formed directly from dilute PDCs (Pedrazzi et al. 2014), could be attributed to small felsic pockets in the crust, reactivated by the mafic intrusion. Such process could be similar but more extended, to the one in the 1971 Teneguia eruption (Araña and Ibarrola 1973) and the 2011–2012 submarine El Hierro eruption (Troll et al. 2012; Meletlidis et al. 2012; Sigmarsson et al. 2013). Besides, phonolitic lavas were dredged on the submarine extension of the southern ridge of El Hierro (Abratis et al.

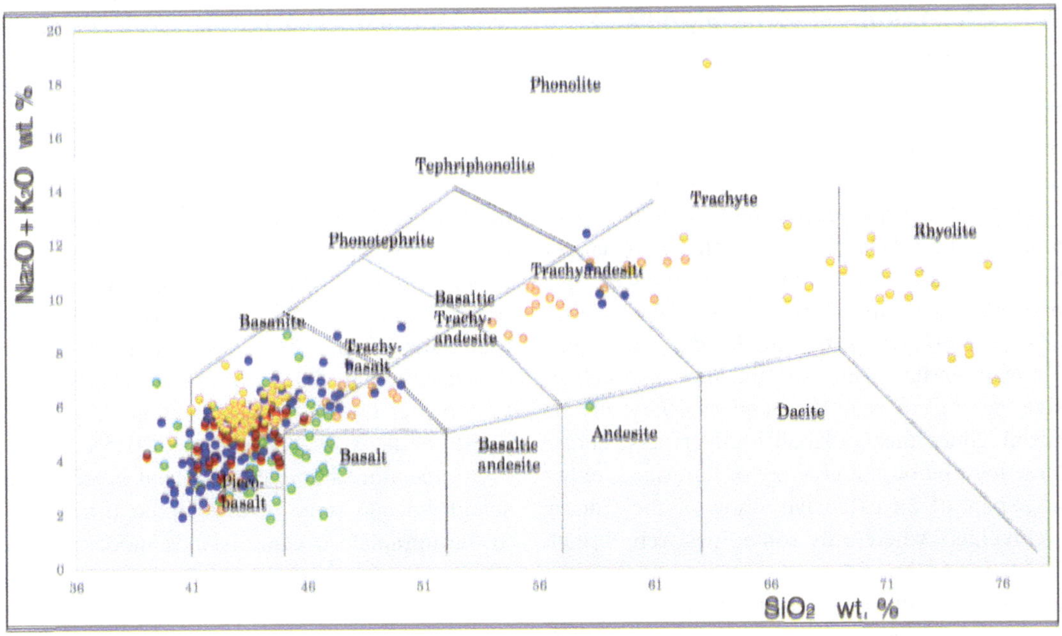

Fig. 2.2 Total alkalis versus silica (TAS) plot showing classification of El Hierro rocks (data represent the existing bibliography uptodate). Yellow dots: 2011–2012 eruptive products. Red dots: submarine products from the rifts. Blue dots: products of the subaerial rifts. Orange dots: products of El Golfo edifice. Green dots: products of Tiñor edifice

2002) but appear to belong to an older seamount structure, overlaid by the younger insular edifice (van den Bogaard 2013).

A special mention should be made on the studies of mantle ultramafic xenoliths—interpreted to reflect metasomatism of lithospheric mantle by young basaltic magmas—in order to increase understanding of the internal structure of the island and inner processes of magma crystallization and depth of stagnation (Neumann 1991, Hansteen et al. 1991; Whitehouse and Neumann 1995; Wulff-Pedersen et al. 1996; Neumann and Wulff-Pedersen 1997; Neumann et al. 2004; Oglialoro et al. 2017).

2.4 Volcanic Activity Through Time

The improvements of the geochronological techniques during the last decades have allowed the determination of more precise ages for the geological formations on the islands, resulting in a significant improvement of our knowledge about the volcanic evolution of the Canary region. In the case of El Hierro, however, on account of a remarkable lack of reliable stratigraphic markers for the whole island, the efforts to date the eruptive products and constrain the age of the volcanic activity and cataclysmic events such as the giant landslides, are particularly challenging.

Back in 1972, Abdel-Monem et al., as a continuation of their 1971 geochronological, stratigraphic and geomagnetic study of the Canary Islands eastern province, extended their studies on the rest of the islands including El Hierro. That was the first attempt to deliver quantitative correlations across the archipelago. They dated only five samples, using K-Ar method, on basalts, with sampling points chosen in accordance with the geology aspects indicated by Hausen (1964) and oriented to the scarp of El Golfo landslide. The obtained ages were between 3 and 0.19 Ma, for the base and the uppermost rim of the El Golfo scarp respectively. Also, they sampled more than forty lava flows, in search of reversed polarity, but with no success. This work, apart from establishing the basis for further works on the geochronology of El Hierro, evidences that a common evolution of the island was not probable.

Pellicer (1975, 1977), in her PhD Thesis, added the first ^{14}C ages, from samples of charcoal localized in deposits of evolved products (trachytic pumices—Malpaso Member) in the central sector of the island and in a cineritc deposit right above them, with an age of 6.8 and 4.2 ka, respectively, assuming that period as the beginning of the recent activity series.

A new ^{14}C data was provided by Hernández-Pacheco (1982), associated to the deposits of Lomo Negro eruption, in the western part of the island. The fresh aspect of the lava flows, in conjunction with some historical documents mentioning an intensive seismic episode in 1793 AD, and the lack of a more comprehensive calibration curve, led to an inaccurate estimation of the period of the eruption.

A total of 10 new K-Ar ages were provided in 1993 (Fúster et al.) concerning the age of the cycles of volcanic activity, concluding that the recent activity was younger than 50 ka and the age of the embayments of El Golfo and Las Playas should be less than 0.5 Ma. Additionally, the authors estimated an average eruptive rate and growth rate of 0.5 km^3 per kyrs.

The advent of oceanographic studies not only improved the knowledge about the submarine deposits and the structure of the island but also permitted a comparison between the ages of the submarine deposits with the mechanism that produced them. Deposits from El Julan landslide were dated older than 60 ka, in relation to a Saharan debris flow (Holcomb and Searle 1991), subsequently confirmed as a > 160 ka deposit, with dates obtained from the lavas on the onshore part (Carracedo et al. 1997; Day et al. 1997). Later, Masson (1996), estimated an age of 10–17 ka for the turbidites generated by the El Golfo collapse, an age that was in agreement with the later work of Guillou et al. (1996) who supported that the landslide took place before 15 and after 45 ka. Research on the giant flank collapse of El Golfo volcano continued and Longpré et al. (2011), based on new ^{40}Ar-^{39}Ar ages on samples of the rim and the base of the embayment, constrained the episode sometime between 87 and 40 ka, with the maximum age being the most confident. Thus far, according to the different works on the embayment, the failure could have occured in two phases, first the collapse of the El Golfo volcanic edifice itself, and then the collapse of the volcanic platform generated from the movement. This approach, according to Carracedo et al. (2001), is supported by observations along excavated sub-horizontal water extraction galleries and the geology of the marine abrasion platform beneath the lava flows that later filled the El Golfo Embayment. In any case, although its age is still a matter of debate, the El Golfo collapse should be considered as the most recent gravitational collapse in the archipelago.

In their extensive study of paleomagnetic polarities and K-Ar dating by Guillou et al. (1996) dated eighteen samples and contributed to the establishing of the main stages of evolution for the island, and most importantly, defined a maximum age of 1.2 Ma for the subaerial volcanism. The age obtained by means of ^{14}C for the Mt. Chamuscada lavas (Fig. 2.1), in the central sector, is still considered one of the youngest eruptions on the island, with an age of 2.5 ka. The same results were used later by Day et al. (1997) to set the age of the aborted slope failure developed by the San Andrés fault system in the eastern sector of the island. Once again, in 1999, a new dating of the scarp of the El Golfo was carried out along with the rifts formation rocks (Szérémèta et al. 1999), providing four K-Ar ages from 0.35 to 0.13 Ma.

In the work of Carracedo et al. (2001), new K-Ar ages and stratigraphic correlations contributed to constrain the temporal evolution of El Hierro and allowed the division of the subaerial rocks in three main units, associating them to cycles of development and destruction. Based on this work, and in an in-depth study of the rifts onshore and offshore products and structures, Acosta et al. (2003), argued that their submarine sections were active earlier than 0.5 Ma, with periods of rejuvenation through the evolution of the island.

For the Holocene age of two eruptive centres, pointed out by Pellicer (1977) and Guillou et al (1996), an age of 5 ka was speculated by Lundstrom et al. (2003) for a sample taken onshore of the southern rift, a couple of km from the point of the upcoming 2011–2012 submarine eruption. Furthermore, new ^{14}C data was obtained by Pérez-Torrado et al. (2011) for volcanoes in the central sector of the island. They dated the same trachytic deposits that Pellicer had done (1975, 1977) and also a deposit below them. The obtained ages range from 3.9 ka for the evolved material to 8 ka for the underneath material, ages that could be considered coherent with those reported in 1977. The third sample, dated on a basaltic fallout deposit, gave an age of almost 5 ka, situating the event between the other two. In 2012, Rodríguez-Gonzalez et al. added a new K-Ar age, ~ 9 ka, for the Holocene eruption of Montaña del Tesoro, located at NE coast of the island.

The age proposed for the Lomo Negro eruption (Hernández-Pacheco 1982), was revised by Villasante-Marcos and Pavón-Carrasco (2014) by means of a paleomagnetic study, with 29 samples in six different locations. They obtained three different age ranges, all of them belonging to the prehistoric period for the island, with the younger one between 1499 and 1602 AD.

A comprehensive geochronological appraisal of the volcano's stratigraphic sequence was published by Becerril et al. (2016a, b). These authors reviewed all previous geochronological information and provided five new ^{40}Ar-^{39}Ar and one ^{14}C ages, and determined new relative ages based on geomorphological criteria. Three new ages came from Tiñor edifice products, and the other three, including a charcoal sample, from materials of the rift volcanism. The results, regarding the samples of the older part of the island, are consistent with the ages proposed by Guillou et al. (1996). In relation to the samples representing the rift activity, they yielded ages between 2.3 and 22 ka, with the former being close to the 2.5 ka eruption dated by Guillou et al. (1996).

Lately, Risica et al. (2022) presented a new study concerning the Holocene activity on the island, where they combined paleomagnetic and C^{14} datation methods. They obtained ages of eleven subaerial eruptions, some of them for the first time, as in the case of the well-known Lajial eruption in the south part of the island, distributed on the three rift arms. Also, they revealed that the recent volcanic activity does not follow a predetermine pattern and that periods of high eruptive frequency could exist, identifying two of them. Finally, the new data obtained in this work of reconstruction, permit to rearrange the ages of known Holocene eruptions, add new ones, and improve our knowledge concerning the rift volcanism.

2.5 Eruptive Styles, Volcano Types and Volcanic Deposits

Eruptive activity in the emerged stage of El Hierro has been dominantly effusive with frequent generation of lava fields. Consequently, the island edifice hosts a considerable number of monogenetic volcanoes ($\sim$ 220) that include cinder and spatter cones, and tuff rings (the latter related to hydrovolcanism), all of them associated with mild explosive activity, generating dominantly basaltic (sensu latu) eruptive products.

Based on the study of the most recent volcanic record (Holocene period), the main eruptive styles on El Hierro include: hawaiian, strombolian and violent strombolian. The first two are the most common eruptive styles observed on the island, resulting in extensive lava flows fields and spatter and cinder cones made of scoria agglutinates and well-bedded lapilli, scoria and ash, respectively. Most of the lava flows on El Hierro, emplaced from cones located on and off the rift zones, reached the sea forming lava-fed deltas.

Violent strombolian activity, i.e., explosive activity that produced sustained eruption columns up to $\sim$ 10 km high, without reaching the tropopause, and with the dominant clast sizes being ash to lapilli (Valentine 1998; Arrighi et al. 2001; Valentine and Gregg 2008), has also been recognized through the presence of several distal ash deposits on the geological record of the island associated to Tanganasoga Volcano (Fig. 2.1).

Hydromagmatic eruptions include the interaction of magmas with both phreatic and sea water (deep and coastal eruptions). Phreatomagmatic episodes have been described for Ventejís, La Caldereta and Hoya de Fileba (Fig. 2.1), generating rhythmic laminated tuff sequences of coarse juvenile ash and lapilli rich beds with accidental lithic fragments, which also occurred at the interior of the island but in less frequency than those mentioned above (IGME 2010). In addition, some hydromagmatic eruptions occurred along the coast producing tuff ring deposits along the west part of the Island (Becerril 2009; Carracedo et al. 2001; IGME 2010) (Fig. 2.1).

Eruptions related to felsic magmas and producing either trachytic lava flows (Guillou et al., 1996) or trachytic pyroclastic deposits (Pellicer 1977; IGME 2010), have also been described. Noteworthy is the occurrence of a base-surge-type explosive eruption that generated dilute pyroclastic surge deposits covering an area of more than 15 km^2 around the Malpaso area, called Malpaso Member (Fig. 2.1) (Pedrazzi et al. 2014).

In addition to the subaerial volcanism, bathymetric studies (Gee et al. 2001) have revealed that a significant number of well-preserved volcanic cones exist on the submarine flanks of the island, in particular on the continuation of the southern rift, which suggests that significant submarine volcanic activity has also occurred in recent times. As a confirmation of this observation, a submarine eruption occurred in 2011–2012 on the southern rift zone, 2 km off the coast of El Hierro (e.g., López et al. 2012; Martí et al. 2013a, b). For more information, see Chap. 12 (Galindo et al.) of this book.

Regarding the size of the El Hierro eruptions, in terms of magnitude (Pyle 2015) or total volume of erupted material (Dense Rock Equivalent —DRE)*, the largest eruptions occurring during the last growing cycle of the island correspond to volumes of the order of 0.042–0.15 km^3 (Tanganasoga and Mt. del Tesoro, the latter calculated by Rodríguez-González et al. 2012). Tanganasoga represents one of the most important eruptive episodes in the last few thousand years on El Hierro (Fig. 2.1). This event occurred inside the depression of El Golfo along a N–S-oriented fissure, no later than 20 kyrs ago (Carracedo et al. 2001). It resulted in the formation of several cones and emission centres, giving rise to one of the largest volcanic edifices on the island via the accumulation of ankaramitic lavas and pyroclastic deposits (Carracedo et al. 2001) (Fig. 2.1). A minimum total volume of erupted material value is estimated for Mt. Los Cascajos, with 0.0016 km^3. This volcano is located at the NE rift of the island, and forms an eruptive fissure formed by a few vents with N52°E direction (Fig. 2.1). Its age was obtained through the ^{14}C method giving the most recent dated age for the island so far (2.28 ka; Becerril et al. 2016a, b). DRE volume of exposed materials in El Hierro is similar to those assigned to other monogenetic fields, which normally have volumes between 0.0001 and a few km^3 for individual eruptions (e.g., Kereszturi et al. 2013). For example, the erupted volume of magma in the Canary Islands typically ranges from 0.001 to 0.2 km^3 (DRE) (Sobradelo et al. 2011). In the Garrotxa Volcanic Field (Spain) the total volume of extruded magma in each eruption ranges from 0.01 to 0.2 km^3 (DRE) (Bolós et al. 2014). The volumes of basaltic eruptions on Terceira (Açores, Portugal) range in size from 0.1 km^3 to less than 0.001 km^3 (DRE) (Self 1976). In the case of Auckland (New Zealand), monogenetic field volumes are in the range of 0.00007–0.698 km^3 (Kereszturi et al. 2013).

Regarding the Volcanic Explosivity Index (VEI) of El Hierro eruptions, most of VEI values are in the range 0–2, whilst the erupted volume of magma, using mean magma density of 2.8 g/cm^3 and average rock density of 2.44 g/cm^3 (obtained from laboratory analysis of El Hierro samples), for most of the recent eruptions on El Hierro lies within the range 0.0001–0.1 km^3(DRE). DRE calculation was based on the volume of exposed materials (lavas and pyroclastic deposits) noting that total volumes are minimum estimates. By comparing pre- and post-eruption high-resolution bathymetries of the Tagoro eruption the total bulk volume erupted during the submarine eruption of 2011–2013 was estimated at 0.33 km^3 (Rivera et al. 2013).

Footnote: The Volcanic Explosivity Index (VEI) (Newhall and Self 1982) and Dense Rock Equivalent (DRE) derived from the volumetric data of the eruptions were also calculated using the equation: DRE (km^3) = volume of volcanic deposit (km^3) * density of volcanic deposit (kg/m^3) / magma density (kg/m^3).

2.6 Volcano-Structural System

First volcano-structural studies on El Hierro were mainly concerned with depicting general tectonic elements and fractures on the island (Hausen 1964; Coello 1971). Pellicer (1977) developed the first schematic volcano-structural map, including eruptive vents and lineaments. This was further refined and added to by Navarro and Soler (1995) who analysed the island's main volcano-structural elements, paying special attention to the distribution of dykes, particularly inside water galleries. Day et al. (1997) focused on the main fault system of the island, which was also mapped by IGME (2010). Other works have mainly focused on exploring the geologic and tectonic evolution of the island (Carracedo 1996; IGME 2010; Carracedo et al. 2001). Münn et al. (2006) used scaled analogue experiments to reproduce the geometry of rift zones and the unstable flanks observed on El Hierro and to associate them with gravitational volcano spreading. Also, Manconi et al. (2009) suggested that the stress perturbation caused by the formation of the largest and most recent of these flank collapses—the El Golfo landslide—likely influenced the magma plumbing system of the island, leading to the eruption of a higher proportion of denser and less evolved magmas.

The superposition of constructive and destructive episodes has led to a complex internal structure in El Hierro that conditioned occasional changes in the local stress fields. It is possible to partially identify this structure along the direct exposures created by erosion and/or landsliding and in the water galleries or tunnels. The uppermost part of the feeding-system forms a sheet intrusion complex in which inclined sheets (mostly dipping 80–90°) and their surficial expression as eruptive fissures, developed preferentially following NE, W and S trends parallel to the rift systems (Carracedo et al. 2001). Eruptions are then fissural fed mostly by subvertical dykes. Nevertheless, a radial strike distribution, which can be related to constructive episodes, is observed in the on-land structures and also in the submarine eruptive fissures, which are radial with respect to the centre of the island suggesting the existence of a rather uniform stress field during the constructive episodes that was maintained during the growth of the whole island (Becerril et al. 2015). Other volcano-structural features such as faults are also recognizable on and offshore the island.

Volcanic vents and eruptive fissures are the most distinguishable elements on the island and can be also identified on the surrounding submarine areas. Subaerial vents are distributed mainly along the rifts, and only a few of them are located inside the landslide valleys (Fig. 2.1). Submarine eruptive vents are common around the island, except within the submarine landslide valleys, where only a few emission centres have been inferred.

Eruptive fissures defined by the alignments of volcanic vents are mainly observed in the rift zones and on the submarine area. Most have the same trend as their associated rift (NE, W and S) and the majority are related to the last cycle of activity of El Hierro (the past 158 ka), these often being partially buried by lava flows representing portions of originally more extended lineaments (Figs. 2.1 and 2.3). By contrast, the submarine fissures display a radial distribution with respect to the centre of the island (Becerril et al. 2015).

Faults are mainly exposed on the island's NE rift, where a graben system has been identified (IGME 2010; Klimeš et al. 2016). These faults are well exposed with steeply dipping planes striking NE–SW (Day et al. 1997). In the southern and western rifts a few outcropping faults show N–S and E–W strikes, respectively. There are also normal faults running parallel to the scarp of El Golfo, some of which are gaping, that is, open fractures. These latter, however, are not related to regional tectonics but rather associated with the more local sector collapses.

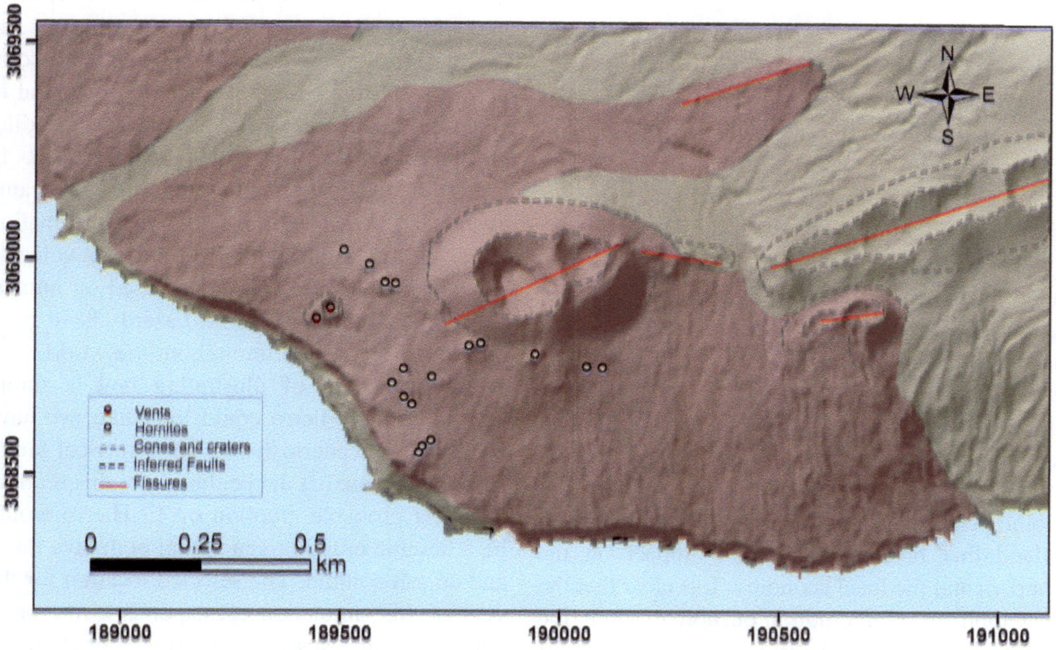

Fig. 2.3 Geomorphological and volcano-tectonic sketch of Orchilla, one of the most recent eruptions on El Hierro

2.7 Future Potentially Hazardous Volcanic Processes

Based on the premise that future eruptions will be similar than past ones, the most likely future volcanic activity on the El Hierro volcanic edifice, both onshore and offshore, will be characterised by mafic eruptions, fed by subvertical dykes that will form eruptive fissures of several hundred to thousand meters length. Their products will be lava flows (pahoehoe or a'a types), with lengths longer than 8 km that most probably will reach the sea, and mean thickness of 3 m. The extension of ballistic projectiles and proximal fallout will be of the order of not more than several km^2, forming small scoria cones (less than 0.1 km^3 DRE). Offshore eruptions are the most expected in the near future, since most of the volcanic edifice is submerged. Another possible scenario, although less probable than the aforementioned off and onshore eruptions, and that is registered in geological times, would be related to a coastal eruption. As it has been described in Sect. 2.5, phreatomagmatic deposits have been identified as phases within strombolian eruptions, being also expected in the future onshore activity.

According to the most common eruptive styles (Hawaiian and Strombolian) and the estimated explosivity of the volcanic activity (VEI: 0–2) on El Hierro, the most likely eruptions on the island are not expected to have a significant impact on a continental or global scale. Nevertheless, the products described by Pedrazzi et al. (2014) on the top of the island were formed by a more vigorous eruption. Therefore, we cannot discard future violent strombolian activity on the island, although these are less probable than the previously-described scenarios.

The volcanic activity of El Hierro, although quite rare at a human time scale, may greatly affect the local socio-economic system, a case in point being the last submarine eruption of the island that brought a negative impact on the tourism and the local economy. It is clear that the development of an eruption onshore or along the coast may have more serious consequences for the local population than the impact of the last submarine eruption.

The repercussions of a possible onshore or coastal eruption could even go beyond the regional scale, affecting the other islands of the Canary Archipelago, as well as the air and maritime traffic between them, and between the islands and the mainland or African Continent. Nevertheless, the intrinsic uncertainty of the volcanic process makes it difficult, in the majority of the volcanic areas, to forecast the exact behaviour of the future eruptions.

2.7.1 Numerical Scenario Simulations of a Likely Future Eruption

In the case of El Hierro, where we have no data about historical activity, all the information about the recent onshore volcanic activity should be based on the geological record. However, this record is affected, since it is a small island with an intensive volcanic activity and its edifices and products suffer the effect of a strong erosion.

El Hierro, like the rest of the archipelago islands, has had a significant growth and progress in terms of population and infrastructures, including an increase of tourism that is the driving force of the economy for the local community. Volcanic eruptions have the potential to generate social or economic impacts, depending on their location, magnitude and duration. In order to reduce the potential impact of a volcanic event, and also to complement the information extracted from the geological record, volcanic risk assessment methodologies based on numerical simulations of volcanic hazards have been developed and tested throughout the world.

With the aim of illustrating how a future eruption on El Hierro could look like, we have computed a scenario based on numerical simulations. As stated at the beginning of this section, the most probable eruption on El Hierro would be a basaltic eruption generating both lava flows and an ash column. The selected location for the vent of this simulation corresponds to the point with the highest susceptibility (i.e., probability of

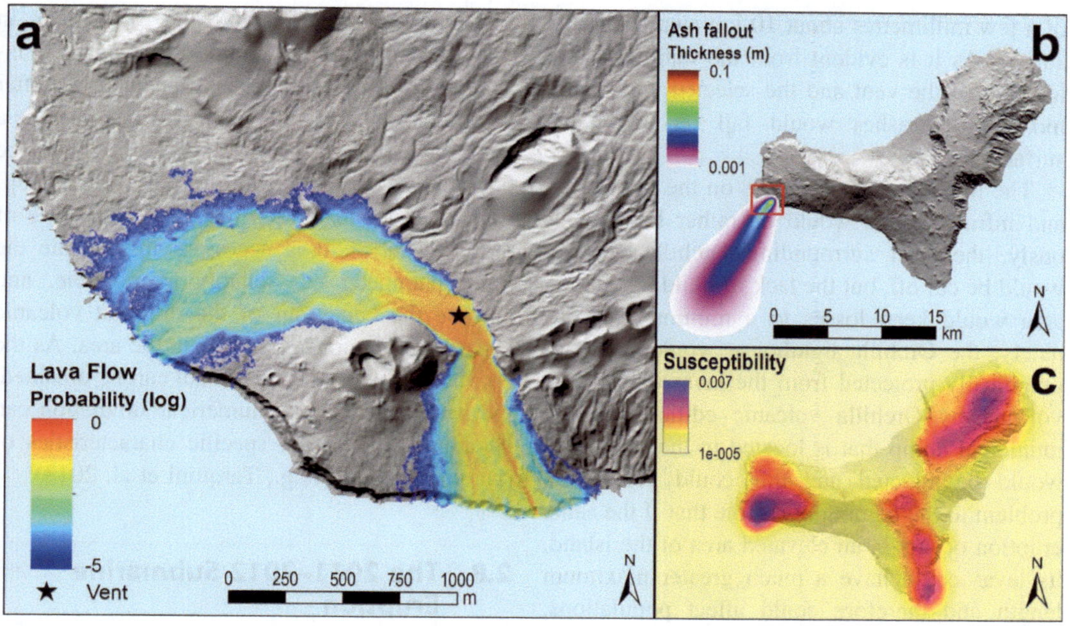

Fig. 2.4 Numerical scenario for a strong strombolian basaltic eruption considering lava flow and ash fallout hazards (see text for details). **a** Probabilistic scenario for lava flow hazard (Note that the probability values are on a logarithmic scale). Black star: location of the vent. **b** Deterministic scenario for ash fallout (see text for details). Red rectangle: area panel (a). **c** Susceptibility map (i.e., probability of hosting a new vent) according to Becerril et al. (2013a, b). Scale of (c) is the same as that of (b)

hosting a future vent) in the island, according to Becerril et al. (2013a, b), which is located on the SW of the island, to the NE of Orchilla volcano (the susceptibility map of El Hierro is shown in Fig. 2.4c). Therefore, we simulated from that vent the areas that could be potentially covered both with lava flows and pyroclastic fallout deposits. For the lava flow simulation, we used the probabilistic model by Felpeto et al. (2001) and included in VORIS tool (Felpeto et al. 2007). The input parameters of this model are the maximum possible length of the lava flow, a height correction parameter, which represents the topographic height that the lava can overflow, a DEM (Digital Elevation model) and the location of the vent(s). Figure 2.4a shows the result obtained after computing 100.000 possible paths, considering a height correction of 3 m (the mean thickness of El Hierro lavas—Sect. 2.7), a maximum length high enough for the lavas to reach the sea and a DEM with a cell size of 5 m. It is observed that two different flows can be developed, one to the S and another to the W, being the former more likely than the latter.

For the ash fallout, we used the deterministic advection-diffusion model included in VORIS tool (Felpeto et al. 2007). The main input parameters of this model are vent location, total volume emitted, grain size distribution, column height and wind field. Figure 2.4b shows the results obtained for the abovementioned vent, a total deposit volume of 0.001 km^3 and a grain size distribution characterized by a mean value of 2ϕ and a standard deviation of 1.5. The value selected for column height was 2.5 km, which can be considered a typical value for a strong strombolian eruption, or slightly higher than M1 eruptions in Mastin et al. (2009). Selected wind field corresponds to a typical spring day, considering the most frequent direction value for that season for heights of 500 m, 1, 2 and 3 km, and their corresponding wind speeds. The results show that the expected deposits have their dispersal axis oriented to the SW, with a thickness of a few millimetres about 10 km away from the source. As it is evident from the map, with the location of the vent and the selected wind field, most of the ashes would fall on the ocean surface.

The impact of this scenario on the population and infrastructures would be rather low. Obviously, the road surrounding Orchilla volcano would be cut off, but the lack of buildings in this area would keep losses to a minimum. Fortunately, the Orchilla lighthouse would be topographically protected from the lavas of the new volcano by Orchilla volcanic edifice, but the municipal dump that is located in the same area would be affected and this could generate a problem to the community. Note that if the same eruption occurs in an elevated area of the island, its lavas could have a much greater maximum length and therefore could affect populations, especially if the vent would be located in the NE Rift. Two recent examples of the destructive forces of these lava flows in the Canaries are the 1706 eruption that destroyed part of the village and isolated the harbour of Garachico in Tenerife, during the first days of the eruption, and the recent 2021 La Palma eruption which buried almost 1676 buildings and 73.8 km of roads. In the same case, the ash fall, depending on the wind direction during the eruption, could affect any of the major towns of the island and also the airport, causing its closure until the removal of the ash.

In the last century there have been similar eruptions in other islands of the archipelago (Chinyero 1909, in Tenerife and San Juan, 1949, Teneguía, 1971 and the 2021 eruption in La Palma) and although they only had short phases of violent activity, it was reported that the pyroclastic fallout products affected the neighbour islands (Olave Lacalle 2014; Di Roberto et al. 2016). Obviously, the activity did not represent a serious threat for the inhabitants, concerning the demographic data of the period, or the facilities and the existing critical structures, but the situation is very different nowadays as it has been proved in the 2021 eruption of La Palma.

Numerical scenarios like the one shown in Fig. 2.4 can be of great importance in the management of a volcanic crisis, especially just after the opening of the vent, when they can provide a rapid assessment of the areas that will be potentially covered by the products of the eruption. In order to have a realistic perspective of those scenarios, it is necessary to evaluate the input parameters as reliably as possible, and analyse the behaviour of the different volcanic hazards in previous eruptions in the area. As the eruption evolves and more data can be obtained, the parameters for the numerical simulation can be adjusted to fit the specific characteristics of ongoing eruption (e.g., Tarquini et al. 2018).

2.8 The 2011–2012 Submarine Eruption

During the historical period of El Hierro (last 600 years) there is no reliable proof of neither onshore nor offshore volcanism, unlike La Palma, where 7 eruptions took place during that period, being the last one the 2021 eruption in Cumbre Vieja volcanic area. Despite this fact, the island was always a potential candidate for further eruptions on the archipelago, due to its volcanic/morphological youth and the well-developed rift system that it presents. Studies on volcanic hazard assessment for the islands (Sobradelo et al. 2011) provided that the probability of a new eruption was not so low for the islands, estimated a return period of 25–30 year.

In the early hours of the 10th of October 2011, after almost 3 months of precursory activity (thousands of seismic events and several centimetres of ground deformation), a submarine eruption started (López et al. 2012). The vent was located 2 km far from the southern coast of the island at a depth of 350 m. This eruption was the first one to be completely monitored by the Geographical Spanish Institute (IGN) network and probably one of the few submarine eruptions in the world that have generated datasets of almost any technique in volcano surveillance. The eruption lasted 5 months, and during that period different phases of activity and diverse products were emitted.

In the opening phase of the eruptive activity, gas emissions generated a green discolouration of seawater, which turned brown a couple of days later (Fig. 2.5b), due to the violent degassing carrying away sediments from the ocean floor, with a continuous bubbling. Concerning the eruptive materials, two types of solid products—apart from the ash emissions—were observed.

During the initial stage of the eruption, and mainly in the course of an intensive degassing episode on the 15th of October (Troll et al. 2012; Sigmarsson et al. 2013; Meletlidis et al. 2012), lava bombs on a decimetre scale, characterised by basanitic crusts and white to grey coloured interiors, were emitted (Fig. 2.5c, d). Due to their high percentage of vesicles contained in these fragments, various days later, they were still floating and drifting on the ocean. At the end of October, a new type of volcanic material was emitted, this time entirely basanitic fragments, also highly vesiculated and often with hollow interiors. These fragments, known as "lava balloons", widely exceeded 1 m in size (Fig. 2.5e, f), and were emitted frequently for the next 4 months of the eruption.

Although both type of materials caused a sensation and controversy in the local community, this fact is not that unusual if we bear in mind the bibliographic references and the proposed models for the Ocean Island Basalts (OIBs) and in particular, those for the young Canary Islands, such as La Palma and El Hierro (Klügel et al. 2000; Galipp et al. 2006; Stroncik et al. 2009). Diverse processes take place, such as crustal assimilation, fractional crystallization or mingling and mixing due to subsequently new intrusions. In any case, these processes lead to a change in the composition of these magmatic bodies, towards more evolved members.

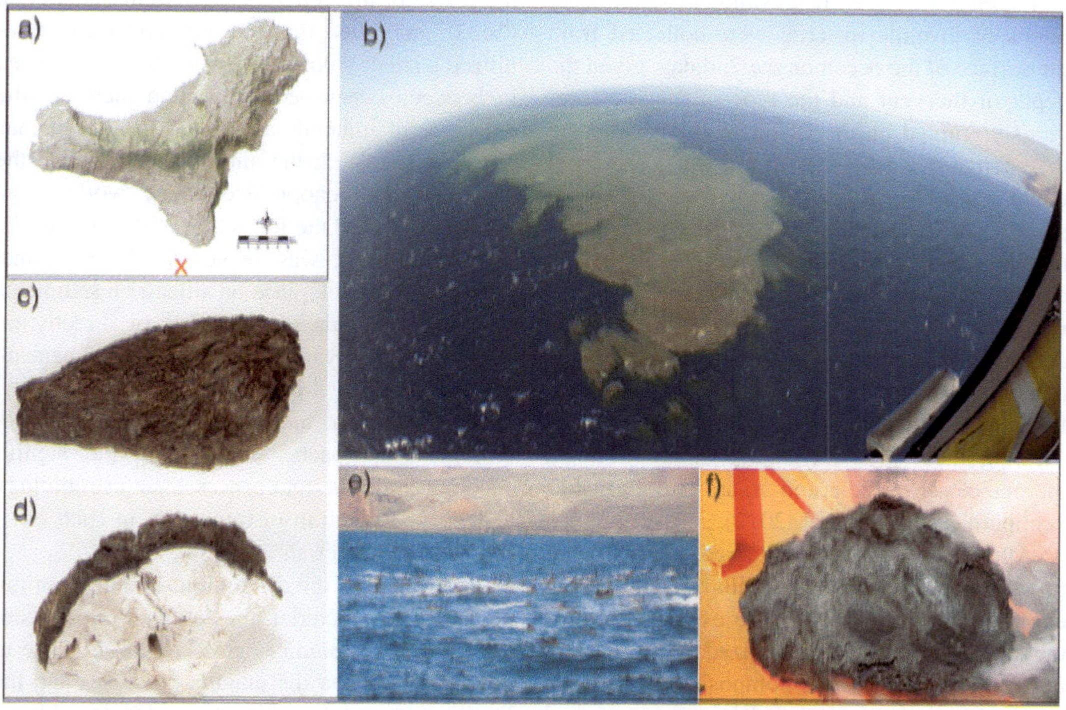

Fig. 2.5 **a** Map of El Hierro with the location of the submarine vent (red cross); **b** aerial photo of the eruption during the first phase; **c** and **d** xenopumice ejected during the 15 October episode (~ 0.20 m); **e** emission of lava balloons and **f** a lava balloon on the vessel (~0.7 m of diameter). All photos courtesy of Instituto Geográfico Nacional

Over the different stages of a volcano evolution, many of the magmatic intrusions beneath the volcanic edifices will not lead to an eruption, converting them to stalled bodies, defining these events as a "failed eruption" (Moran et al. 2011). These episodes occur in almost any tectonic settings, and some of them have been reported in monogenetic volcanism (Gardine et al. 2011; Calais et al. 2008). In El Hierro, following up the 2011–2012 eruption, six post-eruption intrusions took place (Benito-Saz et al. 2017; Domínguez Cerdeña et al. 2018). These magmatic batches migrated from the upper mantle and were trapped in discontinuities in the lower crust, giving birth to sill-like bodies (Stroncik et al. 2009).

2.8.1 On the Origin of the Xenopumices

The emission on the 15th of October took the scientific community by surprise, as the eruption had started 5 days earlier. Although small specimens of juvenile material were collected from the surface of the ocean on the 2nd day, given the depth of the vent and the type of magmas previously erupted on the island, nothing indicated that a massive emission of magmatic material could reach the surface. Furthermore, the composition, texture and shape of these fragments together with the vividness of the degassing episode, produced discussions concerning the magmas involved, the explosivity of the eruption and as a result the level of risk for the habitants (Carey et al. 2014; White et al. 2015). These floating fragments, commonly referred to as "restingolites" after the village of La Restinga, the one closer to the eruption, are also known as xenopumices (Troll et al. 2012).

All the researchers that studied these peculiar products were in agreement with reference to the composition and origin of the external crust of the fragments, and characterized it as juvenile magma, with a basanitic composition. Problems arise when it comes to explain the nature and origin of the glassy, vesicular, and light in colour material, wrapped and melted by the basanitic magma on its way to the surface. While there are a lot of works regarding possible sources and origin, the most plausible hypothesis described them as derived from fragments of marine sediments or from relatively old remobilized volcanic material.

Troll et al. (2012), based on textures, composition, minerals paragenesis, the absence of certain igneous trace element signatures and oxygen isotope values, suggested that this material match the pre-island sedimentary layers. In a subsequent work, Meletlidis et al. 2012, favoured an igneous source, and based also on the textures and chemistry of the material, distinguished two different types of xenopumices, grey and white, depending on the initial composition of the old melted volcanic rock. Additionally, they have employed a thermodynamic model to explain the set of the minerals identified in the material and proposed a model for the magma ascent, including the different events of bubble nucleation and growth, in the total mass of the fragments. As if it was a unified model for the origin of the xenopumices, Sigmarsson et al. (2013), suggested that the basanitic melt remobilized a small volume of a rhyolitic melt trapped in between ocean sediments, and therefore the existence of minerals associated to non-igneous environment. With the aim to shed light on the origin of the xenopumices more works were published, where the formation of the core of the emitted material was related to hydrothermal processes that took place on silica-rich sediments (Rodríguez-Losada et al. 2014) or directly on trachytic to trachyandesitic rocks located beneath the island (Del Moro et al. 2015). Finally, xenopumices also had been studied in an astrobiological approach, emphasizing the significance of hydrothermal processes in astrobiology and in the evolution of future life in such environments (Lalla et al. 2019).

Regardless of the discussions about the origin of this foamy material, the research on the genesis of the xenopumices gave a fresh impulse to our knowledge on the magma processes and the inner structure of the insular volcanic edifices. The interpretations provide clues on the migration of the magma, mingling and mixing processes, interaction with the hosted rocks, and provide a geological frame where the data of the monitoring network could be fitted in. In the two most recent eruptions in the archipelago (La Palma, 1949 and 1971), "exotic" xenoliths were observed (Araña and Ibarrola 1971; Klügel et al. 1999), thus it was not the first time that products with a similar composition were emitted, yet the El Hierro case attracted the attention of the scientists and Civil Protection, as this emission was associated with a higher probability of an explosive event (Carracedo et al. 2012).

2.8.2 Lava Balloons Emissions

From 31 October and throughout the next 4 months, a new type of material started to emerge on the ocean surface. Their size, regularly more than 1 m diameter, didn't prevent them from floating for some minutes before they sank. Once again there was something unusual in the process, but on this occasion, it was the physical properties of the fragments and not the composition. Nor was it the first time of such kind of emission. In the annals of volcanology, emission of lava balloons was reported at least five times before the 2011–2012 El Hierro eruption, during submarine basaltic eruptions; 1877, Hawaii, USA (Moore et al. 1985), 189, Pantelleria island, Italy (Riccò 1892; Conte et al. 2014); 1994, Socorro island, Mexico (Siebe et al. 1995); 1998 and 2001, at Serreta, off Terceira Island, Portugal (Forjaz et al. 2000; Gaspar et al. 2003).

The lava balloons had an almost ellipsoidal shape, up to 2 m long and 1 m in diameter, with an outer porous basanitic shell of a few centimetres in thickness, that enclose one or more gas-filled cavities. Once on the sea surface, although the cracks due to the fast cooling permitted the entrance of the seawater, the high temperature of the basanitic crust (> 400 °C) made the infiltrating water to evaporate and thus maintain the floatability although the gasses had escaped through the same cracks. With the surrounded water cooling down the lava balloon and filling its cavity, the increment of the bulk density made them sink.

It is well known that during an eruption, the petrology and geochemistry of the eruptive products could change, indicating inner processes and interactions with other magmas or rocks and providing valuable information about the depth of the emerging magma. This data could be used as an indicator for the future development of the eruption, its explosivity and durability (Innocenti et al. 2005; Rowe et al. 2011; Albert et al. 2016). Thus, apart from the peculiar type of the eruptive products, the main consideration was to identify any changes in the erupted magma and relate the emission pulses with the data available from the monitoring network, as the direct observation of the eruption, apart from a bathymetric map, was impossible.

Martí et al. (2013a) combined the available data and proposed a model to explain the magma movement and the mechanisms that controlled the eruption. They analysed the temporal evolution of geophysical and petrological data, through the unrest and eruptive episodes, with the aim to obtain possible patterns that would be useful in eruption forecasts in the Canary Islands. In a second work (2013b), Martí and colleagues, by means of petrographic, geochemical and experimental petrology in conjunction with thermodynamic and diffusion modelling on the emitted products, identified major physico-chemical variations which, through correlation with the geophysical changes, allowed them to distinguish two main eruptive episodes and define the characteristics of the erupted magma. Based on the registered seismicity and applying diffusion chronometry to zoned olivine crystals, Longpré et al. (2014), sustained that magma mixing took place during the pre-eruptive seismicity and continued even after the eruption onset. Later, in a comprehensive and multidisciplinary approach to the eruption, improving previous interpretations, Meletlidis et al. (2015), determined 6 pre-eruptive and 3 syn-eruptive stages, using seismic data, ground deformation patterns, gravity data and petrological criteria. They proposed a model of three magmatic environments and employing thermodynamic modelling they defined the path for the magma ascent.

The results obtained from these petrological studies are in good agreement with previous proposed models for the plumbing system beneath El Hierro (Stroncik et al. 2009), although the most relevant additional lesson is the conjointly interpretation of the eruptive event, together with the data acquired by the monitor network. This aspect was also evident in other works, where the geophysical and geodetic data was interpreted taking into account the results of petrological studies (González et al. 2013; Becerril et al. 2013a, b; García et al. 2016; Benito-Saz et al. 2017).

Therefore, the 2011–2012 submarine eruption underlined the need of an interdisciplinary endeavour in monitoring an unrest process, and if any, the subsequent eruption, and analysing any dataset obtained as a whole. Also, an aspect to be considered is the necessity of previous studies of the volcanic areas so any new data could use them as a reference.

2.9 Present and Future Pending Actions on the Island

Effusive and low explosive basaltic monogenetic activity prevails on the island, especially in the current stage of the rift volcanism. This fact together with the lack of documented eruptive events in historical times has led to an underestimation of the potential impacts of volcanism on the island, despite its low frequency and risk.

Volcanological studies on the island began nearly 150 years ago, but the 2011–2012 eruption significatively marked a breakthrough in the knowledge of the island's eruptive activity, promoting the development of several scientific papers regarding almost every field of modern volcanology. However, much work needs to be done in order to characterize, in a comprehensive way, the recent volcanic activity of the whole volcanic edifice, including the subaerial and submarine eruptions.

There is no doubt that the volcanic studies of El Hierro should be addressed and compared in a regional scale. One of the most pending actions on the island would be to widen the database in the fields of petrology, geochemistry and geochronology, including more samples from the submarine structures and deposits to better understand the eruptive dynamics and to accurately calculate the eruptive frequency. New physical volcanology studies would allow us to obtain valuable data of the eruptive dynamics and to better understand the deposit mechanisms and phase changes during an eruption. All these data would be useful as inputs for more refining models of the future activity on El Hierro (and the rest of the islands as this type of eruptions is a common feature), but also, they could be the base for a more accurate monitoring network design.

There are currently some factors that make this task arduous. Firstly, El Hierro is a small island in which to evaluate the extension and volume of volcanic products, such as pyroclasts and ash fallout (especially those associated to the most violent eruption phases) or lava flows, is hard since these are commonly deposited offshore. Additionally, the existence of large landslides on almost any flank of the island have resulted in the loss of high material volumes and geological structures (e.g., vents), making it difficult to obtain a complete picture of the spatial and temporal distribution of the activity. Finally, the long-lasting agricultural tradition of the island and the use of the volcanic material as primary material for the development, contributed to severe landscape alteration.

The 2011–2012 eruption not only had direct effects on the scientific community, providing benefits and pushing forward the research on volcanism, but also on the local community, the authorities and the emergency managers, who could also identify this moment as an inflection point. The impact that a natural phenomenon had on their lives and on the local and regional economy was revised and it became apparent that only a close cooperation of these three involved parties—scientists, emergency managers and community—could further improve the response to such events. The fast densification of the monitoring network and the continuous surveillance in real-time coordination with the first application of the special emergency plan of Civil Protection regarding volcano risk in the

Canary Islands (PEVOLCA), and the collaborative behaviour of the habitants, resulted in the best possible outcome.

There are many future pending tasks that would contribute not only to enhance the scientific knowledge about the island but also to provide a fundamental pillar on which to base the management of the future volcanic activity. These would result in minimizing the physical, economic and social impacts, that is, reducing the volcanic risk. The most important ones would be: (1) to more comprehensively research monogenetic volcanism and compare it with other different ocean volcano islands; (2) to enhance the monitoring networks; (3) to improve the legal framework targeted to respond in an emergency and raise awareness of the population regarding phenomena that would be likely repeated in the future.

References

Abdel-Monem A, Watkins ND, Gast PW (1972) Potassium-argon ages, volcanic stratigraphy and geomagnetic polarity history of the Canary Islands: Tenerife, La Palma and Hierro. Am J Sci 272:805–825

Abratis M, Schmincke H, Hansteen T (2002) Composition and evolution of submarine volcanic rocks from the central and western Canary Islands. Int J Earth Sci 91:562–582

Acosta J, Uchupi E, Muñoz A, Herranz P, Palomo C, Ballesteros M, Group ZEE (2003) Geophysics of the Canary Islands. Mar Geophys Res 24:1–40. https://doi.org/10.1007/s11001-004-1513-3

Albert H, Costa F, Martí J (2016) Years to weeks of seismic unrest and magmatic intrusions precede monogenetic eruptions. Geology 44(G37239):1. https://doi.org/10.1130/G37239.1

Anguita F, Hernán F (2000) The Canary Islands origin: a unifying model. J Volcanol Geotherm Res 103:1–26. ISSN: 0377-0273

Araña V, Ibarrola E (1973) Rhyolitic pumice in the basaltic pyroclasts from the 1971 eruption of Teneguía volcano, Canary Islands. Lithos 6:273–278

Arrighi S, Principe C, Rosi M (2001) Violent Strombolian and subplinian eruptions at Vesuvius during post-1631 activity. Bull Volcanol 63:126–150

Aulinas M, Domínguez D, Rodríguez-González A, Carmona H, Fernández-Turiel JL, Pérez-Torrado FJ, Carracedo JC, Arienzo I, D'Antonio M (2019) The Holocene volcanism at El Hierro: insights from petrology and geochemistry. Geogaceta 65:35–38

Becerril L, Galindo I, Martí J, Gudmundsson A (2015) Three-armed rifts or masked radial pattern of eruptive fissures? The intriguing case of El Hierro volcano (Canary Islands). Tectonophysics 647:33–47

Becerril L, Galve JP, Morales JM, Romero C, Sánchez N, Martí J, Galindo I (2016a) Volcano-structure of El Hierro (Canary Islands). J Maps 12(sup1):43–52. https://doi.org/10.1080/17445647.2016.1157767

Becerril L, Ubide T, Sudo M, Martí J, Galindo I, Gale C, Morales de Francisco J, Yepes J, Lago San José M (2016b) Geochronological constraints on the evolution of El Hierro (Canary Islands). J Afr Earth Sc 113:88–94. https://doi.org/10.1016/j.jafrearsci.2015.10.012

Becerril L, Capello A, Galindo I, Neri M, Del Negro C (2013a) Spatial probability distribution of future volcanic eruptions at El Hierro Island (Canary Islands, Spain). J Volcanol Geotherm Res 257

Becerril L, Galindo I, Gudmundsson A, Morales JM (2013b) Depth of origin of magma in eruptions. Sci Rep 26(3):2762

Becerril L (2009) Approach to volcanic hazard and its effects in coastal areas of the Canary Islands. Master's thesis, Universidad de Las Palmas de Gran Canaria, Spain. Retrieved from http://hdl.handle.net/10553/4595

Benítez Padilla S (1945) Ensayo de síntesis geológica del Archipiélago Canario. Publicaciones de El Museo Canario, Las Palmas

Benito-Saz MA, Parks M, Sigmundsson F, Hooper A, García-Cañada L (2017) Repeated magmatic intrusions at El Hierro Island following the 2011–2012 submarine eruption. J Volcanol Geotherm Res. ISSN: 0377-0273. https://doi.org/10.1016/j.jvolgeores.2017.01.020

Blanco-Montenegro I, Montesinos F, Arnoso J (2018) Aeromagnetic anomalies reveal the link between magmatism and tectonics during the early formation of the Canary Islands. Sci Rep 8. https://doi.org/10.1038/s41598-017-18813-w

Bolós X, Planagumà L, Martí J (2014) Volcanic stratigraphy of the quaternary La Garrotxa volcanic field (NE Iberian Peninsula). J Quat Sci 29(6):547–560

Calais E, d'Oreye N, Albaric J, Deschamps A, Delvaux D, Déverchère J et al (2008) Strain accommodation by slow slip and dyking in a youthful continental rift, East Africa. Nature 456(7223):783–787

Carey S, Katherine B, Sparks S, Stinton A, Ausubel J et al. (2014) Impact of volcanic eruptions on the seafloor around Montserrat, West Indies. Oceanography (Washington D.C.) 27:37–37

Carracedo JC (1996) Morphological and structural evolution of the western Canary Islands: hotspot-induced three-armed rifts or regional tectonics trends? J Volcanol Geotherm Res 72:151–162

Carracedo JC (1999) Growth structure instability and collapse of Canarian volcanoes and comparisons with Hawaiian volcanoes. J Volcanol Geothermal Res 94(1–4):1–19. https://doi.org/10.1016/S0377-0273(99)00095-5

Carracedo JC, Day S, Guillou H, Rodríguez Badiola E, Canas JA, Pérez Torrado FJ (1998) Hotspot volcanism close to a passive continental margin: the Canary Islands. Geol Mag 135(5):591–601i

Carracedo JC, Day S, Guillou H (1999) Giant quaternary landslides and the evolution of La Palma and El Hierro, Canary Islands. J Volcanol Geotherm Res 94:169–190

Carracedo JC, Badiola E, Guillou H, Nuez J, Pérez Torrado FJ (2001) Geology and volcanology of La Palma and El Hierro, W Canaries. Estud Geol 57:1–124

Carracedo JC, Pérez Torrado FJ, González AR, Soler V, Turiel JLF, Troll VR, Wiesmaier S (2012) The 2011 submarine volcanic eruption in El Hierro (Canary Islands). Geol Today 28(2):53–58

Carracedo JC, Day S, Guillou H, Pérez Torrado JF (1997) El Hierro geological excursion guidebook. In: International workshop, volcanism and volcanic hazards in immature intraplate oceanic islands. La Palma, Canary Islands

Carracedo JC, Troll VR, Zaczek K, Rodríguez-González A, Soler V, Deegan FM (2015) The 2011–2012 submarine eruption off El Hierro Canary Islands: New lessons in oceanic island growth and volcanic crisis management. Earth-Sci Rev 150168-200. https://doi.org/10.1016/j.earscirev.2015.06.007

Clarke H, Troll V, Carracedo JC (2009) Phreatomagmatic to Strombolian eruptive activity of basaltic cinder cones: Montaña Los Erales, Tenerife, Canary Islands. J Volcanol Geotherm Res 180:225–245. https://doi.org/10.1016/j.jvolgeores.2008.11.014

Coello J (1971) Contribución a la tectónica de la isla de El Hierro. Estud Geol 27:335–340

Conte AM, Martorelli E, Calarco M, Sposato A, Perinelli C, Coltelli M, Chiocci FL (2014) The 1891 submarine eruption offshore Pantelleria Island (Sicily Channel, Italy): identification of the vent and characterization of products and eruptive style. Geochem Geophys Geosyst 15:2555–2574. https://doi.org/10.1002/2014GC005238

Day S, Carracedo JC, Guillou H (1997) Age and geometry of an aborted rift flank collapse: The San Andrés fault, El Hierro, Canary Islands. Geol Mag 143 (4):523–537

Day S, Pearson G, Macpherson C, Lowry D, Carracedo JC (2010) Evidence for distinct proportions of subducted oceanic crust and lithosphere in HIMU-type mantle beneath El Hierro and La Palma, Canary Islands. Geochim Cosmochim Acta 74:6565–6589. https://doi.org/10.1016/j.gca.2010.08.021

Domínguez Cerdeña I, García-Cañada L, Benito-Saz MA, del Fresno C, Lamolda H, Pereda de Pablo J, Sánchez Sanz C (2018) On the relation between ground surface deformation and seismicity during 2012–2014 successive magmatic intrusions at El Hierro Island. Tectonophysics 744:422–437. ISSN: 0040-1951

Felpeto A, Araña V, Ortiz R, Astiz M, García A (2001) Assessment and modelling of lava flow hazard on Lanzarote (Canary Islands). Nat Hazards 23:247–257. https://doi.org/10.1023/A:1011112330766

Felpeto A, Martí J, Ortiz R (2007) Automatic GIS-based system for volcanic hazard assessment. J Volcanol Geotherm Res 166:106–116

Fernández Navarro L (1908) Observaciones geológicas en la isla de Hierro (Canarias). Mem Sociedad Española de Historia Natural, Tomo V Madrid

Fernández Navarro L (1924) Datos sobre el volcanismo canario. Bull Volcanol (2). Napoli

Fernández Navarro L (1926) Iles Canaries Exc. A7. Congr Géol Int. Madrid

Forjaz VH, Rocha FM, Medeiros JM, Menezes L, Sousa C (2000) Notícias sobre o Vulcão Oceánico da Serreta, Ilha Terceira dos Açores. Observatório Vulcanológico e Geotérmico dos Açores, Publicação, no 10, p 41

Fúster JM, Hernán F, Cendrero A, Coello J, Canragrel JM, Ancochea E, Ibarrola E (1993) Geocronología de la isla de El Hierro (Islas Canarias). Bol R Soc Esp Hist Nat (Sec Geol) 88(1–4):85–97

Gagel C (1910) Die mittelatlantischen Vulkaninseln. In: Handbuch der regionalen Geologie, VI1 Band ,10 Abteilung E. Heidelberg

Galipp K, Klügel A, Hansteen T (2006) Changing depths of magma fractionation and stagnation during the evolution of an oceanic island volcano: La Palma (Canary Islands). J Volcanol Geotherm Res 155:285–300

García A, Cruz-Reyna S, Marrero J, Ortiz R (2016) Short-term volcano-tectonic earthquake forecasts based on a moving mean recurrence time algorithm: the El Hierro seismo-volcanic crisis experience. Nat Haz Earth Syst Sci 16:1135–1144

Gardine M, West ME, Cox T (2011) Dyke emplacement near Parícutin volcano, Mexico in 2006. Bull Volcanol 73:123–132

Gaspar JL, Queiroz G, Pacheco JM, Ferreira T, Wallenstein N, Almeida MH, Coutinho R (2003) Basaltic lava balloons produced during the 1998–2001 Serreta Submarine ridge eruption (Azores). In: White JDL et al. (eds) Explosive subaqueous volcanism. AGU, Washington, pp 205–212

Gee MJR, Watts AB, Masson DG, Mitchell NC (2001) Landslides and the evolution of El Hierro in the Canary Islands. Mar Geol 177(3/4):271–293. https://doi.org/10.1016/S0025-3227(01)00153-0

González PJ, Samsonov SV, Pepe S, Tiampo KF, Tizzani P, Casu F, Fernández J, Camacho AG, Sansosti E (2013) Magma storage and migration associated with the 2011–2012 El Hierro eruption: implications for crustal magmatic systems at oceanic island volcanoes. J Geophys Res: Solid Earth 118. https://doi.org/10.1002/jgrb.50289

Guillou H, Carracedo JC, Pérez Torrado F, Rodríguez Badiola E (1996) K-Ar ages and magnetic stratigraphy of a hotspot-induced, fast grown oceanic island: El Hierro, Canary Islands. J Volcanol Geotherm Res 73:141–155

Hansteen T, Andersen T, Jelsma H (1991) Fluid and silicate glass inclusions in ultramafic and mafic xenoliths from Hierro, Canary Islands: implications for mantle metasomatism. Contrib Miner Petrol 107:242–254. https://doi.org/10.1007/BF00310710

Hausen H (1964) Rasgos geológicos generales de la isla del hierro. Anuario Estud Atlánticos 10:547–593

Hernández-Pacheco A (1982) Sobre una posible erupción en 1793 en la isla de El Hierro (Canarias). Estud Geol 38:15–25

Holcomb RT, Searle RC (1991) Large landslides from oceanic volcanoes. Mar Geotechnol 10(1–2):19–32

IGME (2010) Mapas Geológicos de España, Escala 1:25.000. Isla de El Hierro. Hojas 1105-II (Valverde), 1105-III (Sabinosa), 1105-IV (Frontera), 1108-I/II (La Restinga)

Innocenti S, Furman T, Voight B, Supriyati A (2005) Geochemical, textural and petrographic indicators for eruptive behavior at Merapi Volcano, Indonesia. AGU FallMeeting Abts

Jeffery AJ, Gertisser R (2018) Peralkaline felsic magmatism of the Atlantic Islands. Front Earth Sci 6:145. https://doi.org/10.3389/feart.2018.00145

Jeremine E (1935) Contribution a l'étude des Iles Hierro et Gomera (Archipel Canarien). Bull Soc Fr Minér. París

Kereszturi G, Németh K, Cronin SJ, Agustín-Flores J, Smith IEM, Lindsay J (2013) A model for calculating eruptive volumes for monogenetic volcanoes—implication for the Quatemary Auckland Volcanic Field, New Zealand. J Volcanol Geotherm Res 266:16–33

Klimeš J, Yepes J, Becerril L, Kusák M, Galindo I, Blahut J (2016) Development and recent activity of the San Andres landslide on el Hierro, Canary Islands, Spain. Geomorphology 261:119–131

Klügel A, Schmincke HU, White JD, Hoernle KA (1999) Chronology and volcanology of the 1949 multi-vent rift-zone eruption on La Palma (Canary Islands). J Volcanol Geotherm Res 94(1–4):267–282

Klügel A, Hoernle KA, Schmincke HU, White JDL (2000) The chemically zoned 1949 eruption on La Palma (Canary Islands): Petrologic evolution and magma supply dynamics of a rift zone eruption. J Geophys Res 105:5997–6016. https://doi.org/10.1029/1999JB900334

Klügel A, Hansteen T, Galipp K (2005) Magma storage and underplating beneath Cumbre Vieja volcano, La Palma (Canary Islands). Earth Plan Sci Lett 236(1–2):211

Klügel A, Longpré MA, García-Cañada L, Stix J (2015) Deep intrusions, lateral magma transport and related uplift at ocean island volcanoes. Earth Plan Sci Lett 431:140–149

Von Knebel W (1906) Studien zur Oberflächengestaltung der Inseln Palma und Hierro "Globus". Band 89 Braunschweig

Lalla E, Sanz Arranz A, Lopez-Reyes G, Cote K, Daly M, Konstantinidis M, Rodríguez-Losada JA, Groemer G, Medina J, Martínez-Frías J, Rull F (2019) A micro-Raman and X-ray study of erupted submarine pyroclasts from El Hierro (Spain) and its' astrobiological implications. Life Sci Space Res 21. https://doi.org/10.1016/j.lssr.2019.04.003

Longpré MA, Troll VR, Walter TR, Hansteen TH (2009) Volcanic and geochemical evolution of the Teno massif, Tenerife, Canary Islands: some repercussions of giant landslides on ocean island magmatism. Geochem Geophys Geosyst (G3) 10:Q12017

Longpré MA, Chadwick J, Wijbrans JR, Iping R (2011) Age of the El Golfo debris avalanche, El Hierro (Canary Islands): new constraints from laser and furnace 40Ar/39Ar dating. J Volcanol Geotherm Res 203. https://doi.org/10.1016/j.jvolgeores.2011.04.002

Longpré MA, Klügel A, Diehl A, Stix J (2014) Mixing in mantle magma reservoirs prior to and during the 2011–2012 eruption at El Hierro, Canary Islands. Geology. https://doi.org/10.1130/G35165.1

López C, Blanco MJ et al (2012) Monitoring the volcanic unrest of El Hierro (Canary Islands) before the onset of the 2011–2012 submarine eruption. Geophys Res Lett 39:L13303

Lundstrom CC, Hoernle K, Gill J (2003) U-series disequilibria in volcanic rocks from the Canary Islands: plume versus lithospheric melting. Geochim Cosmochim Acta 67:4153–4177

Maccaferri F, Richter N, Walter TR (2017) The effect of giant lateral collapses on magma pathways and the location of volcanism. Nat Commun 8:1097

Manconi A, Longpré MA. Walter TR, Troll V, Hansteen T (2009) The effects of flank collapses on volcano plumbing systems. Geology 37:1099–1102. https://doi.org/10.1130/G30104A.1

Martí J, Castro A, Rodríguez C, Costa F, Carrasquilla S, Pedreira R, Bolos X (2013a) Correlation of magma evolution and geophysical monitoring during the 2011–2012 El Hierro (Canary Islands) submarine eruption. J Petrol 54:1349–1373

Martí J, Pinel V, López C, Geyer A, Abella R, Tárraga M, Blanco MJ, Castro A, Rodríguez C (2013b) Causes and mechanisms of the 2011-2012 El Hierro (Canary Islands) submarine eruption. J Geophys Res: Solid Earth 118:823–839. https://doi.org/10.1002/jgrb.50087

Masson DG (1996) Catastrophic collapse of the volcanic island of Hierro 15 ka ago. Geology 24:231–234

Masson DG, Watts A, Gee M, Urgeles R, Mitchell N, Le Bas T, Canals M (2002) Slope failure on the flanks of the western Canary Islands. Earth Sci Rev 57(1–35):1

Mastin LG et al (2009) A multidisciplinary effort to assign realistic source parameters to models of volcanic ash-cloud transport and dispersion during eruptions. J Volcanol Geotherm Res 186(1–2):10–21

Meletlidis S, Di Roberto A, Pompilio M, Bertagnini A, Iribarren I, Felpeto A, Torres P, D'Oriano C (2012) Xenopumices from the 2011–2012 submarine eruption of El Hierro (Canary Islands, Spain): constraints on the plumbing system and magma ascent. Geophys Res Lett 39

Meletlidis S, Di Roberto A, Domínguez Cerdeña I, Pompilio M, García-Cañada L, Bertagnini A, Benito-

Saz MA, Del Carlo P, Sainz-Maza Aparicio S (2015) New insight into the 2011–2012 unrest and eruption of El Hierro Island (Canary Islands) based on integrated geophysical, geodetical and petrological data. Ann Geophys 58. https://doi.org/10.4401/ag-6754

Moore JG, Fornari DJ, Clague DA (1985) Basalts from the 1877 submarine eruption of Mauna Loa, Hawaii; new data on the variation of palagonitization rate with temperature (Report)

Moran S, Newhall C, Roman D (2011) Failed magmatic eruptions: late-stage cessation of magma ascent. Bull Volcanol 73:115–122. https://doi.org/10.1007/s00445-010-0444-x

Del Moro S, Di Roberto A, Meletlidis S, Pompilio M, Bertagnini A, Agostini S, Ridolfi F, Renzulli A (2015) Xenopumice erupted on 15 October 2011 offshore of El Hierro (Canary Islands): a subvolcanic snapshot of magmatic, hydrothermal and pyrometamorphic processes. Bull Volcanol 77. https://doi.org/10.1007/s00445-015-0940-0

Münn S, Walter TR, Klügel A (2006) Gravitational spreading controls rift zones and flank instability on El Hierro, Canary Islands. Geol Mag 143:257–268

Navarro A, Soler C (1995) El agua en El Hierro. Cabildo Insular de El Hierro, Valverde, p 97

Neumann ER (1991) Ultramafic and mafic xenoliths from Hierro, Canary Islands: evidence for melt infiltration in the upper mantle. Contrib Miner Petrol 106:236–252

Neumann ER, Wulff-Pedersen E (1997) the origin of highly silicic glass in mantle xenoliths from the Canary Islands. J Petrol 38:1513–1539. https://doi.org/10.1093/petroj/38.11.1513

Neumann ER, Griffin WL, Pearson NJ, O'reilly SY (2004) The Evolution of the Upper Mantle beneath the Canary Islands: Information from Trace Elements and Sr isotope Ratios in Minerals in Mantle Xenoliths Abstract Journal of Petrology 45(12):2573-2612. https://doi.org/10.1093/petrology/egh063

Newhall CG, Self S (1982) The volcanic explosivity index (VEI): an estimate of the explosive magnitude for historical eruptions. J Geophys Res 87:1231–1238

Oglialoro E, Frezzotti ML, Ferrando S, Tirabosch C, Principe C, Groppelli G, Villa I (2017) Lithospheric magma dynamics beneath the El Hierro Volcano, Canary Islands: insights from fluid inclusions. Bull Volcanol 79. https://doi.org/10.1007/s00445-017-1152-6

Olave Lacalle O (2014) Análisis de la dispersión de ceniza de las erupciones de San Juan y El Teneguía en La Palma (Islas Canarias). Relación con las condiciones atmosféricas. Report of Department de Geografía Física, ULL

Pedrazzi D, Becerril L, Martí J, Meletlidis S, Galindo I (2014) Explosive felsic volcanism on El Hierro (Canary Islands). Bull Volcanol 76(10):1–19

Pellicer MJ (1979) Estudio geoquímico del vulcanismo de la isla del Hierro. Archipiélago Canario. Estud Geol 35:15–29

Pellicer MJ (1977) Estudio volcanológico de la Isla de El Hierro (Islas Canarias) (volcanological study of the island of El Hierro (Canary Islands)). Estud Geol 33:181–197

Pellicer MJ (1975) Estudio vulcanológico, petrológico y geoquímico de la isla de El Hierro (Archipiélago canario), Tesis inéditas, Universidad Complutense de Madrid Facultad de Ciencias Geológicas, 332 páginas

Pérez-Torrado F, Carracedo JC, Fernández-Turiel JL, Guillou H, Machín A, González A (2011) Edades C-14 del Rift ONO de El Hierro (Islas Canarias)

Pyle D (2015) Sizes of volcanic eruptions. In: The encyclopedia of volcanoes, pp 263–269. https://doi.org/10.1016/B978-0-12-385938-9.00013-4

Ramalho R (2011) Vertical movements of ocean island volcanoes: insights from a stationary plate. In: Building the Cape Verde Islands. Springer Theses. Springer, Berlin, Heidelberg

Riccò A (1892) Terremoti, sollevamento ed eruzione sottomarina a Pantelleria nella seconda metà dell'ottobre 1891. In: Annali Ufficio Centrale Meteorologogico e Geodinamico, 2nd edn, vol 14

Risica G, Di Roberto A, Speranza F, Del Carlo P, Pompilio M, Meletlidis S, Todrani A (2022) Reconstruction of the subaerial Holocene volcanic activity through paleomagnetic and 14C dating methods: El Hierro (Canary Islands). J Volcanol Geotherm Res 425(107526):0273–0377. https://doi.org/10.1016/j.jvolgeores.2022.107526

Rivera J, Lastras G, Canals M, Acosta J, Arrese B, Hermida N, Micallef A, Tello O, Amblas D (2013) Construction of an oceanic island: Insights from the El Hierro (Canary Islands) 2011–2012 submarine volcanic eruption. Geology 41:355–358

Di Roberto A, Bertagnini A, Del Carlo P, Meletlidis S, Pompilio M (2016) The 1909 Chinyero eruption on Tenerife (Canary Islands): insights from historical accounts, and tephrostratigraphic and geochemical data. Bull Volcanol 78. https://doi.org/10.1007/s00445-016-1083-7

Rodríguez-González A, Pérez-Torrado FJ, Fernandez-Turiel JL, Carracedo JC, Guillou H (2012) GIS-based geomorphological modeling of coastal platform-forming eruptions: Montaña del Tesoro volcano (El Hierro, Canary Islands). In: Geotemas, vol 13. VIII Congreso Geológico de España, Oviedo

Rodríguez-Losada JA, Eff-Darwich A, Hernández-Gutiérrez LE, Viñas R et al. (2014) Petrological and geochemical highlights in the floating fragments of the October 2011 submarine eruption offshore El Hierro (Canary Islands): relevance of submarine hydrothermal processes. J Afr Earth Sci 2015:41–49. https://doi.org/10.1016/j.jafrearsci.2014.11.005

Rowe MC, Peate DW, Peate IU (2011) An investigation into the nature of the magmatic plumbing system at Paricutin Volcano, Mexico. J Petrol 52:2187–2220. https://doi.org/10.1093/petrology/egr044

Schmincke HU, Graf G (2000) DECOS/OMEX II, Cruise N° 43, METEOR-Berichte 20001. University of Hamburg, Alemania, p 99

Self S (1976) The recent volcanology of Terceira, Azores. J Geol Soc London 132:645–666

Siebe C, Komorowski JC, Navarro C, McHone J, Delgado H, Cortes A (1995) Submarine eruption near Socorro Island, México: geochemistry and scanning electron microscopy studies of floating scoria and reticulite. J Volcanol Geotherm Res 68:239–271

Sigmarsson O, Laporte D, Carpentier M et al (2013) Formation of U-depleted rhyolite from a basanite at El Hierro, Canary Islands. Contrib Miner Petrol 165:601–622. https://doi.org/10.1007/s00410-012-0826-5

Simony O (1890) Über eine naturwissenschaftliche Reise nach der westlichen Gruppe der Kanarischen Inseln. Mitteilungen der K. K. Geographischen Gesellschaft in Wien, Wien

Sobradelo R, Martí J, Mendoza-Rosas A, Gomez G (2011) Volcanic hazard assessment for the Canary Islands (Spain) using extreme value theory. Nat Hazards Earth Syst Sci 11:2741–2753. https://doi.org/10.5194/nhess-11-2741-2011

Staudigel H, Féraud G, Giannerini G (1986) The history of intrusive activity on the island of La Palma (Canary Islands). J Volcanol Geotherm Res 27:299–322

Stearns HT (1940) Geology and ground-water resources of the islands of Lanai and Kahoolawe. In: Hawaii (Terr.) division of hydrography bulletin 6, pt 2. Hawaii, pp 117–147

Stroncik N, Klügel A, Hansteen T (2009) The magmatic plumbing system beneath El Hierro (Canary Islands): constraints from phenocrysts and naturally quenched basaltic glasses in submarine rocks. Contrib Miner Petrol 157:593–607. https://doi.org/10.1007/s00410-008-0354-5

Szérémeta N, Laj C, Guillou H, Kissel C, Mazaud A, Carracedo JC (1999) Geomagnetic paleosecular variation in the Brunhes period, from the island of El Hierro (Canary Islands). Earth Planet Sci Lett 165(3–4):241–253

Tarquini S, de' Michieli Vitturi M, Jensen EH, Pedersen GB, Barsotti S, Coppola D, Pfeffer MA (2018) Modeling lava flow propagation over a flat landscape by using MrLavaLoba: the case of the 2014–2015 eruption at Holuhraun, Iceland. Ann Geophys 62:228. https://doi.org/10.4401/ag-7812

Troll V, Klügel A, Longpré MA, Burchardt S et al (2012) Floating stones off El Hierro, Canary Islands: xenoliths of pre-island sedimentary origin in the early products of the October 2011 eruption. Solid Earth 3:97–110. https://doi.org/10.5194/se-3-97-2012

Troll V, Carracedo JC (2016) The geology of El Hierro. In: Troll VR, Carracedo JC (eds) The geology of the Canary Islands. Elsevier, pp 43–99. ISBN: 9780128096635. https://doi.org/10.1016/B978-0-12-809663-5.00002-5.

Urgelés R, Canals M, Baraza J, Alonso B, Masson DG (1997) The last major megalandslides in the Canary Islands: The El Golfo debris avalanche and the Canary debris flow, west Hierro Island. J Geophys Res 102:20305–20323

Urgelés R, Masson DG, Canals M, Watts AB, Le Bas T (1999) Recurrent large-scale landsliding on the west flank of La Palma, Canary Islands. J Geophys Res 104 (2533):1–25348

Valentine GA (1998) Eruption column physics. In: Freundt A, Rosi M (eds) From Magma to Tephra—modelling physical processes of explosive volcanic eruptions. Elsevier, Amsterdam, pp 91–138

Valentine GA, Gregg TKP (2008) Continental basaltic volcanoes—processes and problems. J Geophys Res 177(4):857–873

Van den Bogaard P (2013) The origin of the Canary Island Seamount Province—new ages of old seamounts. Sci Rep 3:2107. https://doi.org/10.1038/srep02107

Villasante-Marcos V, Pavón-Carrasco FJ (2014) Paleomagnetic constraints on the age of Lomo Negro Volcanic Eruption (El Hierro, Canary Islands). Geophys J Int 199:1514. https://doi.org/10.1093/gji/ggu346

Von Fritsch K (1863) Reisebilder von den Kanarischen. Inseln. Petermanns geogr. Mitteilungen. Ergänzungsband V. Gotha

Von Wolff F (1931) Der Vulkanismus. II Band. Spezieller Teil. 2 Teil Die alte Welt. 1 Lieferung. Der Atlantische Ozean Stuttgart

Walker GPL (1990) Geology and volcanology of the Hawaiian Islands. Pacific Sci 44:315–347

Walter O (1894) Petrographische Studien an Gesteinen der Insel Hierro Nebst einer Beigabe. Karl von FRITSCH Geognostische Aufzeichnungen über den Inseln Halle

White J, Schipper C, Kazuhiko K (2015) Submarine explosive eruptions. In: The encyclopedia of volcanoes, pp 553–569. https://doi.org/10.1016/B978-0-12-385938-9.00031-6

Whitehouse M, Neumann ER (1995) Sr-Nd-Pb isotope data for ultramafic xenoliths from Hierro, Canary Islands: melt infiltration processes in the upper mantle. Contrib Miner Petrol 119:239–246. https://doi.org/10.1007/BF00307284

Wulff-Pedersen E, Neumann ER, Jensen B (1996) The upper mantle under La Palma, Canary Islands: formation of Si−K−Na-rich melt and its importance as a metasomatic agent. Contrib Miner Petrol 125:113–139

Zaczek K, Troll V, Cachao M et al (2015) Nannofossils in 2011 El Hierro eruptive products reinstate plume model for Canary Islands. Sci Rep 5:7945. https://doi.org/10.1038/srep07945

Zanon V, Pimentel A, Auxerre M, Marchini G, Stuart FM (2020) Unravelling the magma feeding system of a young basaltic oceanic volcano. Lithos 352. https://doi.org/10.1016/j.lithos.2019.105325

Review of Submarine Eruptions in El Hierro Prior to Tagoro

3

C. Guillén, M. C. Romero, and I. Galindo

Abstract

Hydrovolcanism, resulting from the interaction of magma with water, is a frequent volcanic process found in many geotectonic settings and fundamental in the growth of oceanic islands. Examples of shallow to intermediate-depth hydrovolcanic activity are frequent in the Canary Islands and characterise volcanism from the seamount stage and emergent island stage to historical volcanism. We present a review of the existing information on subaqueous shallow and intermediate-depth hydrovolcanic eruptions on El Hierro prior to the 2011–2012 Tagoro hydrovolcanic eruption. New information is provided based on the study of subaerial and intermediate-depth volcanic deposits, as well as on the analysis of possible historical submarine eruptions. Our findings show that there is evidence of several shallow and intermediate-depth submarine eruptions around El Hierro Island, some having occurred in historical times. Knowledge of these eruptions is essential not only to better understand the evolution of the island, to establish future coastal and submarine eruptive scenarios, but also to improve volcanic risk analysis. This work highlights the fact that, when dealing with volcanic islands, it is essential to include not only the coastline as a boundary, but also to consider the subaerial and submarine part of the islands in a continuous fashion, always taking into account sea level changes in the past as well as in the future.

Keywords

Hydrovolcanism · Phreatomagmatic eruptions · Nearshore deposits · Submarine eruptions · El hierro · Canary islands

3.1 Introduction

Hydrovolcanic eruptions are common in coastal environments, where ascending magma interacts with water in shallow or intermediate-depth environments, depending on the location of the vent (Sheridan and Wohletz 1983; Németh and Kósik 2020). In the Canary Islands, hydrovolcanic deposits related to shallow and intermediate-depth

C. Guillén
Escuela Universitaria de Turismo de Santa Cruz de Tenerife (EUTUR), Santa Cruz de Tenerife, Spain
e-mail: cayetano.guillem@eutur.es

M. C. Romero
Geography Department, Universidad de La Laguna, Campus de Guajara S/N. La Laguna, Tenerife, Spain
e-mail: mcromero@ull.edu.es

I. Galindo (✉)
Geological Extreme Events and Heritage, Spanish Institute of Geology and Mining (IGME-CSIC), Unit of Canary Islands, Alonso Alvarado, 43, 2°A, 35003 Las Palmas de Gran Canaria, Spain
e-mail: ines.galindo@csic.es

P. J. González (ed.), *El Hierro Island*, Active Volcanoes of the World,
https://doi.org/10.1007/978-3-031-35135-8_3

submarine eruptions are abundant in all the islands and islets (Martí and Colombo 1990; De la Nuez et al. 1993; De la Nuez and Alvarez 1997; De la Nuez and Quesada 1999; García-Cacho aand Romero 2000; Romero et al. 2000; Hansen et al. 2008; Carmona et al. 2011; Becerril et al. 2013; Pedrazzi et al. 2013; Becerril 2014; Galindo et al. 2016, 2017, 2019). Explosive hydrovolcanic edifices are found built-in seamounts (e.g. Herrera et al. 2008; Gutiérrez et al. 2002) and in volcanic sequences of the shield-building phase, of Mio-Pliocene age, corresponding to the initial phases of the islands' emersion (Schmidt and Schmincke 2000), as well as in later Pleistocene volcanic rift formations or even forming monogenetic cones in coastal and subaqueous shallow to intermediate-depth sectors (Quesada et al. 1988; De la Nuez et al. 1993; De la Nuez and Alvarez 1997; De la Nuez and Quesada 1999; Clarke et al. 2005). Bathymetric studies revealed that there are numerous well-preserved volcanic cones along the submarine flanks of the islands (Romero et al. 2000; Gee et al. 2001; Krastel and Schmincke 2002; Acosta et al. 2005; Becerril 2014; Becerril et al. 2015; Galindo et al. 2016).

The Tagoro eruption offshore the island of El Hierro in 2011–2012 (López et al. 2012) has shown that subaquatic eruptions are one of the possible eruptive scenarios in the Canaries that must be considered when carrying out volcanic hazard analyses (Becerril et al. 2013; Becerril 2014; Galindo et al. 2016). However, Tagoro was not the first subaquatic eruption in historical times in the Canary Islands. During the eruption of Timanfaya in the island of Lanzarote (1730–36), the eruptive fissure spread from the land into the sea. The underwater vent was shallow enough to produce visible explosions with ejection of pyroclasts, a rising steam-rich plume and an incredible amount of dead fish (Romero 1991). Taking into account this historical events and the fact that the Canary Islands have a coastal perimeter of more than 1583 km, it is crucial for a better understanding of this type of eruptions in the Canaries, as there is the potential for similar hydrovolcanic eruptions to occur in the future.

As a first approach, this work focuses on the study of hydrovolcanic eruptions that have occurred in shallow or intermediate-depth submarine environments throughout the evolution of El Hierro. Based on the study of subaerial and submarine deposits, the interpretation of submerged coastal geomorphology and the review of historical sources, several hydrovolcanic eruptions have been identified on the submarine flanks of the island, demonstrating that in addition to intermediate-depth eruptions, such as the one in 2011–2012, other direct explosive hydrovolcanic eruptions have also taken place on the island.

3.2 Geological Setting

The island of El Hierro, the westernmost island of the Canary Islands (Fig. 3.1) rises from the ocean floor from depths of roughly 5 km, reaching a maximum altitude of 1503 m above sea level and covering a surface area of only 269 km^2. The emerged part of the island consists mainly of alkaline basalts and, to a lesser extent, more differentiated trachybasalts and trachytes (Fuster et al. 1993). The island's geological history can be divided into three phases (Guillou et al. 1996; Carracedo et al. 1997): (1) Tiñor Edifice (1.12–0.88 Ma); (2) El Golfo-Las Playas Edifice (545–176 ka); and (3) rift volcanism (158 ka-present). Along the rift zones, coastal lava deltas include rocks younger than about 20 ka (Carracedo et al. 2001). Rift volcanism is characterised by mafic monogenetic volcanic cones (Aulinas et al. 2019), with marked eruptive fissure alignments, which consists of lava flows and ballistic pyroclasts resulting from typical Hawaiian to Strombolian eruptions and violent Strombolian events (Becerril 2014; Becerril et al. 2013). Volcanic cones are mainly concentrated along three main ridges of NE, WSW and S direction; however, showing a clearly radial distribution below sea level (Becerril et al. 2016).

Available data on the island's historical volcanism are limited and uncertain. Some authors consider that during the last 600 years there has been no volcanic activity on the island, with the exception of the 2011–2012 eruption located in Las Calmas sea, in the submerged extension of

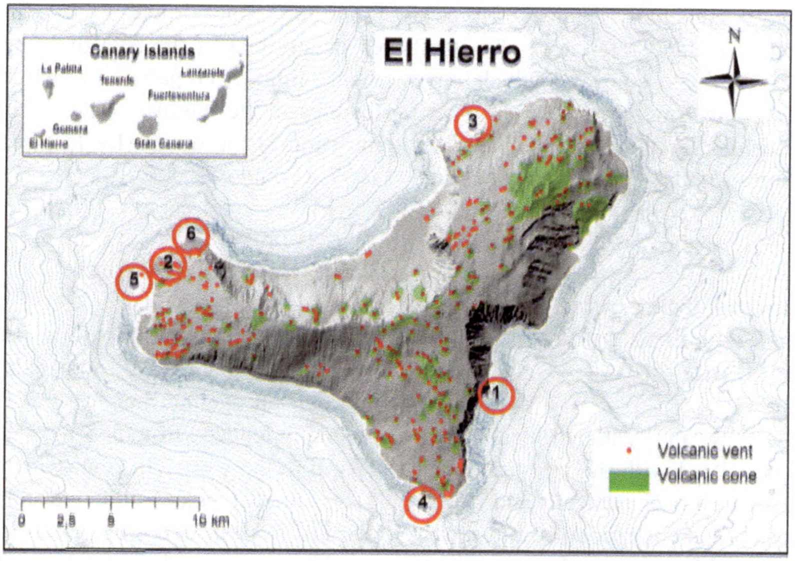

Fig. 3.1 Shaded relief of El Hierro Island and offshore bathymetry showing the location of hydrovolcanic outcrops from submarine eruptions (except Tagoro volcano): (1) Punta del Corral; (2) Hoya del Verodal; (3) Las Bermejas de Tierra; (4) El Bajón; (5) Baja de Anacón: and (6) Punta de la Arena. Inset shows the location of El Hierro in the Canary Islands. Bathymetric Data: Redmic (https://redmic.es/service-ogc-catalog/service-ogc-info/layer-e6ee55c1-d444-48b2-86a9-541b93c067f3)

the S rift (Becerril et al. 2016; Troll and Carracedo 2016). Conversely, there are reports of a seismic crisis towards the end of the eighteenth century, which some authors associate it with the Lomo Negro volcano on the WSW rift (Hernández Pacheco 1982), while others link it to a submarine volcanic event (Bravo 1965–1968).

3.3 Methods

Onshore data was gathered through fieldwork and direct observations made during geological reconnaissance flights carried out in 2008. The inaccessibility and poor exposure of some of the outcrops has prevented a more detailed description of the deposits. The submarine outcrops have been analysed based on bathymetric data from the REDMIC platform (https://redmic.es/atlas) and image analysis of photographs taken by divers.

The review of documentary sources has been fundamental to the study of historical events on the island of El Hierro. Local information is scarce due to the loss of the island's municipal and religious documents from the period between the fifteenth and nineteenth centuries, as the archives were burned in repeated fires in 1553, 1658, 1763 and 1899. For this reason, it has been necessary to resort to indirect and very scattered documentary sources, collected by authors who, in most cases, had not even visited the island. Therefore, records of expeditions conducted between the tenth and fifteenth centuries, as well as chronicles and accounts of the fifteenth and sixteenth centuries, travel books, old dictionaries and geographical compilations of the seventeenth and eighteenth centuries, or catalogues and lists of natural phenomena drawn up from the sixteenth century onwards, have become essential references for this study.

In the absence of primary sources, the compilation of events related to the natural history of the island had to be established from the study of dispersed sources, usually very brief and indirect, which evidently entails notable inaccuracies with respect to specific dates, characteristics of the eruptive processes or, affected places. A review of the existing documentary sources (Romero 2016) has brought to light references of at least three possible eruptions, some of which have already been briefly mentioned by Romero (1991) and included in the Smithsonian GVP catalogue (https://volcano.si.edu/volcano.cfm?vn=383020).

3.4 Submarine Volcanism of the Early Emergence of the Island

There are only a few and scattered hyaloclastite tuff outcrops associated with the construction of Tiñor and El Golfo edifices (Balcells and Gómez 1997). These outcrops correspond to old monogenetic volcanic edifices formed during Surtseyan eruptions that characterise the island early build up stage and are accordingly part of the lower sequences of both edifices. These outcrops are discordantly covered by younger lava flows and pyroclastic density current deposits but can be easily distinguished by their intense yellow-mustard colour of the palagonitised materials. In general, they are located at the base of active cliffs or in palaeo-cliffs (Figs. 3.1, 3.2).

3.4.1 Punta Del Corral

In the southern area of Las Playas amphitheatre, at the base of a 600 m high escarpment, there is an outcrop of palagonitised hyaloclastite tuffs that can only be seen at sea or from the air (Fig. 3.2a, b). Punta del Corral is the lowest sequence of this cliff, corresponding to the base of the Tiñor Edifice, with an age of 1.1 Ma (Troll and Carracedo 2016). A network of subvertical and inclined greyish dykes crosses the edifice trending NW–SE and NE–SW.

Difficulty in accessing the area prevented direct observation in the field; however, thick stratified deposits can be seen at large scale, with alternating materials of different grain size, marked by differential erosion. The general aspect of the volcanic edifice is similar to that of other tuff cones found in the Canary Islands, especially that of the islet of Montaña Clara, in the Chinijo Archipelago, which shows similar features in the internal structure, likewise in the intrusion of the dykes (De la Nuez and Alvarez 1997; Galindo et al. 2019).

3.4.2 Hoya Del Verodal

The entire northeastern tip of the island is crossed by an escarpment with a noticeable arched outline and an average height of 300 m, which is gradually reduced from 535 m in the east to only 175 m in the west. This escarpment corresponds to an ancient palaeo-cliff, partially covered by detrital slopes, exposing two different stratigraphic sequences (Fig. 3.2c): A basal sequence, essentially made up of basaltic lava piles from El Golfo edifice, which disappears towards the southern section; and an upper sequence comprising lavas and pyroclasts belonging to the volcanic rift stage (Carracedo et al. 1997, 2001; Balcells and Gómez 1997). Embedded in the lower sequence, there are two large outcrops of pyroclastic material, clearly discordant with the stratigraphy of the rest of the cliff, which give rise to two protuberances, as shoulders, in the vertical profile of the escarpment and that stand out due to the yellowish colour of their rocks (Fig. 3.2c–e). These two outcrops consist of hydrovolcanic tuffs between 28 and 43 m thick, extensively palagonitised, and of massive wet-surge feature type (Balcells and Gómez 1997). They show, however, some very thin ash-rich layers that give certain lamination to the deposits, which is better expressed in the northern outcrop than in the southern one.

In addition to juvenile pyroclasts, dense lithic clasts with angular to sub-rounded shapes and maximum sizes in the order of 25–30 cm are frequent. The outcrops are approximately 800 m apart and dip 20–25° in opposite directions, which suggests that they are part of a volcanic vent, now covered by younger lavas, corresponding to a tuff-ring type volcanic edifice (Troll and Carracedo 2016).

Fig. 3.2 Palagonitised hyaloclastite tuff deposits from the emergence phase of El Hierro Island: **a**, **b** Aerial view of the cliff of Punta del Corral crossed by dikes; **c** General view of Hoya del Verodal showing the location of the two outcrops exposed in the palaeo-cliff (**d**, **e**). The estimated diameter of the tuff-ring is approximately 800 m. See location in Fig. 3.1

3.5 Surtseyan Deposits on the Rifts

They correspond to small deposits of finely-bedded hyaloclastic tuff arranged in areas close to the coastline of the rift zones: Las Bermejas de Tierra to the NE, Punta del Verodal and Punta de La Arena to the WNW and Punta de La Restinga to the South (Figs. 3.1, 3.3, 3.4, 3.5). These deposits are located: (1) at the base of the cliffs (Las Bermejas de Tierra); (2) underwater, on the submerged flanks of the island, at distances of around 500 m from the coast (Punta del Verodal and Punta de La Restinga); and (3) on a lava platform younger than 20 ka. The deposit located at the base of Las Bermejas de Tierra cliff is part of the lower sequence of the NE volcanic rift and shows many similarities with those described above; the offshore sites are small submarine promontories of irregular morphology, with escarpments and high slopes, made up of hyaloclastic tuffs, rising 70–80 m above the seabed.

3.5.1 Las Bermejas De Tierra

Most of the NE rift is characterised by an active rock cliff less than 100 m high, carved over lava

Fig. 3.3 **a**, **b** Las Bermejas de Tierra, NE rift. **c–f** El Bajón edifice in the submerged area near La Restinga. The red star in C indicates the location of the probable vent based on the strike and dip directions of layers (in red). **d** Surtseyan deposits; **e** Pothole; **f** Detail of cross-stratification. **g**, **h** Baja de Anacón islet, at Punta de El Verodal. The bathymetric maps have been obtained from the REDMIC platform (https://redmic.es/atlas). Source of photos **d**, **e**, and **f**: Tamia Brito, La Restinga-Mar de Las Calmas Marine Reserve, Cabildo de El Hierro Council. See location in Fig. 3.1

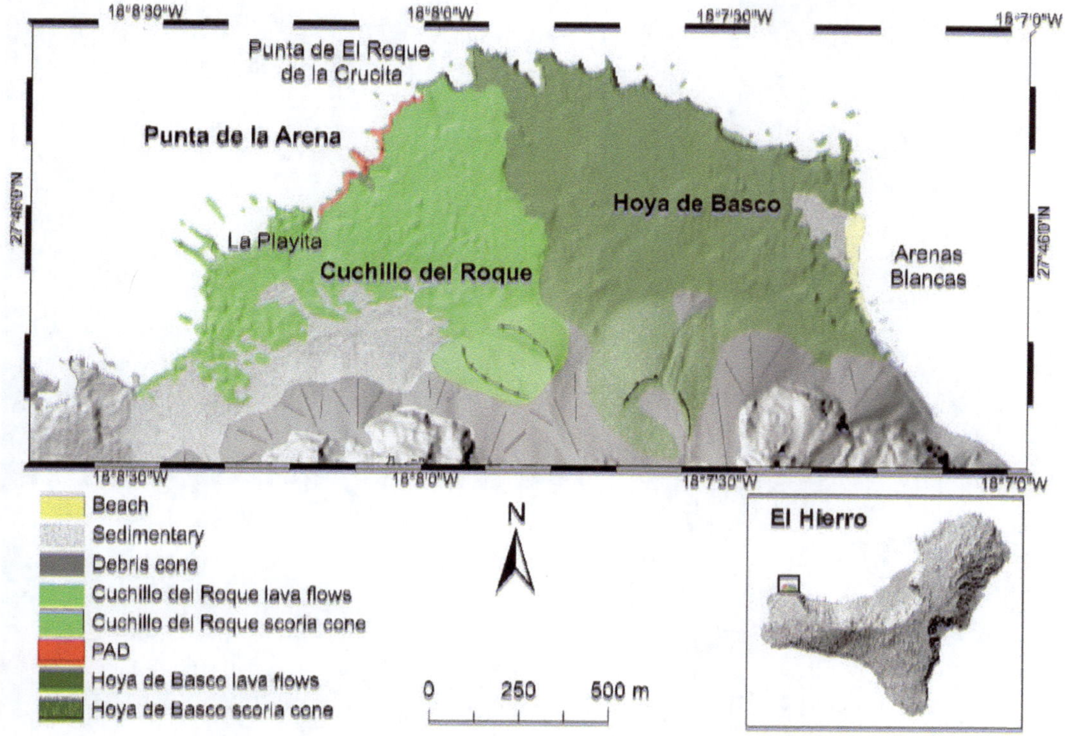

Fig. 3.4 Geomorphological map of El Verodal lava platform showing the location of Punta de la Arena outcrop, that is interbedded between Hoya de Basco and Cuchillo del Roque lava flows

piles and mafic pyroclastic deposits. In the western area of this rift, the cliff reaches heights of up to 400 m. The cliff shows a complex internal structure primarily made up of alternation of scoriaceous and phreatomagmatic deposits and some intercalated lava flows. The base of this escarpment comprises palagonitised tuffs cut by dykes, probably due to slope and marine erosion processes (Fig. 3.3a, b).

Inaccessibility to the cliff has prevented direct observation of this deposit, but the similarities with the submarine deposits of the Tiñor and El Golfo edifices allow us to identify it as the remains of a tuff-cone or tuff-ring formed during the initial phases of the rift volcanism, and therefore, they are considerably younger than 158 ka (Becerril et al. 2013).

3.5.2 El Bajón

In the submarine extension of the SSE rift into the Mar de Las Calmas Marine Reserve, at a distance between 460 and 500 m from the southern tip of La Restinga, there is a submarine promontory that rises from a depth of approximately 80 m in the south and 49 m in the north, with the summit at 9 and 6 m minimum depth respectively (Figs. 3.1, 3.3c–f). The currents around this promontory are very strong, which favours the existence of abundant fauna; this fact, along with the spectacular rock formations, has made it one of the most popular diving spots on the island and one of the places of geological interest in the UNESCO Global Geopark of El Hierro.

Fig. 3.5 Punta de la Arena hydrovolcanic deposits (PAD): **a** General view of the main outcrop showing the four distinct facies. The layers show thickness variations as result of the irregular underlying surface. **b** Detail of the PDA1 and PDA2 facies. **c** Bomb sag structures of PAD2. Loose lithics from the erosion of the upper facies stand out on the deposit; **e** Detail of the PDA4 facies; and **d** Accretionary lapilli of the PDA2 facies

El Bajón rises from a platform with an escarpment of more than 20 m high towards the southernmost part, with a warping path in a parallel direction, although slightly curved towards the SW. This elevation shows an approximate elliptical shape in plane view with the longer axis trending N-S, although with a slight curving towards the NW in its northern sector (Fig. 3.3c). Erosive features cut its eastern and western flanks, like ravines with lengths of around 200 m. This topographical layout might suggest the existence of an open-arched edifice towards the NW. However, there is no morphological evidence of crater rims or volcanic vents. The entire promontory consists of hydrovolcanic tuff deposits, arranged in coarsely laminated packages that dip divergently to the north and south from the saddle between its two main peaks, so this could be interpreted as the location of the possible volcanic vent (Fig. 3.3c–f). In some areas, the characteristic cross laminations of pyroclastic surges can be observed (Fig. 3.3f).

These are the eroded remains of a tuff-cone, with vertical walls of almost 70 m, and frequent cornices and shoulders that result from the differential erosion of materials with different grain sizes. Potholes formed by rock abrasion are observed around 17 m deep (Fig. 3.3e), similar to those described in hydrovolcanic deposits in Lanzarote and the Chinijo Archipelago (Galindo et al. 2017, 2019). These structures are typical of shallow littoral environments suggesting that this Surtseyan-type volcanic edifice was already formed when the sea level was lower than it is today.

3.5.3 Baja De Anacón

At the south-western end of the El Verodal coastal shelf, at the end of the WNW rift, about 360 m W of the coast, there is a small islet 80 m long with a NW–SE orientation (Figs. 3.1, 3.3g, h), which is crescent-shaped, open to the SW and barely protrudes 1.7 m from the sea surface. This islet, known as Baja de Anacón, has been previously interpreted as the remains of a volcanic edifice, without determining its genesis (Navarro and Soler 1991).

The geomorphological analysis of the bathymetry and the examination of photographs allow drawing the conclusion that it corresponds to the southern flank remains of a tuff cone rising from a depth of 58 m. It is deeply eroded and consists of consolidated tuffs, finely laminated and of greyish-yellowish colour. The islet is usually hit by heavy swell, which has prevented direct in situ observation of the proximal deposits. However, the arched morphology of the edifice and the dip of the exposed tuff levels, gently dipping NE, suggests that the vent of this Surtseyan cone was located to the SW.

3.5.4 Punta De La Arena Deposits (PAD)

El Verodal littoral platform was formed by the stacking of lavas from different eruptions, the youngest lavas being issued from Hoya de Basco (HB) and Cuchillo del Roque (CR) volcanic cones (Fig. 3.4). Interbedded between these lava flows, a sequence of pyroclastic deposits is exposed at Punta de La Arena (Figs. 3.1, 3.5). This outcrop extends along 620 m in the cliff, between 6 and 19 m above sea level, and tops small, isolated rocks formed by the retreat of the coastline. Four main facies can be recognized: (1) stratified ash facies; (2) stratified ash with bomb sag facies; (3) massive facies; and (4) lithic-rich facies.

The base of the sequence consists of a stratified ash facies (PAD1), a matrix-supported deposit of ~20 cm maximum thickness, gently undulating and mostly formed by fine ash, with very few accidental lithics (Fig. 3.5a, b). Parallel laminations are marked by changes in grain size (ranging from fine ash up to lapilli dimensions). Some occasional lithics of size smaller than 10 cm are observed at this layer. This facies is extensively palagonitised.

The stratified ash with bomb sag facies (PAD2) has a maximum thickness of 30 cm. This facies is matrix-supported, with a fine to medium ash matrix. It is laminated with alternating thin layers of ash and lapilli that show general reverse grading. There are some clast-supported layers at the top of the sequence. The deposit is rich in accidental lithic fragments and juvenile fragments, including cauliflower bombs and accretionary lapilli (Fig. 3.5e). Some lithic and juvenile fragments are wrapped up by a thin coating of ash. Bomb sag structures are very frequent and cause deformation of PAD1 layers (Fig. 3.5a–c). Generally, the lithic fragments are mafic, dense, prismatic and angular, showing major axes up to 25 cm. This facies is less palagonitised than PAD1.

The massive facies (PAD3) has a maximum thickness of 60 cm (Fig. 3.5a). It is matrix-supported at the base with clast-supported layers at the top. It is massive with an incipient lamination. The matrix consists of coarse ash and lapilli sized fragments, showing reverse gradation towards the top. Juvenile fragments are scarce. Lithic fragments are similar to those of PAD2 but with smaller size (<10 cm).

At the top of the sequence is the lithic-rich facies (PAD4), a clast-supported deposit (Fig. 3.5a, d) with a maximum thickness of 25 cm. The top is very irregular because it was clearly eroded before the emplacement of the Cuchillo del Roque lavas and some clasts are reworked. This facies is characterized by the abundance of accidental lithic fragments and subordinate juvenile clasts. Both types of fragments have a mean size of 5–10 cm, with some lithic fragments up to 20 cm (major axis). PAD4 is unconformably overlain by the Cuchillo del Roque lavas, which incorporated some lithics at the base during its movement over PAD4 deposits. The lava flow caused thermal alteration of this deposit showing reddish colours in some areas.

The facies association of PAD suggests it corresponds to deposits of a Surtseyan-type eruption, emplaced in a subaerial environment. The general upward coarsening of the levels towards the top indicates a decrease in the efficiency of magma fragmentation and thus, in the water/magma ratio towards the final phases of the eruption. The PAD1 and PAD2 facies suggest high-energy explosions with abundant water content and emplacement of ballistic projectiles (PAD2). The palagonitisation affecting the deposit is due to the presence of hot water during emplacement, being more abundant in the basal part of the outcrop. The PAD progressively decrease in thickness and clast size towards the NE and onshore (Fig. 3.5a), and the accidental lithic clasts decrease both in size and in number in the same direction, suggesting that the vent was located offshore and SW of this location. Thus, it is likely that these deposits are related to the tuff ring of Baja de Anacón. However, further studies are needed to resolve uncertainties about the age and formation of these deposits.

3.6 Historical Surtseyan and Submarine Volcanism

A review of historical chronicles provides information on possible historical coastal eruptions on El Hierro prior to the 2011–2012 Tagoro eruption. There are only documentary references in the scientific literature of a seismic crisis in 1793 (Darias Padrón 1929; Bethencourt Massieu 1982; Romero 1991, 2016), which some authors associate to the Lomo Negro Volcano (Hernández Pacheco 1982), while others link it to a possible submarine volcanic event (Bravo 1965–1968). Historical sources, however, not only record the progress of the 1793 seismic crisis, but also of five different volcanic episodes (a pre-Hispanic eruption, 1690, 1677, 1692, 1777) some of them categorised as probable, uncertain or doubtful (Romero 2016). The supposed volcanic events of 1677 and 1692, slightly cited by Romero (1991, 2007), have gone unnoticed and do not appear in the bibliography, even though they are listed as uncertain in the Smithsonian catalogue (GPV). Only references to possible eighteenth century events seem to correspond to submarine volcanic episodes.

3.6.1 June-November 1721 Surtseyan Eruption: A Legend?

Between June and November 1721, many people observed an island in the sea, west of El Hierro Island. Based on a legend, the inhabitants of El Hierro identified it as the island of San

Borondón. It is one of the most popular legends of the Canary Islands running from the 16th to nineteenth centuries, and still present today. It narrates the mythical existence of an eighth island that appears and disappears in the sea (Poggio and Regueira 2009) and can be seen from Tenerife, La Gomera, La Palma and, mainly, from El Hierro. This tale is linked to the legendary journey of Saint Brandan of Clonfert, an Irish friar of the sixth century, who travelled across the ocean for seven years, accompanied by seventeen monks, in search of Paradise. Among the islands they travelled through, there is a rocky island with lava that they call Hell Island, and they find another one where the monks disembark to celebrate Easter, fleeing from it when the island trembles and begins to move, as it was not solid ground but a whale (Corbellá Díaz and Medina López 1997).

This kind of "whale island" myth or legend is the one that has survived the most in the Canary Islands, becoming the island of San Borondón. This island is so deeply rooted in the Atlantic culture, including that of the Canary Islands, that from the middle of the fourteenth century until the end of the eighteenth century it used to appear mapped at various points in the ocean, and even next to the rest of the Canary Islands (Tous Meliá 1996; Poggio and Regueira 2009). The sighting of the island of San Borondón from the western islands led to four expeditions in 1526, 1570, 1604 and 1721 (Ruano 1979; Corbellá Díaz and Medina López 1997; Poggio and Regueira 2009). Some of these expeditions were even recorded in notarial acts, which include the statements of the witnesses who observed the island (Ruano 1979).

The reiteration of the phenomenon over several months in 1721 led the authorities to draw up a series of notarial records, in which witness statements were recorded. These official proceedings carried out in the year 1721 to ascertain the existence of the island of San Borondón includes the information gathered on El Hierro regarding these appearances (Corbellá Díaz and Medina López 1997). Most of the witnesses described the presence of an island to the west of El Hierro usually covered with mist. In this dossier, along with accounts that seem to be more focused on the predefined idea of the mythical and legendary image of San Borondón, there are some testimonies that deserve to be considered for the information they provide. Thus, Juan Padrón Morales, a resident of Temuica on El Hierro, in his statement of 7th November 1721, states (Corbellá Díaz and Medina López 1997):

> *"...they saw opposite the defence of this island (El Hierro) an island clearly and distinctly that they all recognised, and that over the ten or twelve previous days they saw two fires on the same side where they had seen the island, one that lasted a short time, and the other longer, and they saw them with all certainty and that they judged that these fires were on the island...".*

Although there is no reference to volcanic events in the Canary Islands in 1721, the ten to twelve days sighting and the observation of fires in the sea, one of shorter duration than the other, could suggest that there was an eruption in the sea. The appearance of the island of San Borondón has usually been attributed to optical or meteorological phenomena or to the presence of floating plant debris (Viera and Clavijo 1772–1783; Fritsch 1867). Nevertheless, the sight of an island appearing and disappearing has also been explained as a consequence of volcanic processes (Barrow 1807), and it is not surprising if we take into account some of the witness statements in 1721 where the presence of fires in the sea was reported. The mythical legend of San Borondón may be the sum of all these phenomena, so it could be likely that some of the descriptions of the sighting of the island of San Borondón may have been recounting the development of a submarine volcanic episode, for which there is no historical, geographical or geological evidence.

3.6.2 Thirteenth September 1777 Submarine Eruption

The first record of a volcanic eruption on the island of El Hierro in 1777 is contained in a list of earthquakes, famines, inundations, storms, tempest, frost, and accidental fires in the fifth edition of an English book, *The tablet of memory*

(1783), published just 6 years after the eruptive event, and it only states:

> *"Volcano, in the Isle of Ferro, broke out Sept. 13, 1777, which threw out an immense amount of red water that discoloured the sea for several leagues".*

This description, however, does not mention whether the island named Ferro corresponds to one of the Canary Islands. This place name to designate the island of El Hierro was commonly used from the time immediately prior to the conquest. It appears for the first time in maps and cartouches from the mid-fourteenth century, where it is labelled as Fero, which Trapero (2008) interprets as Ferro. Despite its secular isolation, the island of Ferro was well known as the site of the 0 meridian. However, in order to verify whether these references are related to one of the Canary Islands and not to another volcanic island with a similar name, such as the Faroe Islands for example, we have traced in the current bibliography the eruptions that took place during that year. Only 11 volcanoes were active during the year 1777, according to the worldwide eruption list (Siebert et al. 2010). None of them occurred on an island with a similar name or whose spelling could be confused with "Ferro", and there is no data on a submarine eruption happening at that time on the Faroe Islands.

The information referring to this episode of 1777 would be repeatedly incorporated without further data by other authors, in other lists and catalogues throughout the nineteenth century (Young 1807; Tegg 1811). Only Harmonville (1845) adds to this information:

> *"Volcano of the Ferro Island. The first eruption of this island's volcano took place in 1662; it was renewed in 1677 after a violent tremor. After an interval of a century, in 1777, the crater opened again, and since then no further eruptions have been reported. It has been recorded as an extraordinary event that at the time of the eruption of 13 September 1777, reddish water came out of the crater at such an intensity that it splashed into the sea many leagues away".*

And Webster (1758–1843) highlights:

> *"A volcano in Ferro discharged discoloured water, but no lava".*

It could be thought that, given the temporal proximity of this event to the 1793 seismic crisis, there may be a possible confusion or error on dates between the two events, and that the references to these two years correspond, in reality, to a single volcanic episode. However, the first news concerning the 1777 eruption was published in 1783, i.e., six years after the eruption, but ten years before the seismic crisis of 1793. This clearly shows that there is no confusion on dates between these two events in the late eighteenth century.

It is striking that there is not a single record of this eruptive event in the local and regional literature, even though it took place during a period of strong historical production in the Canary Islands. However, it is worth considering that these types of phenomena were recorded according to the losses caused and that traditionally pirates and corsairs maintained the island's population inland, due to fear of attacks. On the other hand, it is no coincidence that news referring to the Canary Islands is found in English texts, since during the eighteenth century the archipelago maintained strong trade relations with England.

Although the location of this eruption is unknown, there is no doubt it took place in the sea and close to the coast, in an environment and under conditions very similar to those observed in the Mar de Las Calmas in 2011–2012, and therefore with no damage to the island territory or harm to its population. On the other hand, while the lack of contemporary local sources could be interpreted against the authenticity of the 1777 eruption, the analysis of these indirect sources supports the veracity of this eruption for several reasons: (1) The appearance of discolored water has been described for other submarine intermediate depth eruptions such as the 2011–2012 eruption in El Hierro (López et al. 2012), Serreta (Casas et al. 2018) or Monowai (Chadwick et al. 2008) is evidence of a similar eruptive scenario for the 1777 eruption; (2) This type of eruptive scenario was little known in the late 18th and early nineteenth centuries and attracted much attention from chroniclers and compilers of natural phenomena, hence its inclusion in the catalogues; and (3) The virtual absence of population on the coast throughout the island's history, due to constant fear of barbaric corsair and pirate attacks. For instance, the town settlement of La Restinga, located at the southernmost tip of the island, only dates back to the 1960s.

3.6.3 The Seismic Crisis of 27th March–27th June 1793: A Submarine Eruption?

From 27th March to 27th June 1793 the island of El Hierro was affected by a rather long seismic crisis. Local authorities recorded these events (Darías Padrón 1929; Bethencourt Massieu 1982; Romero 1991) and a dossier was drawn up (AHN 1793; Bethencourt 1982). Most of the earthquakes in this series correspond to a swarm of moderate intensity, yet felt by the population, having been assigned intensities of IV to VI (Rueda Núñez et al. 2020). On 8th May, at 11:00 pm, one of the strongest earthquakes shook the island making the population flee their homes to sleep in the shelter of the countryside. The following days (9th–13th May) the earthquakes were so recurrent that made many at odds reconcile. Even the island's mayor said: "*if it were necessary to hold a formal rehearsal for the hour of reckoning, the islanders have already done it and lived it*". The strongest earthquake occurred on 15th June, causing major rockfalls all over the island, but especially in El Golfo area. Although some buildings in Valverde, the capital of the island, were affected by this seismic event, it was in El Golfo area that the earthquake caused the total or partial collapse of some houses. After this last event, the number and intensity of earthquakes decreased, although Cosme de Burós reports that in El Golfo "*the earth is always with some moving, and as if boiling*". From 27th June onwards, there are no more references to seismic events on the island. The experience lived by the population and the authorities in previous volcanic events during the eighteenth century (eruptions of 1704–05, 1706, 1712 and 1730–36 in the Canary Islands) meant that these were interpreted as precursors of a new volcanic event, even establishing possible eruptive scenarios. However, although the evacuation of the island was planned, there is no mention in the existing documentation of it ever taking place.

Several months of precursor earthquakes, an earthquake of greater intensity, and then, the continuous ground-shake made some authors link this seismic crisis to a possible volcanic eruption. Bravo (1965–1968), in view of the lack of geological evidence and the inability of assigning an eruptive centre to this period, suggests that it could have been related to a submarine volcanic event, probably in the vicinity of El Golfo.

Other authors consider that the seismic sequence could have also been associated with a subaerial eruption in the northwestern area of the island. The most solid arguments regarding the development of an eruption at that time were provided by Hernández Pacheco (1982), who considers that the seismic crisis could have also been associated with a subaerial eruption located at Lomo Negro, based on a C^{14} dating of plant remains found under the lavas of the Lomo Negro eruptive complex, which gave an age of around 1800 AD. However, recent studies suggest (that Lomo Negro deposits could correspond to an eruption occurred in the XVI century CE (1498–1595 CE) based on paleomagnetic (Villasante-Marcos and Pavón-Carrasco 2014) and geochronological data (Risica et al. 2022).

The Lomo Negro volcanic complex described by Hernández Pacheco (1982), resulting from the 1793 eruption, is an eruptive fissure 4–6 m wide and approximately 53 m long. The fissure is closed to the north by a spatter hornito about 3 m high, which contains a 2 m in diameter and 14.6 m deep crater. The lava flows emitted from these vents, "aa" and "pahoehoe", cover an area of only 0.54 km^2. However, Lomo Negro is part of a larger eruptive fissure that includes Hoya del Verodal cinder cone. Such volcanic episode would have been definitely noticed by the inhabitants of the island. The Hernández Pacheco's Lomo Negro eruption does not describe the volcanological and morphological characteristics that define the most recent mafic monogenetic volcanism on El Hierro. These eruptions are characterised by the opening of long eruptive fissures, usually arranged transversally to the hillsides on which they open (Becerril et al. 2013). This arrangement is common in the historical volcanism of the Canary Islands (Romero 1991) where volcanic cones resulting mainly from Strombolian activity are found on the highest part of the eruptive fissure. However, effusive vents and the construction of small spatter cones are located at the base of the edifices.

The absence of documentary sources for all the dates mentioned by Villasante-Marcos and Pavón-Carrasco (2014) makes it impossible to link this volcanic complex with a documented eruption in the historical period. Nevertheless, these conclusions not only allow us to rule out a subaerial eruption on the island during the 1793 crisis, but also to link the processes that occurred to either an intrusive magmatic event or a submarine eruption (Carracedo et al. 1997; Villasante-Marcos and Pavón-Carrasco 2014). In this case, the sequence of events is similar to that observed during the 2011–2012 Tagoro eruption. The documentary evidence of continuous tremor towards the end of the 1793 seismic sequence and the reduction in the number of earthquakes felt onwards can be interpreted as corresponding to the initial phase of a submarine eruption. It is difficult to know the location of the eruption; as the Tagoro eruption showed, the location of the precursory seismicity is not clearly related to the location of the eruptive centres.

The hypothesis of a submarine eruption is also supported by the fact that fishermen of El Hierro were affected by a crisis on 1793, similar to what happened in 2011–2012 when tourist diving companies and fishermen were the most affected economic sectors, as sailing, diving and fishing was forbidden in the waters near the island, due to the spread of a brownish, reddish and greenish stain in the sea water, contaminated by volcanic gases. A similar situation was experienced on the island at the end of the eighteenth century (Ruiz de la Serna and Cruz Quintana 1973). Towards the middle of the eighteenth century, the fishermen of Las Palmas formed a confraternity under the patronage of Saint Elmo, the patron saint of sailors. For its operation, each of its members had to contribute with a 3% of the number of fish caught in their boat. Out of this amount, 1.5% went to the Patron's worship and the adornment of the hermitage; the other 1.5% was used to attend to social matters, medical-pharmaceutical care, widows' or orphans' pensions, help for the elderly, burial expenses, etc. The confreres still found ways to set aside a part of their income for donations to charitable works that went beyond their confraternity. Béthencourt Massieu (1989) pointed out as an example of this the donation made on the occasion of the 1793 earthquakes: *"...on one occasion, they donated fifty pesos to alleviate the evils caused by an earthquake on the island of El Hierro..."*. It is obvious, therefore, that it was the fishermen of the island of El Hierro who were most affected by the crisis of 1793. This would also support the development of another submarine volcanic eruption in the last decades of the eighteenth century.

3.7 Discussion and Conclusions

Intermediate-depth and shallow (Surtseyan) submarine eruptions are characteristics of the seamount and emergent island stages of a volcanic island, since it begins to grow on the ocean floor until it emerges above sea level and the volcanic conduits are isolated from seawater. This type of volcanism coexists throughout the island's eruptive history with subaerial volcanism; however, most of the littoral volcanic deposits are overlaid by subaerial deposits as the island grows and increases its surface. This study provides a first approach to the knowledge of the volcanic submarine events that have occurred in El Hierro Island. Only a few examples of Surtseyan activity are found at the base of the main volcanic formations, corresponding to extensively palagonitised tuffs. More recent and located near the coast, in the submarine extension of the S and WNW rifts, evidence of at least two Surtseyan eruptions has been located, one of them younger than 20 ka. In addition, based on historical documents we suggest that a possible Surtseyan eruption took place in 1721 and two intermediate-depth submarine eruptions occurred in 1777 and 1793 with similar characteristics to the Tagoro eruption.

The occurrence of Surtseyan eruptions has been classified as a hazardous phenomenon (Verolino et al. 2019), and therefore represent a considerable risk for the island of El Hierro. Data collected about eruptions in shallow marine environments on the island of El Hierro show that, although their frequency can be considered as low, they cannot be ruled out as future eruptive scenarios, given that several volcanic episodes with these features have been recorded during the Holocene. Documentary evidence also seems to suggest that the seismic crisis of 1793 could also correspond to a submarine eruptive episode. Although it did not cause direct damage to the population, it strongly affected the island's fishing activity (Béthencourt Massieu 1989). In any case, the island of El Hierro has not been the first in the Canaries to record historical submarine volcanic events. The first historical record of this type of eruption in the Canary Islands is found in the priest of Yaiza's description of the great eruption of Timanfaya, during which he mentions the appearance on the coast of dead deep-sea fish. It must be considered that the historical period in the Canary Islands begins in the twentieth century and that the inhabitants of the islands lived far from the coast in order to protect themselves from attacks by corsairs and pirates (De Armas 1947–50). Therefore, the scarcity of submarine eruptions versus subaerial eruptions in historical times could also be related to the lack of observations instead of the lack of events. Although submarine hydrovolcanism is not the most common eruptive style in the Canary Islands in historical times, this work reveals that explosive hydrovolcanic eruptions of Surtseyan, Satral and Phreatomagmatic type have occurred. Thus, more studies are needed to better understand this type of eruptive scenarios.

Despite the likelihood of this kind of event recurring on El Hierro, as well as in any other Canary Islands, prior to the 2011–2012 Tagoro eruption, no volcanic hazard studies carried out in the Canary Islands had considered submarine eruptions (Felpeto et al. 2007; Laín et al. 2008; Martí and Felpeto 2010; Sobradelo and Martí 2010). Since Tagoro eruption, the possibility of submarine eruptions started to be included in some volcanic susceptibility studies (Becerril et al. 2013; Becerril 2014; Galindo et al. 2016) but the scarcity of data makes it difficult to create possible scenarios based on local data. It is worth emphasising that the Canary Government through the Canary Islands Volcanic Emergency Plan (PEVOLCA; Gobierno de Canarias 2010, 2018), in the versions prior and after the Tagoro eruption, only mention hydrovolcanic eruptions in a general way and do not consider specific scenarios of explosive hydrovolcanic eruptions of Surtseyan, Satral or Phreatomagmatic types neither in the event trees, nor in the volcanic risk maps. The analysis of the eruptive scenarios and volcanic hazards of El Hierro must, therefore, consider any volcanic phenomenon that may originate at any insular point, including the submarine flanks of the island, both intermediate-depth and shallow, as other authors on other islands have suggested (Queiroz et al. 2008).

The expected rise of sea level in the coming decades can lead to an increase in future Surtseyan-style eruptions; This prediction, together with the growth of population, the fact that their social and economic activities and tourist infrastructures are located in coastal areas, determines that the risk associated with these volcanic events will increase. With this work we emphasise the need to improve the knowledge of this type of eruptions in volcanic islands, working first on the identification of the associated deposits, such as those presented in this work, as well as reviewing the historical documentation that could provide useful information on these volcanic events and their effects.

Acknowledgements This research has been funded by the Canarian Agency for Research, Innovation, and Society of Information (ACIISI) under the Government of the Canary Islands through the project LIGCANARIAS (ProID2017010159). The authors would also like to highlight the collaboration of the staff of the La Restinga-Mar de Las Calmas Marine Reserve, El Hierro UNESCO Global Geopark and El Hierro Man and Biosphere Reserve (Cabildo de El Hierro). We especially appreciate the collaboration of Tamia Brito, who has contributed with her knowledge of submarine areas and has provided the photos of El Bajón. We thank the reviewers Adriano Pimentel and Ulrich Kueppers for their constructive comments.

References

Acosta J, Uchupi E, Smith D, Muñoz A, Herranz P, Palomo C, ZEE Working Group (2005) Comparison of volcanic rifts on La Palma and El Hierro, Canary Islands and the island of Hawaii. Geophys Canary Islands 14:59–90

AHN (1793) Expediente formado a representación del Regente de la Real Audiencia de Canaria, en que se remite la carta original que le ha dirigido el S Alcalde Mayor de la Ysla del Yerro, dándole cuenta de los terremotos que se han experimentado en aquella Ysla y aflicción en que se hallan sus moradores. In: En Romero Ruiz C, (1991a) Las manifestaciones volcánicas históricas del Archipiélago Canario. Santa Cruz de Tenerife. Gobierno Autónomo de Canarias. Vol I

Aulinas M, Domínguez D, Rodríguez-González A, Carmona H, Fernández-Turiel JL, Pérez-Torrado FJ, D'Antonio M (2019) The Holocene volcanism at El Hierro: insights from petrology and geochemistry. Geogaceta 65:35–38

Balcells R, Gómez JA (1997) Memorias y mapas geológicos del Plan MAGNA a escala 1: 25.000. El Hierro Island: Hoja 1105-III, Sabinosa. Spanish Geological Survey, Madrid, p 71

Barrow J (1807) Voyage a la Cochinchine, par les iles de Madère, de Ténériffe et du Cap Verd. Traduit par Malte-Brun. Chez F. Buisson, Paris

Becerril L, Cappello A, Galindo I, Neri M, Del Negro C (2013) Spatial probability distribution of future volcanic eruptions at El Hierro Island (Canary Islands, Spain). J Volcanol Geotherm Res 257:21–30

Becerril L, Galindo I, Martí J, Gudmundsson A (2015) Three-armed rifts or masked radial pattern of eruptive fissures? The intriguing case of El Hierro volcano (Canary Islands). Tectonophysics 647:33–47

Becerril L, Ubide T, Sudo M, Martí J, Galindo I, Galé C, Lago M (2016) Geochrono-logical constraints on the evolution of El Hierro (Canary Islands). J Afr Earth Sci 113:88–94

Becerril L (2014) Volcano-Structural Study and Long-Term Volcanic Hazard Assessment on El Hierro Island (Canary Islands), Tesis Doctoral. Universidad de Zaragoza

Bethencourt Massieu A (1982) Los terremotos de 1793 en El Hierro. Homenaje a Alfonso Trujillo. Tomo II. Aula de Cultura del Excmo. Cabildo Insular de Tenerife, Santa Cruz de Tenerife, pp 15–28

Béthencourt Massieu A (1989) La cofradía de Mareantes de San Telmo en Las Palmas de Gran Canaria: proyecto de un montepío textil (1781–1805). Espacio Tiempo y Forma. Serie IV, Historia Moderna, vol 2

Bravo T (1965–1968) Hidrogeología de la isla de El Hierro. Anuario del Instituto de Estudios Canarios, Estudios Canarios, pp 11–13

Carmona J, Romero C, Dóniz J, García A (2011) Characterization and facies analysis of the hydrovolcanic deposits of Montaña Pelada tuff ring: Tenerife, Canary Islands. J Afr Earth Sci 59(1):41–50

Carracedo JC, Rodríguez-Badiola E, Guillou H, Nuez Pestana JD, Pérez Torrado FJ (2001) Geology and volcanology of La Palma and El Hierro, Western Canaries. Estud Geol 57:175–265

Carracedo JC, Day S, Guillou H, Pérez Torrado FJ (1997) Geology of the island of El Hierro, Canary Islands: stratigraphy, structure and tectonism. Excursion guidebook. In: International workshop on volcanism and volcanic hazard in immature intraplate oceanic islands. Tene-rife/Gran Canaria. Estación Volcanológica de Canarias and the Universidad de Las Palmas

Clarke H, Troll V, Carracedo J, Byrne K, Gould R (2005) Changing eruptive styles and textural features from phreatomagmatic to strombolian activity of basaltic littoral cones: Los Erales cinder cone, Tenerife, Canary Islands. Estud Geol 61:121–134

Corbellá Díaz D, Medina López J (1997) Noticias de la isla de San Borondón. Instituto de Es-tudios Canarios. Documentos para la historia lingüística de Canarias. San Cristóbal de La Laguna.

Darías Padrón DV (1929) Noticias generales históricas sobre la isla del Hierro, una de las Canarias. San Cristóbal de La Laguna. Imprenta Curbelo

De Armas AR (1947–50) Piraterías y ataques navales contra las Islas Canarias (Vol. I, II y III) Consejo Superior de Investigaciones Científicas, Instituto Jerónimo Zurita, Madrid

De la Nuez J, Quesada ML (1999) El edificio hidromagmático de Montaña Goteras en la Palma (Islas Canarias). Bol Geol Min 110:19–24

De la Nuez J, Alonso J, Quesada ML, Macau M (1993) Edificios hidromagmáticos costeros de Tenerife (Islas Canarias). Revista De La Sociedad Geológica De España 57:47–59

De la Nuez Pestana J, Álvarez MLQ, Blanco JJA (1997) Los volcanes de los islotes al norte de Lanzarote: Islas Canarias. Fundación César Manrique

Felpeto A, Martí J, Ortiz R (2007) Automatic GIS-based system for volcanic hazard assessment. J Volcanol Geotherm Res 166:106–116

Fritsch KV (1867) Las islas Canarias. Cuadros de viaje. Traducción, Estudio Introductorio y notas de José Juan Batista Rodríguez y Encarnación Tabares Plasencia (2006). Tenerife, Gran Canaria. Taller de Historia. Gobierno de Canarias. Dirección general de Patrimonio Histórico. Cabildo de Lanzarote y La Gomera. Centro de la Cultura Popular Canaria

Fúster JM, Hernán F, Cendrero A, Coello J, Cantagrel JM, Ancochea E, Ibarrola E (1993) Geocronologia de la isla de El Hierro (Islas Canarias). Bol R Soc Esp Hist Nat 88(1–4):85–97

Galindo I, Romero C, Sánchez N, Morales JM (2016) Quantitative volcanic susceptibility analysis of Lanzarote and Chinijo Islands based on kernel density estimation via a linear diffusion process. Sci Rep 6:27381. https://doi.org/10.1038/srep27381

Galindo I, Romero C, Llorente M, Rubio JC, Vegas J, Sánchez N, Díaz A (2017) Resultados preliminares del inventario de lugares de interés geológico submarinos del Geoparque Mundial UNESCO de Lanzarote y archipiélago Chinijo. In: Investigando el mar: viaje al planeta agua. Actas XII Semana Científica Telesforo Bravo. Instituto de Estudios Hispánicos de Canarias, pp 15–40

Galindo I, Vegas J, Romero C, Llorente M, Martín-González E, Ruio JC, Díaz A, Mangas J, Mateo E, Sánchez N (2019) Geoheritage in the shallow submarine slopes of an oceanic volcanic edifice: a new option for diving geotourism. In: Mateo E, Martínez Frías J, Vegas J (eds) Lanzarote and Chinijo Islands geopark: from earth to space. Geoheritage, Geoparks and Geotourism. Springer Nature, New York, pp 85–98. https://doi.org/10.1007/978-3-030-13130-2_6

García-Cacho L, Romero C (2000) Fenómenos Hidromagmáticos en Lanzarote. In: Astiz M, García A (eds) Curso Internacional de Volcanología y Geofísica Volcánica. Serie Casa de los Volcanes 7, Excmo. Cabildo Insular de Lanzarote, Madrid, pp 153–162

Gee MJR, Masson DG, Watts AB, Mitchell NC (2001) Offshore continuation of volcanic rift zones, El Hierro. Canary Islands. J. Volcanol Geoth Res 105:107–119

Gobierno de Canarias (2010) Plan Especial de Protección Civil y Atención de Emergencias por riesgo volcánico en la Comunidad Autónoma de Canarias (PEVOLCA). Boletín Oficial De Canarias 140:18723–19064

Gobierno de Canarias (2018) Plan especial de protección civil y atención de emergencias por riesgo volcánico en la comunidad autónoma de canarias (PEVOLCA). Boletín Oficial De Canarias 158:26312–26542

Guillou H, Carracedo JC, Pérez-Torrado FJ, Rodríguez Badiola E (1996) K-Ar ages and magnetic stratigraphy of a hotspot-induced, fast grown oceanic island: El Hierro, Canary Islands. J Volcanol Geotherm Res 73:141–155

Gutiérrez M, Casillas Fernández R, Balogh K, Ahijado A, Castillo C (2002) La serie volcánica submarina del complejo basal de Fuerteventura: Crecimiento submarino y emersión de la isla

Hansen A, Pérez Torrado FJ, Rodríguez González A, Gimeno D, Aulinas M, Carracedo JC (2008) Evolución volcánica de La Isleta (Gran Canaria, Islas Canarias). Geogaceta 10:1293–1296

Harmonville L (1845) Dizionario delle date. Venezia. Tomo terzo

Hernández Pacheco A (1982) Sobre una posible erupción en 1793 en la isla del Hierro (Cana-rias). Estudios Geológicos nº 38. Madrid, Facultad de Geológicas. Universidad Complutense de Madrid

Herrera R, Huertas Coronel MJ, Ancochea Soto E (2008) Edades 40Ar-39Ar del Complejo Basal de la isla de La Gomera. Geogaceta 44:7–10

Krastel S, Schmincke HU (2002) The channel between Gran Canaria and Tenerife: constructive processes and destructive events during the evolution of volcanic islands. Int J Earth Sci 91(4):629–641

Laín L, Bellido F, Galindo I, Pérez F, Mancebo MJ, Llorente M (2008) La cartografía de peligrosidad volcánica de Tenerife. In: Galindo I, Laín L, Llorente M (eds) El estudio y la gestión de los riesgos geológicos. Instituto Geológico y Minero de España, Madrid, pp 175–186

López C, Blanco MJ, Abella R, Brenes B, Cabrera Rodríguez VM, Casas B, Domínguez Cerdeña I, Felpeto A, de Villalta MF, del Fresno C, García O, García-Arias MJ, García-Cañada L, Gomis Moreno A, González-Alonso E, Guzmán Pérez J, Iribarren I, López-Díaz R, Luengo-Oroz N, Meletlidis S, Moreno M, Moure D, de Pablo JP, Rodero C, Romero E, Sainz-Maza S, Sentre Domingo MA, Torres PA, Trigo P, Villasante-Marcos V (2012) Monitoring the volcanic unrest of El Hierro (Canary Islands) before the onset of the 2011–2012 sub-marine eruption: Monitoring El Hierro Prior 2001 eruption. Geophys Res Lett 39:1846. https://doi.org/10.1029/2012GL051846

Martí J, Colombo F (1990) Estratigrafía, sedimentoíogia y mecanismos eruptivos del edificio hidromagmático de El Golfo (Lanzarote). Bol Geol Min 101:560–579

Martí J, Felpeto A (2010) Methodology for the computation of volcanic susceptibility: an example for mafic and felsic eruptions on Tenerife (Canary Islands). J Volcanol Geotherm Res 195:69–77

Navarro JM, Soler C (1991). El agua en el Hierro. Resumen del Avance del Plan Hidrológico. Excmo. Cabildo Insular de El Hierro y Consejería de Obras Públicas, Vivienda y Aguas, p 49

Németh K, Kósik S (2020) Review of explosive hydrovolcanism. Geosciences 10:44

Pedrazzi D, Martí J, Geyer A (2013) Stratigraphy, sedimentology and eruptive mechanisms in the tuff cone of El Golfo (Lanzarote, Canary Islands). Bull Volcanol 75(7):1–17

Poggio Capote M, Regueira Benítez L (2009) La isla perdida. Memorias de San Borondón desde La Palma. Cartas Diferentes Ediciones. La Palma, Islas Canarias.

Queiroz G, Pacheco JM, Gaspar JL, Aspinall WP, Guest JE, Ferreira T (2008) The last 5000 years of activity at Sete Cidades volcano (São Miguel Island, Azores): implications for hazard assessment. J Volcanol Geoth Res 178(3):562–573

Quesada M, Alonso J, De La Nuez J (1988) Evolución submarino-subaérea del edificio hidromagmático de la Caldereta (La Palma, Canarias). In: II Congr. Geol., España, 377–386.

Risica G, Roberto AD, Speranza F, Carlo PD, Pompilio M, Meletlidis S, Todrani A (2022) Reconstruction of the subaerial Holocene volcanic activity through paleomagnetic and 14C dating methods: El Hierro (Canary Islands). J Volcanol Geotherm Res 425:107526

Romero C, García-Cacho L, Araña V, Luque A, Felpeto A (2000) Submarine volcanism surrounding Tenerife, Canary Islands: implications for tectonic controls, and oceanic shield forming processes. J Volcanol Geotherm Res 103(1–4):105–119

Romero C (1991) Las Manifestaciones volcánicas históricas del Archipiélago Canario. Consejería de Política Territorial Gobierno de Canarias. Sta. Cruz de Tenerife. p 2

Romero C (2007) La visión geográfica de Dámaso de Quesada y Chaves. En Quesada y Chávez, Dámaso de (2007) Canarias ilustrada y puente americano [1770–1779]. Edición a cargo de Paz Fernández Palomeque, Carmen Gómez-Pablos Calvo y Rafael Padrón Fernández. La Laguna. Instituto de Estudios Canarios

Romero C (2016) Historias de volcanes. Isla de El Hierro. En Arozena y Romero (Coord) Temas y Lugares. Homenaje a Eduardo Martínez de Pisón. Serie Homenajes 7. Secretariado de Publicaciones, Universidad de La Laguna, pp 327–373

Ruano EB (1979) San Borondón» octava isla canaria. Seminario de Historia de América de la Universidad de Valladolid, Casa-Museo de Colón

Rueda Nuñez JJ, Abella Meléndez R, Blanco Sánchez MJ, Díaz Suárez EA, Domínguez Cerdeña IF, Domínguez Valbuena J, Fernández de Villalta Compagni M, del Fresno Rodríguez-Portugal C, López Díaz R, López Moreno C, López Muga M, Muñoz Santamaría A (2020). Revisión del catálogo sísmico de las Islas Canarias (1341–2000). Centro Nacional de Información Geográfica. p 230

Ruiz de la Serna E, Cruz Quintana S (1973) Prehistoria y protohistoria de Benito Pérez Galdós. Contribución a una biografía. Los Mareantes de San Telmo. Gran Canaria. Ediciones del Excmo. Cabildo insular de Gran Canaria

Schmidt RALF, Schmincke HU (2000) Seamounts and Island Building. Encyc Volcan 14:383–402

Sheridan MF, Wohletz KH (1983) Hydrovolcanism: basic considerations and review. J Volcanol Geotherm Res 17(1–4):1–29

Siebert L, Simkin T, Kimberly P (2010) Volcanoes of the world, 3rd edn. Univ of California Press, California

Sobradelo R, Martí J (2010) Bayesian event tree for long-term volcanic hazard assessment: application to Teide-Pico Viejo stratovolcanoes, Tenerife, Canary Islands. J Geophys Res Solid Earth 63:115

Tegg T (1811) Chronology or the historian's companion; being an authentic register of events, from the earliest period to the present time. Printed for Thomas Tegg, London

Tous Melía J (1996) El Plan de las Afortunadas Islas del Reyno de Canarias y la isla de San Borondón. Colección Las islas Canarias a través de la Cartografía. Juan Tous Meliá

Trapero M (2008) Sobre los nombres antiguos y modernos que tuvieron y tienen Las Islas de Canarias Estudios de traducción, cultura, lengua y literatura». In memoriam Virgilio Moya Jiménez (ed. Isabel Pascua, Bernardette Rey- Jouvin, Marcos Sarmiento). Universidad de Las Palmas de Gran Canaria. Servicio de Publicaciones, pp 71–100

Troll VR, Carracedo JC (2016) The geology of the Canary Islands. Elsevier, Amsterdam

Verolino A, White JDL, Dürig T, Cappuccio F (2019) Black Point-Pyroclasts of a Surtseyan eruption show no change during edifice growth to the surface from 100 m water depth. J Volcanol Geotherm Res 384:85–102

Viera y Clavijo J (1772–1783) Noticias de la historia general de las Islas de Canaria: contienen la descripción geográfica de todas... con los principales sucesos de los últimos siglos. Santa Cruz de Tenerife: Idea; Valladolid: Maxtor, 2004. 4 v. Reproducción facsímil de la edición de Madrid: Imprenta de Blas Román

Villasante-Marcos V, Pavón-Carrasco FJ (2014) Palaeomagnetic constraints on the age of Lomo Negro volcanic eruption (El Hierro, Canary Islands). Geophys J Int 199:1497–1514

Webster N (1758–1843) A brief history of epidemic and pestilential diseases; with the principal phenomena of the physical world, which precede and accompany them, and observations deduced from the facts stated, vol I–II. https://quod.lib.umich.edu/e/evans/N27531.0001.001/1:12?rgn=div1;view=toc

Young T (1807) A course of lectures on natural philosophy and the mechanical arts. A Catalogue of works relating to natural philosophy and mechanical arts. With references to particular passages and occasional abstracs and remarks, vol 2. Printed for Johnson, London

Part II

The 2011–2014 Volcanic Unrest and Submarine Eruption: Geology

From Magma Source to Volcanic Sink Under Tagoro Volcano (El Hierro, Canary Islands): Petrologic, Geochemical and Physiographic Evolution of the 2011–2012 Submarine Eruption

4

Antonio M. Álvarez-Valero, Olga Sánchez-Guillamón, Irene Navarro, Helena Albert, Antonio Polo Sánchez, José A. Lozano Rodríguez, Adelina Geyer, Joan Martí, Masao Ban, María Gómez-Ballesteros, Manuel Catalán, Natalia García, Eugenio Fraile-Nuez, Ramón Casillas, María C. Martín-Luis, Desirée Palomino, Juan T. Vázquez, Nieves López-González, Daniel Hernández-Barreña, and Elena Núñez-Guerrero

A. M. Álvarez-Valero (✉) · I. Navarro · A. P. Sánchez · N. García · D. Hernández-Barreña · E. Núñez-Guerrero
Departamento de Geología, Universidad de Salamanca, 37008 Salamanca, Spain
e-mail: aav@usal.es

O. Sánchez-Guillamón · D. Palomino · J. T. Vázquez · N. López-González
Instituto Español de Oceanografía (IEO-CSIC), Centro Oceanográfico de Málaga, 29640 Fuengirola, Spain

H. Albert
Departamento de Mineralogía, Petrología y Geología Aplicada, Universidad de Barcelona, 08028 Barcelona, Spain

J. A. L. Rodríguez · E. Fraile-Nuez
Instituto Español de Oceanografía (IEO-CSIC), Centro Oceanográfico de Canarias, 38180 Santa Cruz de Tenerife, Spain

A. Geyer · J. Martí
Geosciences Barcelona—CSIC, 08028 Barcelona, Spain

M. Ban
Faculty of Science, Yamagata University, Yamagata, Japan

M. Gómez-Ballesteros
Instituto Español de Oceanografía (IEO-CSIC), Servicios Centrales de Madrid, 28080 Madrid, Spain

M. Catalán
Real Instituto Y Observatorio de La Armada, 11110 San Fernando, Cádiz, Spain

R. Casillas · M. C. Martín-Luis
Departamento de Biología Animal, Edafología Y Geología, Universidad de La Laguna, Tenerife, Spain

P. J. González (ed.), *El Hierro Island*, Active Volcanoes of the World,
https://doi.org/10.1007/978-3-031-35135-8_4

Abstract

Active volcanoes are key laboratories to carry out detailed research -and monitoring- about the history of magmas before, during and after eruptions. Tagoro, the submarine active volcano at El Hierro Island (Canary archipelago), is a highly favorable case to assess and monitor its daily ongoing behaviour, as well as to study the links between the processes of magma genesis occurring at depth and their derived eruptive events at the surface. In this interdisciplinary research we combine new results of classical petrology (petrography, geochemistry, and thermodynamics) on the volcanic products expelled by Tagoro during the 2011–2012 eruption, with a high-resolution (5 m grid) bathymetry model carried out during 2017, and recent data from magnetometry, to refine the current knowledge of this eruption. Our results mainly reveal (i) slight magma differentiation and mixing processes at c. 12 km depth during a continuous eruptive pulse; (ii) a similar magmatic evolution and residence times at depth between previous and 2011–2012 eruptions on the island; (iii) an insignificant interaction of external fluids with the magma at depth or within the ascent conduit; (iv) a present-day magnetometric anomaly under the Tagoro's area; (v) a minimum volume estimate for the magma withdrawn from the plumbing system at depth.

Keywords

Magmatic systems · Ocean Island volcanoes · Petrology · Geobarometry · El Hierro

4.1 Introduction

The scientific interest to advance our knowledge of how magma reservoirs behave and evolve is a fundamental and long-lasting target in Earth Sciences in general and in volcanology-petrology-geochemistry in particular, as it is directly related to the type of hazards that volcanoes may generate. When an eruption is already inevitable, the magmatic evolution under the volcano strongly depends on how the plumbing system behaves at depth. Yet, direct access to this information in nature is challenging and infrequent. Hence, combining exhaustive petrological studies of erupted products with volcano monitoring data recorded prior to and during an eruption becomes extremely useful to achieve such purpose (e.g. Scandone and Malone 1985; Saunders et al. 2012; Martí et al. 2013a,b; Tárraga et al. 2014; Albert et al. 2016, 2019). Tagoro submarine volcano, formed 1.8 km away from the southern coast of the El Hierro Island (Canary archipelago) during the 2011–2012 eruption, is an exceptional natural laboratory of an active volcanic system that allows us to study the magmatic conditions that led to eruption with direct implications to forecast the possible occurrence of the next eruption. In fact, in a recent work at Tagoro, it has been demonstrated that once the magma source is triggered at depth, the geochemical signal (of helium isotopic ratios) in an already erupting system, can be received well in advance of the geophysical signals (Álvarez-Valero et al. 2018). This has also been observed in e.g. Etna volcano (Paonita et al. 2016, 2021).

In addition to the considerably increasing scientific interest on El Hierro, the 2011–2012 eruption and the consequent formation of Tagoro volcano has provided a unique opportunity to combine monitoring and petrological data to infer how an intraplate oceanic system prepared for a new eruptive episode (e.g. Stroncik et al. 2009; López et al. 2012; Fraile-Nuez et al. 2012; Carracedo et al. 2012; Pérez-Torrado et al. 2012; Padrón et al. 2013; Santana-Casiano et al. 2013, 2016; Martí et al. 2013a,b; Domínguez et al. 2014; Longpré et al. 2014, 2017; Álvarez-Valero et al. 2018; Taracsák et al. 2019). However, despite this intense research, there are essential aspects of the pre-, syn-, and post-eruption behaviour of the magmatic system that still remain elusive, such as (i) assessing the evolution of the El Hierro subvolcanic system since its previous status before the 2011–2012 submarine eruption to the present day, as well as (ii) estimating the minimum volume of magma available in the deep reservoir(s). This interdisciplinary study combines results of petrology (petrography, mineral chemistry and thermodynamic modelling), geochemistry (bulk rock composition and stable isotopes), and geophysics (high-resolution bathymetry and magnetometry) with the aim of contributing to these questions.

4.2 Overview of Tagoro and El Hierro's Geology

El Hierro is the most occidental (27°5′–27°37′ N; 18°14′–18°16′W) of the volcanic islands of the Canary archipelago in the Atlantic Ocean. All the islands in the Canary archipelago except La Gomera, record volcanic activity during the Holocene (Martí et al. 2013a). According to historic records, in the last 600 years there have been more than ten volcanic eruptions in Lanzarote (1730–1736, 1824), Tenerife (1704, 1705, 1706, 1798, 1909) and La Palma islands (1585, 1646, 1677, 1712, 1949, 1971) (Romero 1991). Typically, the eruptions were characterised by a short period of activity (from few weeks to months), and classified as Hawaiian and Strombolian types building cones with different sizes and lava flow volumes (e.g. Romero 1991; Dóniz et al. 2008). Phonolitic volcanism is also present on the archipelago, although subordinated to mafic volcanism (Pellicer 1975, 1977; Fuster 1993; Balcells and Gomez 1997a, 1997b; Carracedo et al. 2001; Pedrazzi et al. 2014).

El Hierro, with a surface of 280 km^2, has been traditionally assumed to be the youngest island of the archipelago, as the oldest dated subaerial rock comes from 1.12 Ma, but the volcanic edifice is also composed of a 4000 m submarine succession of unknown age (Guillou et al. 1996). Its base displays a circular shape, yet from an aerial view it is more similar to a tetrahedron due to the occurrence of three large volcanic-derived landslides (NW, NE and S) (Carracedo 2011). Becerril et al. (2015, 2016) state that the volcano-structural evolution of the island was controlled by a predominant NE–SW strike, which coincides with the main regional trend of the Canary archipelago as whole, and two other dominant structural directions (N–S and WNW–ESE).

The island registered four main eruptive events that formed and shaped the current morphology of the island, at 1.1 Ma; 600–176; 176–134; and 21–2.5 ka (Carracedo et al. 2001).

In October 2011, after three months of volcanic unrest characterised by more than 10,000 earthquakes (magnitude up to 4.3), 5 cm of ground deformation, and significant changes in gas emissions at El Hierro Island, an underwater eruption gave rise to a new shallow submarine volcano (Tagoro) c. 1.8 km southward of La Restinga village at the south of the island (Fig. 4.1). The eruption released large quantities of mantle-derived gases, solutes and heat into the surrounding waters, CO_2 and He fluxes intensified few days before the seismic events and eruption (Padrón et al. 2013; Melián et al. 2014). Periodic bathymetric mapping carried out by the Spanish Institute of Oceanography (IEO) located the main vent and indicated a major growth of the volcano from an initial 300 m depth to 88 m below the sea surface (Fig. 4.1; e.g. Fraile-Nuez et al. 2012, 2016, 2022, Chap. 8 in this book).

The stratigraphic, petrological and geochemical data obtained from this eruption suggest that Tagoro is a submarine monogenetic volcano developed on the offshore slope of El Hierro Island, characterized by the eruption of mafic alkaline magmas, mainly basanites (e.g. Martí et al. 2013a, b; Longprè et al. 2014; Álvarez-Valero et al. 2018). Recent detailed studies of the petrologic and geochemical features of other offshore products previous to Tagoro (Stroncik et al. 2009), as well as for the 2011–2012 event (e.g. Martí et al. 2013a,b; Longprè et al. 2014) described its plumbing system at depth, suggesting a combination of magma mixing events with subsequent fractionation at uppermost mantle depths.

Fig. 4.1 **a** Location of the submarine eruption and El Hierro Island within the Canary archipelago; **b** Bathymetric map of Tagoro volcano area with the location of the rock samples dredged during the VULCANO-1013 oceanographic campaign. Bathymetric data acquired at high resolution of 5 × 5 m during the VULCANA-0417 oceanographic campaign on board of R/V Ramón Margalef (Spanish Institute of Oceanography—IEO); **c–h** Hand-specimen examples of some dredged basanites during VULCANO-1013 campaign. Samples DA06 **e**, DA07 **f**, DA09 **h** within the main edifice are from 360 to 110 m b.s.l.; DA01 **c**, DA02 **b** in the lava flow deposits of the SW flank from 850 to 400 m b.s.l., and DA08 **g** from 250 to 170 m b.s.l. 1.8 km northwest of the main edifice

4.3 Methods

To improve our understanding of the magmatic evolution that gave rise to the formation of Tagoro volcano we combined existing data with new geophysical and petrological data. Since 2011 to present, the IEO has carried out systematic oceanographic campaigns of bathymetry (Bimbache, VULCANO and VULCANA projects), as well as sporadic trawl-dredging and magnetometry tasks, to monitor Tagoro's evolution.

The bathymetric data used in this chapter were obtained during the VULCANA-0417 campaign yielding a digital elevation model of Tagoro volcano at 5 m resolution. The dredged samples used in this study were obtained during the VULCANO-1013 oceanographic campaign on board of the Ramón Margalef (IEO) oceanographic vessel from 29 October to 12 November 2013 at different locations on the Tagoro edifice. According to the sea-floor description provided by Somoza et al. (2017), the samples were collected both in the oldest part of the volcanic edifice (Jan–Feb 2012) which is the main exposed one (DA01, DA02, DA06, DA07, DA09), and on the collapsed (and youngest) sector of the edifice (DA08a) formed in October 2011 (Fig. 4.1; Table 4.1). These samples were selected among the entire dredge-load material according to their representativity and freshness (Fig. 4.1c–h). In addition, we studied a pre-2011 eruption sample (DA08m; Álvarez-Valero et al. 2018) to be compared with the 2011–2012 episode.

Table 4.1 Main features summary of the dredges collected at Tagoro

Rock stations	Collecting date	Depth b.s.l. (m)	Eruption area	Hand-specimen description	Techniques
DA01	29-10-2013	792–579	Proximal apron	Angular with prismatic shape composed of alternating levels of coarse sand-fine gravel	EMP, XRF, P–T, D/H, $\delta^{18}O$, NG*
				To fine sand-silt (scoria-lapilli and ash)	
DA02	28-10-2013	643–343	Proximal apron	Well sorted fragmentary material of highly vesiculated vitreous-aphanitic pyroclastics	EMP, XRF, P–T, D/H, $\delta^{18}O$
DA06	28-10-2013	205–196	Main cone	Fragments have red or orange patina of iron oxides and hydroxides that may be due to hydrothermal alterations	
DA07	29-10-2013	358–222	Upper slope	Poorly sorted fragmentary material with angular fragments between 0.5 and 20 cm	EMP, GS
DA08	29-10-2013	269–165	NW outer cone	Massive fragments of lapilli (DA08a); rounded dense bomb rich in olivine crystals (DA8m)	EMP, XRF, D/H, $\delta^{18}O$, NG*, GS
DA09	29-10-2013	225–173	NW main cone	Vesiculated aphanitic and vitreous material (50% vesicles, up to 90%)	EMP

EMPA, electron microprobe analyses; XRF, X-ray fluorescence; P–T, geothermobarometric calculations; NG*, noble gases from Álvarez-Valero et al. 2018; GS, Fe–Mg diffusion geospeedometry in olivine crystals; b.s.l.: below sea level.

Whole-rock major and trace element concentrations were determined by X-ray fluorescence (XRF) analysis with a Rigaku RIX2000 spectrometer at Yamagata University (Japan). Operating conditions were 50 kV accelerating voltage and 50 mA current. The preparation method of the glass disks and the calibration methods for major and trace elements based on Yamada et al. (1995). The standards used in the analyses are the GSJ (Geological Survey of Japan) igneous rocks series. Analytical uncertainties are 5% for Nb, Zr, Y, Sr, Rb and Ni, 10% for V and Cr, and 5–15% for Ba. The range of uncertainties for a single element is based on the concentration range observed in standards.

Major element contents in glass and minerals were analysed on thin sections using a JEOL JXA-8230 electron microprobe at the Scientific and Technological Centre of Barcelona University (CCiTUB). Operating conditions were 20 kV acceleration voltage, 15 nA beam current, 2 μm beam diameter (using a defocused beam of up to 20 μm to minimise sodium mobility) and counting time of 10 s per element. The estimated accuracy and standards used were: SiO_2 (±0.13%, wollastonite); Al_2O_3 (±0.09%, corundum); TiO_2 (±0.13%, rutile); Cr_2O_3 (±0.14%, Cr_2O_3); Na_2O (±0.68%, albite); MgO (±0.11%, periclase); MnO (±0.16%, rhodonite); FeO (±0.11%, Fe_2O_3); ZnO (±0.11%, sphalerite); K_2O (±0.19%, feldspar); and CaO (±0.02%, calcium standard) and VO_3 (±0.20%, vanadium metal).

Crystallographic orientations of the compositional profiles were estimated through SHAPE sofware (Dowty 1987) by matching the BSE images with the 2D section view. Crystal DA07 was modelled employing a forsterite model based on Roubault et al. (1963) (Miller Indexes: [010], [001], [021], [110] with central distances 1, 2, 1.4, 1.05, respectively). Whereas, for crystal DA08m we applied 1.8 for the central distance of [001] and 0.4 for the section height.

For the stable isotope analysis of bulk-rock we cut and discarded several centimeters of material from the surface to avoid any potential late isotopic modification by weathering. The hydrogen and oxygen isotopic analyses were carried out at the Servicio General de Análisis de Isótopos Estables (NUCLEUS—University of

Salamanca, Spain). Oxygen in glass and phenocrysts was extracted by fluorination (Clayton and Mayeda 1963) employing a Synrad 25W CO_2 laser (Sharp 1990) and ClF_3 as reagent (e.g. Borthwick and Harmon 1982), and oxygen isotope ratios were measured on a VG-Isotech SIRA-II dual-inlet mass spectrometer. Both internal and international reference standards (NBS-28, NBS-30) were run to check accuracy and precision. Long-term reproducibility for repeated determination of reference samples was better than $\pm$ 0.2‰ (1σ). Results are reported in $\delta^{18}O$ notation relative to the Vienna Standard Mean Ocean Water (V-SMOW) standard, using a $\delta^{18}O$ value of 9.6‰ for NBS-28 (quartz) for the mass spectrometer calibration.

D/H ratios were determined on another SIRA-II mass spectrometer on H_2 gas obtained by reduction over hot depleted-U of the water released by induction heating of samples. A vacuum line (Bigeleisen et al. 1952), following the procedures described by Godfrey (1962) with modifications (Jenkin 1988), was used for gas extraction. Samples were loaded into degassed platinum crucibles that were placed in quartz reaction tubes and heated under vacuum to 125° C overnight to remove any adsorbed H_2O. Results are reported in δD notation relative to the V-SMOW standard, using a $\delta D = -66.7$‰ for NBS-30 (biotite) for the mass spectrometer calibration. Long-term reproducibility for repeated determination of reference samples was better than $\pm$ 2‰ (1σ). The amount of H_2 recovery is known by a baratron gauge reading, which measures the total hydrogen (non-condensables) derived from water according to the ideal gas law. Then, we calculated the water content (% H_2O), as a function of the amount of H_2 obtained and the sample weight (wt%).

For the magnetometric measurements, we employed a Marine Seaspy magnetometer. This equipment was placed 200 m astern of the ship to guarantee an accuracy of more than 1 nT. The rate of acquisition of the data was 0.166 Hz, which would imply a spatial resolution along the line of 31 m at a speed of 10 knots (i.e. 18.52 km/h). The position of the vessel was obtained from ship's own computer system. These locations were affected by a correction that correctly positioned the measurement.

We used Güimar (Tenerife Island) geomagnetic observatory data to extract the contribution by external fields. The internal field contribution was extracted using the IGRF-12 model (Thébault et al. 2015). To determine the depth to the top of magnetic basement of magnetic bodies we have used two different approaches. First, we have applied Werner Deconvolution (Hartman et al. 1971; Ku and Sharp 1983). It is a powerful tool for the interpretation of magnetic profiles. To get a set of depth solutions using this method we have to define the size of a window. The mathematical procedure started using the smallest window size selected. The window moves along the profile to the end. Then the window size is incremented and the entire profile is processed again until the maximum size is reached. As minimum and maximum size we set 100 and 3000 m respectively, and the window expansion and shift increments were set at 100 and 500 m respectively. We have obtained a set of solutions that show a homogenous picture. Most of solutions fall between 700 m and 2 km below the sea level (b.s.l.). To cross-check this, we have obtained the radial average spectrum of the magnetic anomaly map showing a reliable solution. Spector and Grant (1970) stated that the shape of the power-density spectrum of the magnetic anomaly is strongly controlled by the average depth of the magnetic causative body. Specifically, the spectrum decays exponentially with wavenumber, at a rate of decay proportional to the average depth to the top of the magnetic source.

4.4 Results

4.4.1 Bathymetry and Dredged Samples Description

Two large morphological units can be distinguished in the volcanic materials erupted by the 2011–2012 eruption (Fig. 4.1b), namely, the first one that built the main edifice and the second one constituting the flow of material that mainly extends to the southwest (Vázquez et al. 2022, Chap. 7 in this book). The main edifice is at 88 m b.s.l., with slopes dipping 20°–34°, extending up to a depth of c. 400 m b.s.l., where the slope decreases below 20°. The base of this edifice is pseudo-circular with diameter ranging from 0.9 to 1.2 km, and irregularities in shape due to the presence of two sets of morphological elements: (i) four ridges, two of them with a NE–SW direction located on the west flank of the main edifice, a third with an ENE–WSW trend that crosses the proximal secondary cone, and another one with a NNW–SSE orientation located on the eastern flank of the main edifice (Vázquez et al. 2016); (ii) several secondary cones from NNW to SSE decreasing in height (from 88 to 247 m b.s.l.; Fig. 4.1b) and size, showing at least 13 old vents (see also Vázquez et al. 2016).

In hand-specimen, the dredged materials show in general the typical features of basaltic products of a submarine explosive eruption (Fig. 4.1c–h). Samples in dredge DA01 are fragmentary to a large extent. These fragments range from (i) very vesiculated vitreous-aphanitic material with vesicles and quenched glass oriented in parallel and incipient radial disjunction similar to pillow-lava fragments or the so-called "pieces of cake" (sample DA01b), coming from the pillow-lava flow; to (ii) fragments of consolidated lapilli (sample DA01a), likely related to a previous collapsed pyroclastic deposit of the first edifice. In dredge DA02, the samples are fragments of lava flows, vitreous-aphanitic with vesicle contents becoming progressively higher towards the centre of each fragment, coming from the lava sliding of the last edifice. Materials in dredges DA03 and DA04 are glassy, highly vesiculated pyroclastic fragments (c. 70 and 50% vesicles, respectively). The former are similar to the amphora-like floating lava balloons described by Somoza et al. (2017). Sample DA06 is a heterogeneous fragmentary material of mostly vitreous-aphanitic fragments, with a variable range of vesicle amounts and sizes, and most probably represents pyroclastic material of lapilli and ash after a more highly explosive phase of activity. Sample DA07 consists of fairly homogeneous pyroclastic fragments mixed with lava fragments of various different morphologies. Samples in dredge DA08 are heterogeneous with fragments of lapilli (DA08a), and a rounded basanite bomb (DA08m) with numerous colonies of black corals and other organisms. Samples in dredge DA09 are pyroclastic fragments (scoria and lapilli) from explosive eruptions during the last edifice building episode. The studied sample from dredge DA09 is highly vesiculated (50–90%).

4.4.2 Petrographic Analysis

The bubble-rich hypidiomorphic basanites erupted in 2011–2012 generally contain (micro)phenocrysts of mainly olivine from subhedral to anhedral, clinopyroxene (augite), ulvospinel (titanomagnetite), and locally plagioclase (mostly as microlites within the mesostasis). Olivine phenocrysts are generally characterized by reverse forsterite (Fo) zoning patterns from core to rim followed by a Fo decrease towards the edge. Groundmass textures range from hypocrystalline, aphanitic, microcrystalline to porphyric with a wide range of bubble volumes and shapes (Fig. 4.2; see also Martí et al. 2013a, b). Sample DA08m shows the highest modal proportion of phenocrysts, as well as two generations of phenocrysts with local undulated olivine and pyroxene, while samples DA07, DA02 and DA01b display the higher volumetric proportion of vesicles. DA01a and DA08a are more massive than the latter, thus showing a lower vesicular volume. DA01a also shows local undulated olivine phenocrysts (Fig. 4.2). According to the representative hand-specimen and petrographic features of all the collected samples, we selected (i) DA01b, DA02, DA07, DA08a, DA08m, DA09 for the mineral chemistry study in order to analytically cover all the crystals' variability. We also focused on the glasses aiming to track any potential variation compared to their respective bulk rock, thus assessing the compositional evolution of the residual melts; and (ii) samples DA01a, DA01b, DA02, DA08a, DA08m covering the whole range of macroscopic features of all the dredged samples, for the analysis of bulk composition.

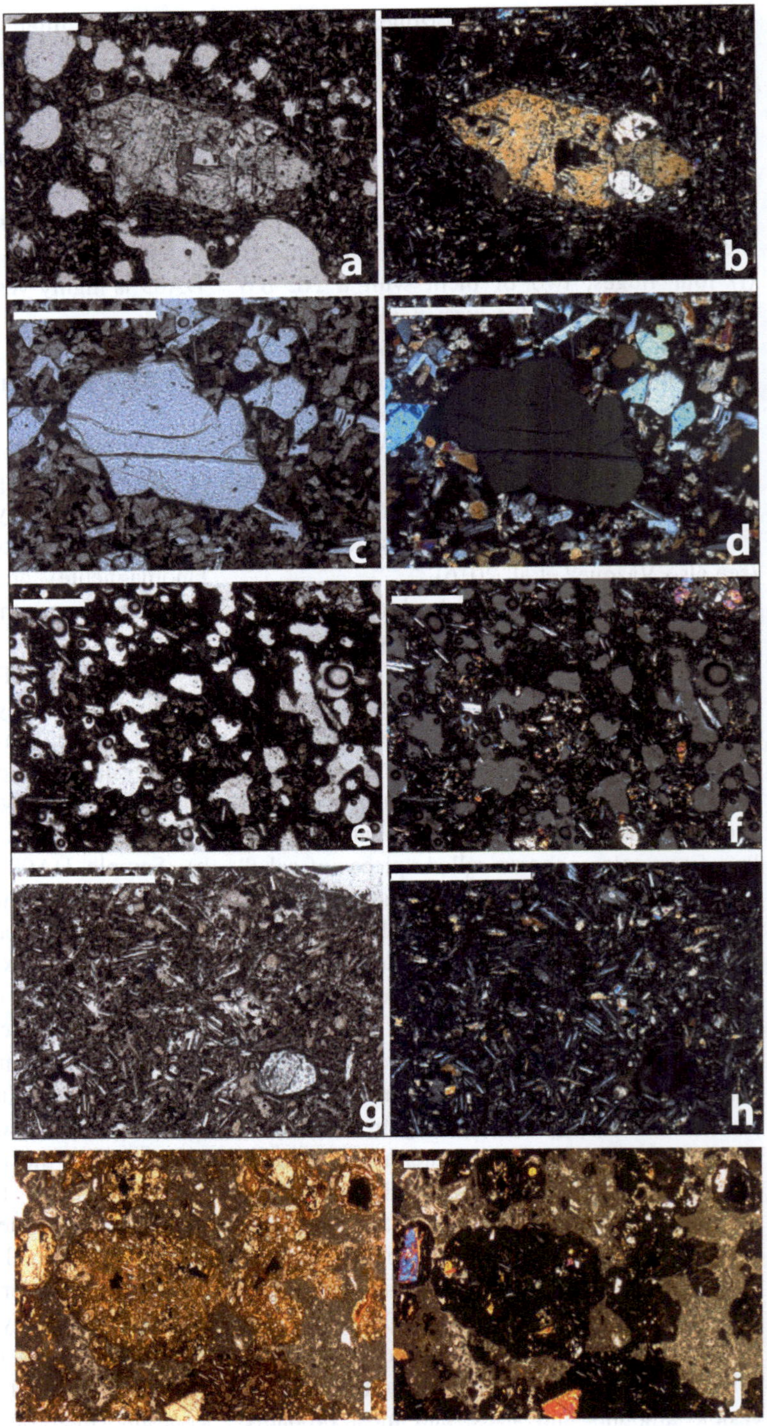

Fig. 4.2 Photomicrographs of the studied samples under plane-polarized light (left column) and cross-polarized light (right column) highlighting: **a**, **b** euhedral olivine phenocryst with uniform optical extinction (DA01b) within a glassy and bubbly groundmass; **c**, **d** subeuhedral olivine phenocryst with undulatory extinction (DA08m) within a microcrystalline groundmass. Similar olivines are also locally observed in DA01a. **e**, **f** Sample DA07 with mostly glassy groundmass and abundant vesicles. **g**, **h** Massive DA01a poor in vesicles and rich in (micro) phenocrystals of Ol, Pl, Cpx and secondary oxides. **i**, **j** Different glass microdomains in sample DA08a. In each image the scale bar is 500 microns

4.4.3 Whole-Rock Geochemistry (X-Ray Fluorescence, XRF)

With the exception of sample DA08m, all samples i.e. DA01a, DA01b, DA02, DA08a are compositionally very similar (Table 4.2; Figs. 4.3, 4.4). All samples are alkaline basanites with MgO and SiO_2 contents of 7.18–8.11 wt% and 42.45–42.94 wt%, respectively, whereas sample DA08m contains 12.56 wt% MgO and 41.24 wt% SiO_2. Sample DA08m also displays different abundances of some trace elements such as Cr (467 ppm) and Ni (299 ppm), with respect to the rest of samples that contain 147–225 ppm Cr, and 94–139 ppm Ni.

4.4.4 Glass and Mineral Chemistry (Electron Microprobe)

Glasses are characterized by a constant composition for each particular eruptive period, yet different compositional ranges as a function of their eruption chronology (along the five months event) and their location at the surface. Only samples belonging to the first edifice (i.e. DA01a and DA08a from the first eruption period starting in October 2011) have more evolved compositions (> 50% SiO_2 wt%) phono-tephrite and phonolite, respect to the subsequently erupted samples that formed the last edifice (ca. 45–48 SiO_2 wt%) (Figs. 4.3, 4.4; Table 4.2). The latter are also compositionally very similar to the glasses of the pre-2011 eruption (i.e. sample DA08m) (45–46 SiO_2 wt%).

Olivine phenocrysts have display a slight reverse zoning patterns and do not vary systematically with their host whole-rock composition. Only the pre-2011 sample DA08m, shows a lower Fo content (Fo_{78}, Fig. 4.2; Table 4.2). Clinopyroxene phenocrysts (classified as diopside) are also similar in composition for all samples with average values of 7 FeO wt%, 22 CaO wt% and 13 MgO wt%. Plagioclase (bytownites) show a fairly constant composition for all samples of c. $An_{78\text{-}86}$. Local oxide minerals are also present as secondary phases (Fig. 4.2).

4.4.5 Geothermobarometry

Pressure (P) and temperature (T) estimates were constrained by applying and combining classical geothermobarometers and thermodynamical modelling through the thermodynamic database of rhyolite-MELTS software v.1.2.0 (Ghiorso and Sack 1995; Asimow and Ghiorso 1998). Concerning the former (e.g. Albarede 1992; Nimis 1999; Putirka 2005, 2008), we applied them between pairs of minerals (rim versus rim) or mineral-glass at equilibrium, i.e. when the equilibrium constant ($K_D = X_{FeO}/X_{MgO}$) mineral (X_{FeO}/X_{MgO}) melt were ca. 0.3 for olivine and clinopyroxene versus glass (e.g. Roeder and Emslie 1970; Putirka et al. 1996; Stroncik et al. 2009). Pressure results of 1.2–2.3 GPa are however locally not in line with the depths from previous studies (e.g. Stroncik et al. 2009; Martí et al. 2013a,b; Longprè et al. 2014). Hence, we are cautious because this inconsistency is related to the differences between the input parameters used in the mentioned experiments and the real conditions of the Tagoro sub-volcanic system: for instance, orthopyroxene is normally assumed in the equilibrium system of the experiments (e.g. Putirka 2008), yet it is absent in the Tagoro samples; or the silica activity experiments carried out under conditions of peridotites (e.g. Albarede 1992) but not properly basanites as the Tagoro rocks. Regarding the thermodynamical modelling, it allows to constrain P and T by considering equilibrium among the phases involved in the main paragenesis, as well as the observed versus computed modal proportions of glass and crystals within the groundmass. For the modelling calculations we used the whole-rock compositions (Table 4.2a) as input "parental" magmas into the rhyolite-MELTS software and retrieved phenocrystic assemblages and abundances with chemical compositions for various T–P–H_2O conditions (an initial 1–1. 5 wt% H_2O) under oxygen fugacity (fO_2) conditions of 1 log units above the Niquel-Niquel-Oxide (NNO) buffer (Ban et al. 2008) at pressures from 0.1 to 0.5 GPa until equilibria were achieved; i.e. we finally compared the best fit of SiO_2–X_{Mg}–T–P conditions between the calculated and the

Table 4.2 Geochemical results of the studied samples

(a)

	SiO_2	Al_2O_3	TiO_2	FeO	Fe_2O_3	MgO	MnO	CaO	Na_2O	K_2O	P_2O_5
D1a	42.94	13.45	4.74	11.24	2.21	7.18	0.18	10.72	3.65	1.49	0.89
D1b	42.57	12.82	4.75	11.3	2.22	8.07	0.18	10.95	3.57	1.44	0.81
D2	42.45	12.78	4.76	11.24	2.2	8.11	0.17	10.93	3.66	1.43	0.81
D8a	42.79	13.34	4.68	11.34	2.22	7.29	0.18	10.62	3.71	1.51	0.89
Pre-2011 eruption											
D8m	41.24	10.74	4.13	11.69	2.29	12.56	0.18	11.32	2.85	1.04	0.87

(a)

	Rb	Ba	Sr	Zr	Y	Nb	Cr	Ni	V	Zn	Cu
D1a	33	359	975	402	35.5	78.4	151	94.3	333	131	75.5
D1b	31.5	337	906	365	32.9	70.9	225	138	343	126	97.1
D2	30.3	334	905	365	32.7	70.4	224	139	335	125	88.4
D8a	33.8	366	979	406	36.1	78.7	147	96.4	321	134	74
Pre-2011 eruption											
D8m	19.8	285	823	287	28.5	58.3	467	299	323	118	89.8

(b)

	SiO_2	Al_2O_3	TiO_2	FeO	MgO	MnO	CaO	Na_2O	K_2O	Total
DA01b	48.47	15.51	2.92	10.04	2.27	0.28	5.67	5.69	3.41	94.26
DA01b	47.97	16.08	3.09	9.78	2.40	0.21	5.93	6.61	3.24	95.33
DA01b	48.26	16.90	2.99	10.24	1.77	0.24	4.38	6.34	3.87	94.99
DA01b	46.79	16.04	3.55	11.20	2.62	0.30	5.91	7.10	3.51	97.05
DA01b	46.56	16.12	3.43	10.43	2.78	0.19	6.83	7.02	3.49	96.88
DA01b	46.77	15.98	3.58	10.76	2.51	0.26	6.30	7.23	3.38	96.81
DA01b	46.62	16.21	3.39	10.14	2.79	0.23	6.77	6.65	3.34	96.15
DA01b	47.18	16.51	3.33	10.15	2.76	0.19	6.79	6.70	3.24	96.87
DA01b	47.79	16.50	3.06	10.36	2.71	0.20	6.17	6.71	3.58	97.13
DA01b	49.83	16.87	2.65	8.49	2.37	0.21	5.85	5.48	3.45	95.21
DA02	46.08	14.91	4.51	12.19	3.65	0.30	8.67	4.04	2.83	97.24
DA02	46.29	15.41	4.55	12.21	3.31	0.25	7.55	4.67	2.96	97.27
DA02	45.41	15.04	4.62	11.99	3.40	0.23	8.38	4.53	2.81	96.46
DA08a	52.90	19.63	1.00	5.51	0.45	0.04	0.10	7.36	6.82	93.84
DA08a	54.41	21.64	0.87	5.17	0.54	0.09	0.34	10.59	5.56	99.21
DA08a	52.15	20.86	0.54	5.09	0.61	0.10	1.94	9.49	6.55	97.37
DA08a	51.83	20.66	0.68	6.34	0.95	0.17	0.70	9.87	7.11	98.32
DA08a	51.24	21.02	0.61	5.95	0.94	0.16	1.03	9.39	6.69	97.03
DA08a	52.56	20.62	0.58	5.88	0.77	0.11	0.70	9.82	7.21	98.26
DA08a	53.36	21.35	0.94	5.54	0.73	0.11	0.21	9.72	7.21	99.20
DA08a	50.91	20.09	0.66	4.78	0.41	0.09	1.50	8.46	6.62	93.53
DA08a	51.23	19.68	0.64	5.66	0.54	0.11	0.05	8.13	7.10	93.14
DA09	44.81	14.96	4.71	11.50	3.86	0.22	9.35	4.99	2.19	96.64

(continued)

Table 4.2 (continued)

(b)

	SiO_2	Al_2O_3	TiO_2	FeO	MgO	MnO	CaO	Na_2O	K_2O	Total
Pre-2011 eruption										
DA08m	45.40	15.13	4.15	10.96	4.44	0.20	8.85	4.77	2.43	96.43
DA08m	45.05	15.82	4.38	11.01	4.54	0.26	8.86	4.95	2.20	97.09
DA08m	45.57	15.77	4.34	10.60	4.48	0.22	8.73	4.78	2.25	96.81
DA08m	45.64	15.76	4.33	10.71	4.58	0.26	8.78	4.68	2.31	97.11
DA08m	45.56	15.70	4.25	10.68	4.48	0.19	8.93	4.77	2.25	96.86
DA08m	45.83	15.62	4.26	10.75	4.70	0.16	8.92	4.78	2.27	97.37
DA08m	46.45	15.83	4.31	10.59	4.31	0.18	8.83	4.96	2.35	97.85
DA08m	45.35	15.61	4.37	10.77	4.53	0.21	8.84	4.94	2.14	96.81
DA08m	45.78	15.79	4.34	10.81	4.61	0.23	8.72	4.96	2.34	97.67
DA08m	45.28	15.51	4.30	10.84	4.42	0.16	8.79	4.93	2.21	96.47
DA08m	45.75	15.61	4.29	10.75	4.27	0.16	8.91	5.02	2.25	97.09

(c)

	SiO_2	Al_2O_3	TiO_2	FeO	MgO	MnO	CaO	Na_2O	K_2O	Total
DA01b	40.03	0.02	0.03	17.42	41.79	0.24	0.27	0.02	0.01	99.87
DA01b	40.40	0.00	0.03	16.71	42.46	0.25	0.26	0.04	0.00	100.17
DA01b	40.04	0.02	0.04	16.98	42.41	0.23	0.26	0.00	0.01	100.00
DA01b	40.09	0.02	0.01	17.61	42.18	0.24	0.28	0.03	0.01	100.51
DA01b	40.76	0.00	0.03	15.29	43.75	0.21	0.27	0.00	0.01	100.34
DA01b	40.55	0.04	0.03	15.23	43.89	0.21	0.27	0.00	0.01	100.27
DA01b	40.87	0.04	0.05	14.20	44.46	0.21	0.24	0.02	0.00	100.16
DA01b	41.01	0.01	0.02	16.80	42.97	0.20	0.19	0.02	0.00	101.24
DA01b	40.94	0.03	0.03	17.21	41.92	0.26	0.29	0.00	0.01	100.71
DA01b	40.01	0.02	0.05	17.73	41.92	0.24	0.23	0.03	0.00	100.22
DA01b	40.53	0.04	0.03	16.45	42.33	0.22	0.28	0.02	0.01	99.95
DA01b	40.30	0.04	0.03	16.42	42.98	0.26	0.28	0.03	0.00	100.40
DA01b	40.23	0.00	0.05	17.73	42.21	0.23	0.26	0.00	0.00	100.73
DA01b	40.21	0.05	0.02	17.21	42.25	0.28	0.26	0.00	0.01	100.29
DA01b	40.29	0.06	0.03	17.55	41.47	0.21	0.28	0.00	0.00	99.90
DA01b	41.36	0.03	0.04	17.36	41.68	0.25	0.29	0.03	0.01	101.09
DA01b	40.47	0.03	0.03	16.92	42.09	0.24	0.29	0.06	0.02	100.18
DA02	39.82	0.03	0.03	16.59	43.35	0.23	0.27	0.00	0.01	100.38
DA02	40.47	0.04	0.03	17.31	42.16	0.26	0.28	0.04	0.00	100.60
DA02	40.47	0.04	0.04	15.93	43.09	0.23	0.24	0.02	0.00	100.09
DA02	40.33	0.05	0.02	16.15	42.81	0.24	0.24	0.00	0.00	99.86
DA02	40.51	0.02	0.04	17.34	41.80	0.21	0.27	0.01	0.01	100.22
DA02	40.63	0.00	0.04	17.10	42.79	0.26	0.29	0.00	0.02	101.14
DA02	40.68	0.05	0.04	17.87	41.59	0.26	0.32	0.00	0.03	100.84
DA02	40.19	0.04	0.02	17.20	42.45	0.26	0.26	0.01	0.00	100.45

(continued)

Table 4.2 (continued)

(c)

	SiO_2	Al_2O_3	TiO_2	FeO	MgO	MnO	CaO	Na_2O	K_2O	Total
DA02	40.97	0.06	0.03	15.79	43.15	0.22	0.28	0.00	0.01	100.53
DA02	40.19	0.01	0.02	15.52	43.60	0.19	0.28	0.00	0.01	99.87
DA02	40.39	0.04	0.01	16.85	42.58	0.25	0.26	0.04	0.01	100.48
DA02	40.30	0.02	0.03	16.69	43.02	0.22	0.27	0.02	0.02	100.67
DA02	40.59	0.04	0.07	17.31	41.54	0.25	0.28	0.00	0.03	100.15
DA02	40.29	0.05	0.09	18.29	41.61	0.24	0.32	0.01	0.01	100.93
DA07	40.37	0.06	0.04	17.19	42.80	0.25	0.18	0.00	0.00	100.91
DA07	40.68	0.05	0.02	14.47	44.92	0.21	0.20	0.01	0.02	100.63
DA07	40.43	0.01	0.04	17.39	42.43	0.28	0.27	0.00	0.01	100.91
DA07	40.35	0.07	0.06	17.60	42.70	0.25	0.26	0.03	0.01	101.36
DA07	40.25	0.05	0.04	17.24	42.71	0.24	0.27	0.00	0.01	100.83
DA07	40.26	0.03	0.06	18.79	40.92	0.26	0.35	0.01	0.00	100.71
DA08a	39.99	0.04	0.02	19.64	40.80	0.27	0.22	0.00	0.02	101.02
DA08a	39.78	0.00	0.03	19.61	41.36	0.28	0.22	0.00	0.04	101.33
DA08a	40.24	0.00	0.07	18.82	41.04	0.29	0.30	0.02	0.01	100.79
DA09	39.81	0.03	0.03	16.38	42.29	0.20	0.25	0.00	0.02	99.04
DA09	40.59	0.07	0.03	14.67	43.93	0.19	0.26	0.00	0.02	99.80
DA09	40.84	0.02	0.02	13.13	44.59	0.17	0.28	0.01	0.00	99.11
DA09	39.93	0.04	0.07	17.65	41.68	0.25	0.27	0.01	0.03	99.94
Pre-2011 eruption										
DA08m	39.85	0.04	0.03	20.08	40.84	0.30	0.21	0.00	0.00	101.36
DA08m	39.12	0.02	0.01	19.71	40.55	0.31	0.20	0.00	0.01	99.94
DA08m	40.22	0.02	0.02	18.30	42.17	0.28	0.13	0.00	0.00	101.15
DA08m	39.89	0.04	0.04	18.83	41.57	0.26	0.14	0.00	0.02	100.82
DA08m	40.38	0.02	0.01	18.08	42.29	0.26	0.11	0.03	0.03	101.23
DA08m	40.37	0.03	0.00	17.31	42.35	0.27	0.18	0.04	0.00	100.56
DA08m	40.73	0.03	0.00	15.42	44.23	0.24	0.29	0.00	0.02	100.98
DA08m	40.11	0.02	0.01	18.91	41.39	0.24	0.13	0.00	0.01	100.81
DA08m	40.60	0.05	0.03	17.35	42.85	0.24	0.16	0.02	0.01	101.35
DA08m	40.54	0.00	0.04	16.12	43.57	0.24	0.33	0.01	0.01	100.88
DA08m	40.26	0.06	0.04	16.16	43.70	0.19	0.32	0.00	0.02	100.81
DA08m	40.60	0.06	0.04	14.88	44.93	0.21	0.23	0.00	0.02	100.98
DA08m	41.11	0.03	0.02	15.56	44.18	0.21	0.27	0.00	0.00	101.40
DA08m	40.52	0.01	0.03	15.49	44.33	0.26	0.23	0.06	0.01	100.95

(d)

	SiO_2	TiO_2	Al_2O_3	FeO	MnO	MgO	CaO	Na_2O
DA01b	47.71	2.12	6.81	8.56	0.22	11.42	21.86	0.79
DA01b	47.33	2.33	7.24	8.34	0.15	11.70	21.79	0.83
DA01b	48.46	2.75	4.44	7.16	0.09	13.56	22.25	0.34

(continued)

Table 4.2 (continued)

(d)

	SiO_2	TiO_2	Al_2O_3	FeO	MnO	MgO	CaO	Na_2O
DA01b	50.40	1.86	4.02	6.01	0.10	14.41	22.23	0.45
DA01b	44.87	4.17	7.97	7.21	0.12	11.73	22.53	0.52
DA01b	48.67	2.90	4.86	7.05	0.11	13.34	22.37	0.45
DA01b	47.84	2.59	6.17	6.44	0.08	13.07	22.35	0.46
DA01b	50.42	1.87	4.14	5.87	0.09	14.16	22.35	0.45
DA01b	46.18	3.22	8.36	6.93	0.14	12.50	21.63	0.60
DA01b	49.79	2.35	4.33	6.82	0.14	14.10	22.18	0.37
DA01b	42.94	4.98	8.85	7.87	0.09	11.01	22.37	0.56
DA01b	48.73	2.95	4.99	6.98	0.12	13.19	22.42	0.44
DA01b	44.34	4.80	8.55	7.85	0.10	11.10	22.29	0.57
DA01b	46.38	3.36	6.96	7.50	0.08	12.23	22.35	0.50
DA01b	48.92	2.25	6.29	6.94	0.13	12.66	21.64	0.63
DA01b	44.98	4.37	7.90	7.60	0.11	11.42	22.35	0.47
DA01b	44.83	4.39	8.26	7.24	0.09	11.39	22.41	0.53
DA01b	49.10	2.81	4.73	7.11	0.13	13.65	22.46	0.42
DA01b	50.34	2.06	3.86	6.50	0.07	14.25	22.25	0.46
DA01b	49.17	2.68	4.87	6.82	0.13	13.43	22.49	0.44
DA02	50.25	2.44	4.21	6.86	0.13	14.04	22.40	0.35
DA02	44.41	4.64	8.49	7.62	0.10	11.31	22.34	0.44
DA02	50.75	2.01	4.15	6.51	0.09	14.26	22.20	0.48
DA02	50.91	1.94	4.09	6.37	0.09	14.22	22.35	0.52
DA02	50.08	2.14	4.13	6.48	0.11	13.93	22.27	0.36
DA02	50.05	2.01	4.10	6.40	0.11	14.10	22.20	0.40
DA07	48.17	2.40	6.25	7.50	0.19	13.04	21.49	0.74
DA07	49.80	2.36	4.41	6.89	0.10	13.93	22.02	0.40
DA07	49.25	2.78	4.60	7.05	0.14	13.50	22.12	0.43
DA07	50.89	1.74	4.10	5.82	0.10	14.33	22.18	0.48
DA07	50.80	1.75	4.14	5.82	0.12	14.35	22.19	0.50
DA07	49.26	2.56	4.28	7.17	0.12	13.90	22.01	0.42
DA08a	47.88	2.50	6.81	7.03	0.12	13.16	21.36	0.83
DA08a	48.78	2.46	4.81	6.75	0.14	14.22	22.23	0.53
DA08a	50.24	2.28	4.15	6.48	0.12	14.02	22.40	0.43
DA08a	48.30	2.47	6.75	6.94	0.12	12.96	21.44	0.71
DA08a	50.26	1.79	4.70	6.92	0.13	13.68	21.56	0.67
DA08a	47.94	2.68	6.67	6.88	0.11	13.04	21.93	0.64
DA08a	49.30	2.34	4.54	6.46	0.13	13.86	22.66	0.50
DA08a	47.26	3.02	6.24	6.90	0.12	12.97	22.35	0.54
DA08a	45.87	3.83	7.55	7.30	0.10	11.90	22.30	0.57

(continued)

Table 4.2 (continued)

(d)

	SiO_2	TiO_2	Al_2O_3	FeO	MnO	MgO	CaO	Na_2O
DA08a	48.11	2.97	5.39	7.32	0.13	13.30	22.03	0.48
DA08a	49.28	2.63	5.04	7.12	0.14	13.81	21.76	0.46
DA08a	48.09	2.49	6.17	7.10	0.13	13.12	22.35	0.66
DA08a	47.01	2.92	7.36	8.09	0.12	11.76	22.15	0.69
DA08a	45.62	4.08	8.34	7.42	0.10	11.91	22.06	0.61
DA08a	45.10	4.34	8.12	7.77	0.12	11.78	21.95	0.58
DA09	49.84	2.13	4.04	6.67	0.13	13.99	21.85	0.42
DA09	49.79	1.96	3.82	6.41	0.11	14.24	22.08	0.35
DA09	42.37	5.38	9.69	8.16	0.09	10.86	22.06	0.58
DA09	49.76	2.14	4.01	6.50	0.11	13.99	22.22	0.40
DA09	44.10	4.55	8.41	7.77	0.13	11.57	21.89	0.56
DA09	49.33	2.24	4.11	6.69	0.11	14.11	22.06	0.44
DA09	44.17	4.76	8.60	7.48	0.09	11.23	22.03	0.51
Pre-2011 eruption								
DA08m	48.30	2.74	5.14	6.87	0.11	13.27	22.59	0.42
DA08m	44.82	4.11	8.36	7.50	0.10	11.31	22.63	0.53
DA08m	48.25	2.87	5.24	7.05	0.14	13.03	22.59	0.53
DA08m	48.31	2.87	5.48	7.04	0.13	12.87	22.66	0.54
DA08m	43.95	4.88	9.15	7.90	0.10	11.11	22.40	0.52
DA08m	48.73	2.82	5.50	7.03	0.15	13.11	22.56	0.42

(d)

	K_2O	Total	Mg/(Mg + Fe_2)	Fe_2/(Fetot)	Al/(Al + Fe_3 + Cr)	Jadeite	Acmite	Diopside
DA01b	0.01	99.56	0.753	0.779	0.835	0.048	0.009	0.943
DA01b	0.00	99.76	0.779	0.711	0.809	0.048	0.011	0.940
DA01b	0.02	99.13	0.800	0.844	0.848	0.021	0.004	0.976
DA01b	0.00	99.75	0.822	0.924	0.926	0.030	0.002	0.967
DA01b	0.02	99.24	0.806	0.696	0.837	0.032	0.006	0.962
DA01b	0.01	99.86	0.795	0.868	0.880	0.029	0.004	0.967
DA01b	0.01	99.45	0.813	0.830	0.888	0.030	0.004	0.966
DA01b	0.02	99.71	0.812	0.997	0.997	0.032	0.000	0.968
DA01b	0.00	99.70	0.811	0.751	0.872	0.038	0.006	0.957
DA01b	0.01	100.16	0.802	0.912	0.911	0.024	0.002	0.974
DA01b	0.01	98.78	0.804	0.610	0.802	0.033	0.008	0.959
DA01b	0.00	99.86	0.784	0.930	0.935	0.030	0.002	0.968
DA01b	0.00	99.71	0.759	0.798	0.884	0.036	0.005	0.959
DA01b	0.01	99.48	0.796	0.745	0.837	0.031	0.006	0.963
DA01b	0.03	99.56	0.762	1.018	1.014	0.046	-0.001	0.954
DA01b	0.00	99.30	0.760	0.845	0.904	0.031	0.003	0.966

(continued)

Table 4.2 (continued)

(d)

	K_2O	Total	Mg/(Mg + Fe_2)	Fe_2/(Fetot)	Al/(Al + Fe_3 + Cr)	Jadeite	Acmite	Diopside
DA01b	0.00	99.25	0.772	0.828	0.904	0.035	0.004	0.962
DA01b	0.01	100.45	0.802	0.844	0.858	0.026	0.004	0.970
DA01b	0.01	99.94	0.812	0.907	0.900	0.030	0.003	0.967
DA01b	0.02	100.11	0.794	0.912	0.920	0.029	0.003	0.968
DA02	0.03	100.80	0.789	0.978	0.975	0.025	0.001	0.975
DA02	0.00	99.45	0.760	0.835	0.905	0.029	0.003	0.968
DA02	0.03	100.61	0.803	0.959	0.956	0.032	0.001	0.966
DA02	0.02	100.62	0.807	0.952	0.949	0.035	0.002	0.963
DA02	0.03	99.63	0.791	1.012	1.013	0.027	0.000	0.974
DA02	0.02	99.52	0.806	0.945	0.942	0.027	0.002	0.971
DA07	0.01	99.85	0.814	0.709	0.801	0.043	0.011	0.946
DA07	0.00	99.98	0.788	0.968	0.965	0.028	0.001	0.971
DA07	0.00	99.95	0.779	0.969	0.968	0.030	0.001	0.969
DA07	0.02	100.04	0.808	1.041	1.043	0.036	-0.001	0.965
DA07	0.01	100.00	0.814	1.004	1.004	0.036	0.000	0.964
DA07	0.01	99.79	0.801	0.856	0.854	0.026	0.004	0.970
DA08a	0.00	99.83	0.833	0.669	0.805	0.048	0.012	0.940
DA08a	0.03	100.00	0.863	0.597	0.714	0.027	0.011	0.962
DA08a	0.01	100.19	0.799	0.972	0.970	0.030	0.001	0.969
DA08a	0.00	99.86	0.795	0.860	0.907	0.047	0.005	0.949
DA08a	0.00	99.76	0.790	0.936	0.938	0.045	0.003	0.952
DA08a	0.02	100.01	0.815	0.765	0.853	0.039	0.007	0.954
DA08a	0.00	99.87	0.835	0.753	0.801	0.029	0.007	0.964
DA08a	0.01	99.51	0.828	0.694	0.807	0.032	0.008	0.961
DA08a	0.02	99.52	0.787	0.786	0.872	0.036	0.005	0.959
DA08a	0.00	99.79	0.800	0.808	0.844	0.029	0.005	0.965
DA08a	0.00	100.31	0.791	0.914	0.921	0.030	0.003	0.967
DA08a	0.02	100.16	0.839	0.631	0.768	0.036	0.011	0.953
DA08a	0.01	100.16	0.774	0.758	0.841	0.042	0.008	0.950
DA08a	0.02	100.21	0.788	0.771	0.873	0.039	0.006	0.956
DA08a	0.01	99.86	0.782	0.754	0.857	0.036	0.006	0.958
DA09	0.00	99.16	0.794	0.973	0.969	0.030	0.001	0.969
DA09	0.01	98.96	0.812	0.916	0.909	0.023	0.002	0.975
DA09	0.01	99.30	0.793	0.621	0.815	0.035	0.008	0.957
DA09	0.02	99.25	0.806	0.921	0.917	0.027	0.002	0.971
DA09	0.00	99.07	0.792	0.698	0.835	0.034	0.007	0.959
DA09	0.02	99.23	0.824	0.804	0.815	0.026	0.006	0.968
DA09	0.05	99.04	0.762	0.835	0.907	0.034	0.003	0.962

(continued)

Table 4.2 (continued)

(d)

	K_2O	Total	Mg/(Mg + Fe_2)	Fe_2/(Fetot)	Al/(Al + Fe_3 + Cr)	Jadeite	Acmite	Diopside
Pre-2011 eruption								
DA08m	0.02	99.60	0.811	0.802	0.842	0.026	0.005	0.970
DA08m	0.02	99.51	0.785	0.738	0.857	0.033	0.005	0.962
DA08m	0.00	99.81	0.809	0.780	0.827	0.032	0.007	0.962
DA08m	0.01	100.01	0.800	0.813	0.855	0.033	0.006	0.961
DA08m	0.03	100.19	0.773	0.737	0.861	0.033	0.005	0.962
DA08m	0.01	100.45	0.785	0.911	0.925	0.028	0.002	0.970

(e)

	SiO_2	TiO_2	Al_2O_3	FeO	MnO	MgO	CaO	Na_2O	K_2O	Total	X_{Ab}	X_{An}	X_{Or}
DA07	52.86	0.25	29.22	0.66	0.02	0.12	12.47	4.26	0.32	100.18	0.375	0.607	0.018
DA07	51.26	0.18	30.44	0.66	0.00	0.11	13.99	3.31	0.22	100.21	0.296	0.691	0.013
DA08a	52.27	0.23	29.98	0.70	0.00	0.12	13.03	4.02	0.25	100.63	0.353	0.633	0.014
DA09	52.06	0.24	29.37	0.69	0.01	0.12	12.96	3.90	0.35	99.69	0.345	0.634	0.020
DA09	52.30	0.23	29.46	0.73	0.01	0.09	12.86	3.83	0.29	99.80	0.344	0.639	0.017
Pre-2011 eruption													
DA08m	53.15	0.20	28.94	0.55	0.00	0.08	12.39	4.24	0.32	99.88	0.375	0.606	0.019
DA08m	53.75	0.20	29.05	0.47	0.00	0.10	11.70	4.67	0.40	100.35	0.410	0.567	0.023
DA08m	52.23	0.25	29.65	0.60	0.01	0.06	12.76	4.04	0.33	99.93	0.357	0.624	0.019
DA08m	53.31	0.23	29.09	0.52	0.00	0.09	12.09	4.43	0.38	100.15	0.390	0.588	0.022
DA08m	52.99	0.24	29.05	0.50	0.03	0.08	12.09	4.43	0.40	99.81	0.390	0.587	0.023
DA08m	54.05	0.20	28.57	0.52	0.02	0.04	11.42	4.94	0.43	100.19	0.428	0.547	0.024
DA08m	54.20	0.22	28.73	0.59	0.00	0.10	11.71	4.70	0.37	100.62	0.412	0.567	0.021
DA08m	53.53	0.21	29.46	0.51	0.00	0.09	12.32	4.44	0.39	100.99	0.386	0.592	0.022
DA08m	53.34	0.22	29.04	0.59	0.00	0.07	12.28	4.29	0.35	100.19	0.379	0.600	0.021

(f)

	DA01a	DA01b	DA02	DA08m	Lava-restingolite	Restingolite
δD%	− 89.8	− 96.4	− 96.6	− 96.3	− 87.3	− 72.2
$\delta^{18}O$%	6.2	6.1	6.2	6.4	6.8	11.1

(a) Bulk-rock (XRF) analysis (oxides in wt%; elements in ppm). (b–e) EPMA (wt%) of glass, olivine, clinopyroxene and plagioclase, respectively. (f) δD and δ18O isotope results of the Tagoro basanites and restingolite

observed results for the glass (i.e. residual liquid) and the main cotectic phases (i.e. Ol, Cpx and Pl, with total amount up to 15% for the 2011 eruption samples, and up to 30% for sample DA08m; Table 4.3). Our modelling results indicate that the assemblage olivine or clinopyroxene + plagioclase + residual glass reaches equilibrium at c. 4 kbar (i.e. c. 12 km depth assuming an average oceanic crustal density of 3000 kg/m^3), and 1130 °C, in general for all samples (Table 4.3), except for sample DA08m at slightly lower T of 1100 °C. The "silica activity barometer" of Putirka (2008, Eq. 42) for the bulk composition of sample DA08m, indicates deeper

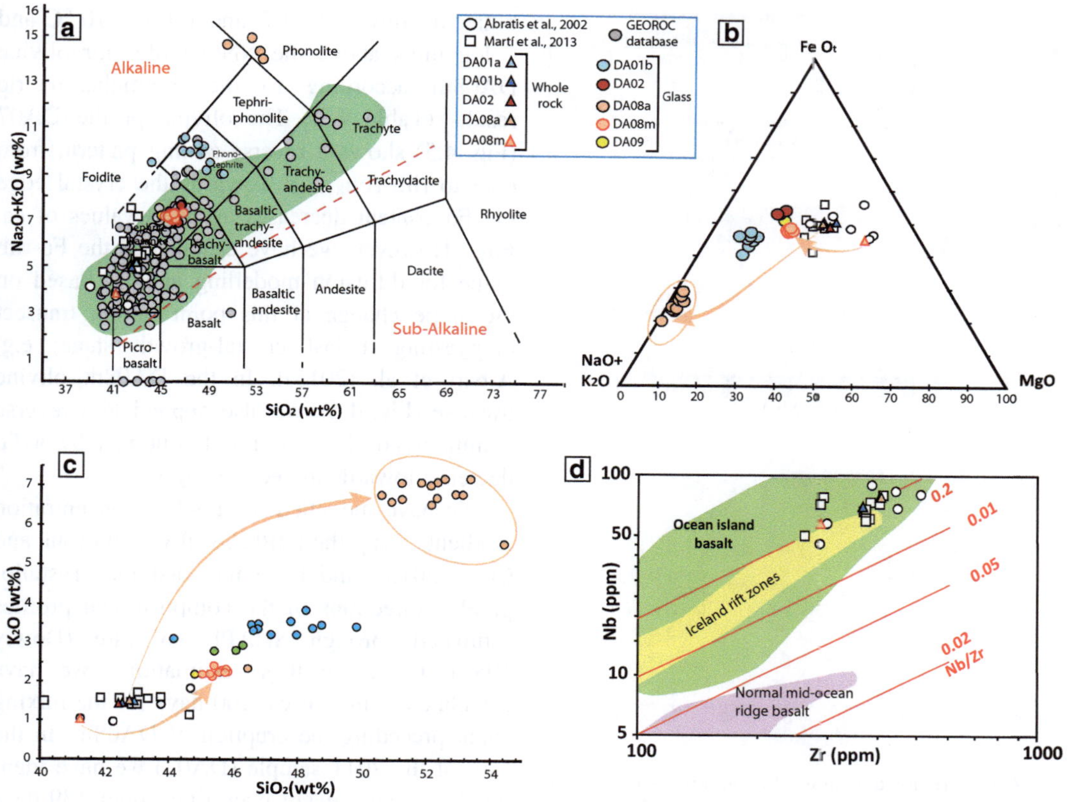

Fig. 4.3 **a** Total Alkalis versus Silica (TAS) diagram (Le Bas et al. 1986) displaying the alkaline geochemical affinity of the Tagoro rocks. Gray circles represent data for Canary Islands basanites compiled from the GEOROC database (http://georoc.mpch-mainz.gwdg.de/georoc/). The green field represents the magmatic whole-rocks of El Hierro described in Troll et al. (2012); **b** Glass compositional variation shown on a $Na_2O + K_2O$, FeOt, MgO (AFM) ternary projection (Irvine and Baragar 1971), and **c** K_2O versus SiO_2 diagram (Le Maitre 1984). Orange arrows highlight the compositional variation from the whole-rock to the glass data within the same sample; **d** Nb versus Zr comparative projection of the studied basanites with basalts from Iceland, oceanic islands and normal segments of mid-ocean ridge (Fitton and Godard 2004, for MgO > 5 wt%)

values (c. 20–24 km; López et al. 2012). The inclusion of residual glass in the thermodynamical model locally yields to inconsistent results of the output composition of the cotectic phases, as glass is equilibrating at the latest stage of cooling and might not be fully related with the first crystallised phases (i.e. Ol and/or Cpx). These results are also in line with the two main magma storage regions (10–12 and 22–36 km depth) beneath El Hierro island estimated through fluid inclusions studies (Oglialoro et al. 2017), and multifrequency, multisensor interferometric analysis of spaceborne radar images (González et al. 2013).

4.4.6 Olivine Diffusion Geospeedometry

We have acquired two compositional profiles (points every 20–30 µm) from two olivine crystals (samples DA07 and DA08m) to carry out a preliminary comparison between the 2011 early erupted products (DA07; Fig. 4.5) and the previous 2011 erupted products (DA08m; Fig. 4.6). As mentioned, olivine crystals erupted during the 2011 eruption display a variety of core and rim compositions suggesting the occurrence of multiple open-system processes before the triggering of the eruption (Martí et al. 2013b; Longpré et al.

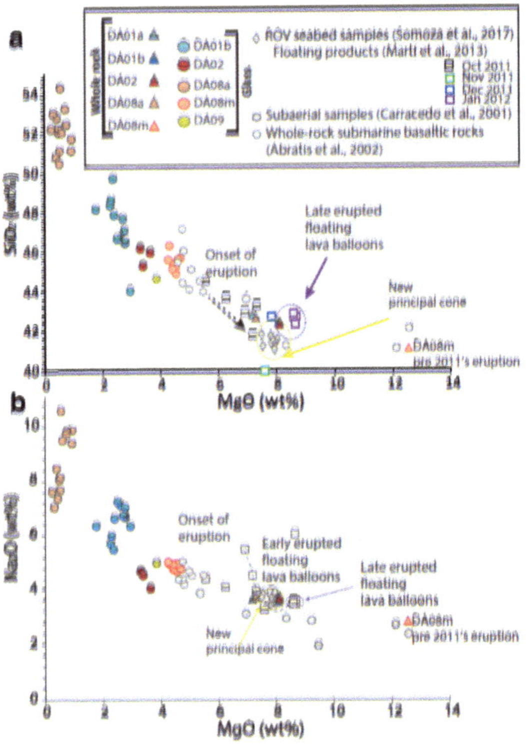

Fig. 4.4 Variation diagrams (after Somoza et al. 2017) of whole-rock major elements SiO_2 (**a**) and Na_2O (**b**) versus MgO of the studied basanites compared to other Tagoro rocks from literature (see legend). Samples DA01a and DA08a match the data for the first edifice, whereas DA02 and DA01b coincide with the last erupted products

2014). The detailed study of the olivine patterns from different eruptions is beyond the scope of this book chapter, but the comparison between these two profiles (Figs. 4.5, 4.6) highlights the importance of understanding the processes occurring at depth aiming to build a consistent model of the volcanic plumbing system.

The T and fO_2 used for modelling are 1130 ± 10 °C and 1 log units above the NNO buffer for olivine DA07, and 1100 ± 10 °C and 1 log units above the NNO buffer for olivine DA08m according to the thermobarometric results (Table 4.3). The olivine profile DA07 (Fig. 4.5) shows a reverse zoning pattern, from core to rim (Fo_{81-84}). Towards the crystal edge the Fo content decreases again to values of ca. Fo_{79}. However, we have established the Fo-rim value for diffusion modelling at Fo_{82} based on the slope change at this point of the transect (suggesting a last crystal-growth stage; e.g. Albert et al. (2015)). In the DA08m olivine traverse (Fig. 4.6), we also appreciate a reverse zoning trend (Fo_{78} to Fo_{82}) followed by a Fo decrease towards the edge (Fo_{80}).

We have modelled the Fe–Mg concentration gradients using the DIPRA software (Girona and Costa 2013), and have included the crystallographic directions of the compositional profiles estimated through SHAPE software (Dowty 1987). Based on these estimations, we have calculated a time of ca. 200 days for the mixing event preceding the eruption of DA08m. In the case of the 2011 sample (DA07) we have identified a mixing event occurring about 189 days followed by a resting time of 86 days prior to eruption.

4.4.7 Stable Isotopes

The results of oxygen and hydrogen stable isotope analysis (Table 4.2; Fig. 4.6) show $\delta^{18}O$ values (6.2–6.4 ‰) matching MORB values (5–6‰) and arc volcanoes (5–8‰) (Bindeman 2008). D/H results display similar values within the field of primary magmatic fluids for all samples (including the pre 2011 eruption DA08m; c. −96‰). Only sample DA08a from

Table 4.3 Summary of the P–T calculations obtained for the studied basanites

	Sample		DA01a	DA01b	DA02	DA08a	DA08m
MELTS	Ol + Cpx + Glass	P (kbar)(± 0.6)	4	4	4	4	4
		T (°C) (± 40)	1130	1120	1120	1130	1100
	% residual melt; % crystals		90; 10	85; 15	85; 15	90; 10	70; 30
	Putirka (2008) Eq. 42 at 1100 °C	P (kbar)					7.0 ± 0.5

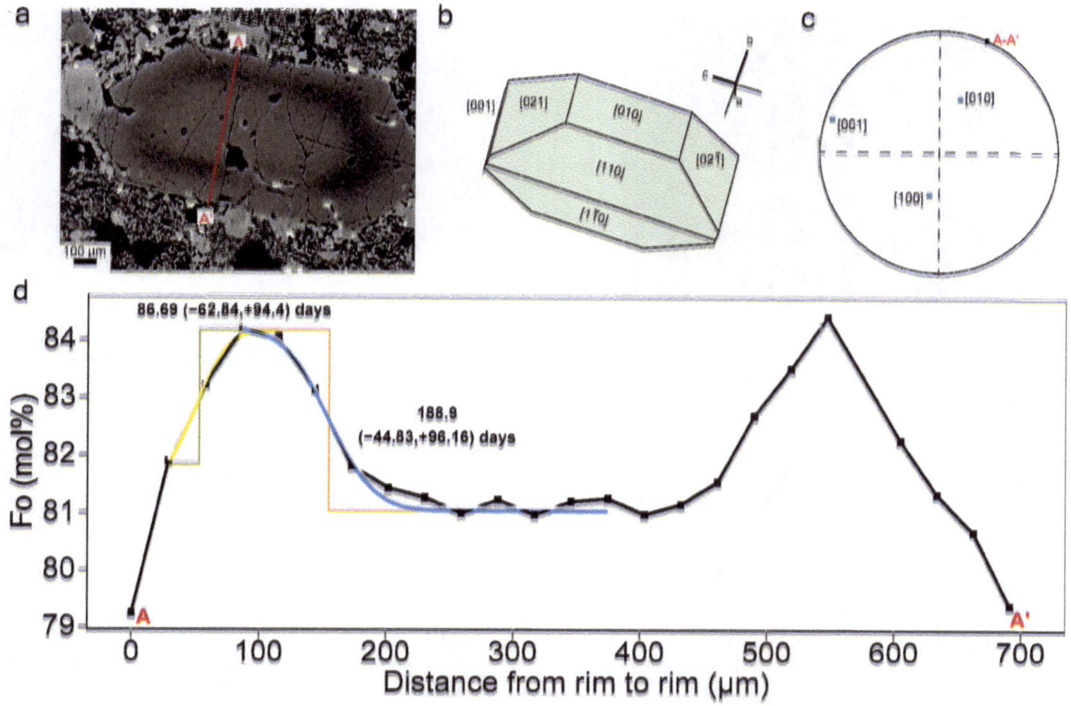

Fig. 4.5 Olivine of sample DA07. **a** Back-scattered electron (BSE) image of the crystal and position of the compositional traverse (A–A'); **b** Crystal orientation defined by SHAPE; and **c** Stereographic poles and profile orientation projection (dark green squares); **d** Olivine Fo (mol%) profile from rim to rim (A–A') and best-fit diffusion model. See the main text for the details of the calculated timescales

the earlier eruptive stage in October 2011 shows a slightly lighter value of c. −89%, which is consistent with the presence of different glass microdomains (Figs. 4.1j, 4.2i) from the magma interaction with the surrounding sediments there present prior to eruption.

Attempting to assess the potential sediments assimilation by the first products emitted in October 2011, we also analysed a white-floating sample (the so-called restingolite; e.g. Carracedo et al. 2012) and its basaltic host material. The results evidence the isotopic fractionation of the bulk values by the assimilation process of the two distinctive endmembers values. This is confirmed by the restingolite showing lighter D/H value of c. −72%, with a higher $\delta^{18}O$ value (up to 11.2%) matching the values range of sedimentary fluids (Fig. 4.6). Its host basaltic in-contact sample show also certain variation of the primary magmatic values (6.8% for $\delta^{18}O$, and −80% for D/H), which is also consistent with the major and trace elements data (Fig. 4.4, Table 4.2a–f).

4.4.8 Magnetometry

We explored whether the magnetic response represents a proxy for a magma source at depth, and it can be integrated with the rest of the petrologic and geochemical database as an informative tool of the magmatic scenario beneath Tagoro. Based on the exponential spectrum decay assumption we can estimate the depth to the top of magnetic source(s). Results show two boundaries of the magnetic anomaly (inset at Fig. 4.7): one corresponding to superficial magnetic contribution at 1 km depth, and a second one at 3 km depth. Our results match with Werner Deconvolution results. To resolve the

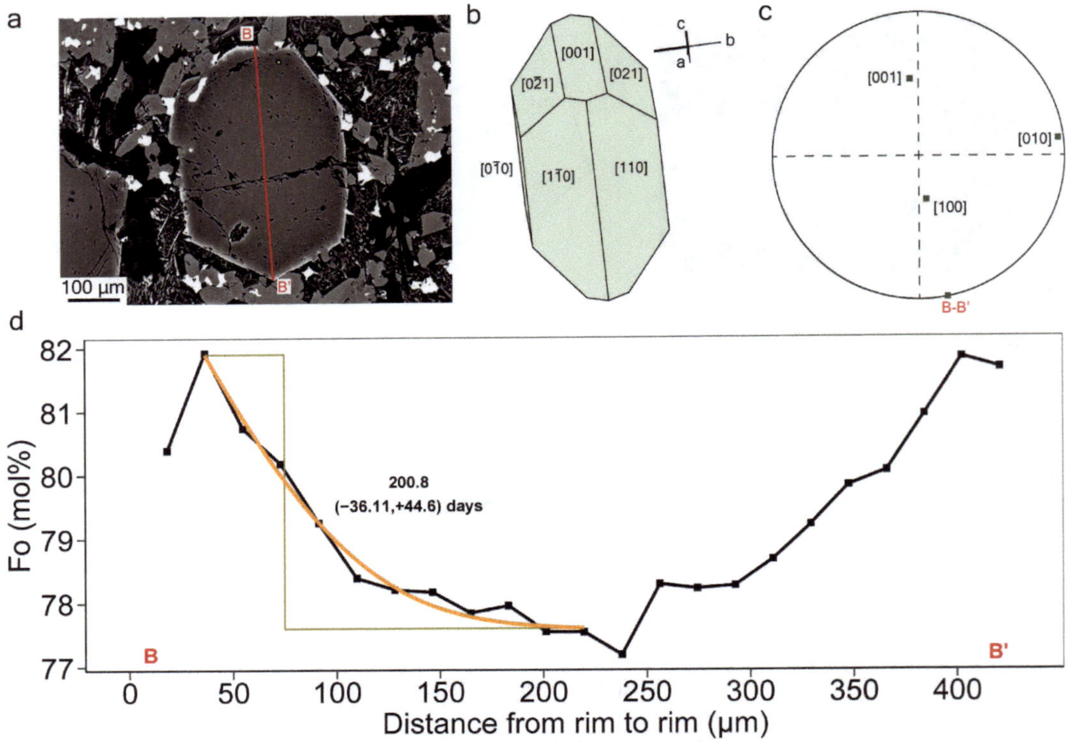

Fig. 4.6 Olivine of sample DA08m. **a** Back-scattered electron (BSE) image of the crystal and position of the compositional traverse (B–B'); **b** Crystal orientation defined by SHAPE; and **c** Stereographic poles and profile orientation projection (dark green squares); **d** Olivine Fo (mol%) profile from rim to rim (B–B') and best-fit diffusion model. See the main text for the details of the calculated timescales

depth extension of possible magnetic contributions we have used the approach proposed by Blakely (1988, 1995) and Schuey et al. (1977). Blanco-Montenegro et al. (2003) applied the method based on the position of a maximum value along the layer axis (k_{max}) to a grid centred in Gran Canaria Island, suggesting the presence of rocks located at mantle-like depths, which could behave as magnetic sources. We followed a similar technique. In order to solve the peak in the spectrum, we used 1 km as the depth of the layer's top and 10 km as depth to the Moho in the area (see also Carracedo et al. 2015). It leads to a 50 km-size for the window. Thus, using the magnetic anomaly map (Fig. 4.7) we got the radial-average spectrum of the magnetic anomaly map, and obtained a k_{max} of 0.0297 cycles/km, as well as the depth to the bottom of this magnetic source lying at 15 km b.s.l.

4.5 Discussion

4.5.1 Magmatic Evolution Up to Eruption: Mixing and Differentiation Processes

The compositional comparison (major and trace elements) between the bulk-rock (i.e. assumed as parental magma) and the evolved residual glasses (Fig. 4.3; Table 4.2) confirm the minor variation during the magmatic evolution related to the Tagoro 2011 eruption (see also Martí et al. 2013a,b). In this study we selected sample DA08m as the reference magma composition of a previous submarine eruption in the region (Álvarez-Valero et al. 2018) to be compared with the 2011–2012 products. The petrologic features

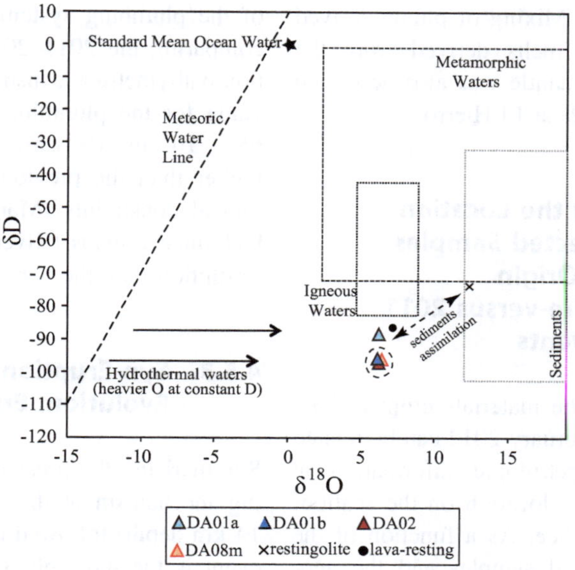

Fig. 4.7 Representation of δD versus $\delta^{18}O$ values (%) of the Tagoro samples in Taylor's (1967) diagram of water isotopic composition (Table 4.2). The temperature of the magmas (ca. 1100°C) did not allow for oxygen equilibration or fractionation (typical at such high T) during rapid eruption, and water δD values are therefore comparable with solid samples. Isotopic variation is only evident after sediment assimilation during the first eruption period starting in October 2011

of sample DA08m with respect to the other samples erupted in 2011–2012, show (i) significantly higher MgO, Cr and Ni contents; (ii) two different P estimates of c. 24 km and c. 10–14 km depth, in line with the presence of two magma sources; and (iii) similar residence times and Fo compositional plateaus compared to the 2011–2012 samples (Longprè et al. 2014).

All this combined with the zoning patterns of the olivine crystals (Figs. 4.5, 4.6) showing mixing-derived zoning patterns (c. ≈190 and ≈90 days for the 2011–2012 eruption, and ≈200 days for the pre-2011 episode), supports the hypotheses that (i) the two submarine eruptions had similar subsurface magmatic evolution processes; and (ii) a mixing mechanism of two compositionally similar mafic magmas, one from a deeper source ascending and filling up shallower reservoirs. The deeper ascent allowed interaction between the magmabatches at c. 20–24 and 10–14 km depth. These depths range are in line with those retrieved by studies of e.g. fluid inclusions (Oglialoro et al. 2017) and geodetic (González et al. 2013). In the latter, a shallowest magma reservoir at ca. 4–5 km depth is also described, which is consistent with the depth of the reservoir's roof identified by our magnetometry survey (Fig. 4.7). The presence of two sets of olivine and pyroxene phenocrysts with and without optical undulatory extinction correlated to different deformation degrees. This in turn may be related to the two main different depths of the magmabatches; i.e. a mantle-derived source and a shallower's one, respectively (see also Longprè et al. 2014; Taracsák et al. 2019). This is also consistent with the $^3He/^4He$ results of sample DA08m with typical values of pristine magmas c. 8.7–9.7 (R_A) (Álvarez-Valero et al. 2018), whereas usually more evolved magmatic values of 2.6–5.0 (R_A) correspond to the 2011 early eruption samples DA01b and DA08a.

Therefore, this magmatic evolution is remarkable along the geological record of El Hierro, where magma mixing mechanisms at depth have characterised most of the historical eruptions on the island (e.g. Stroncik et al. 2009;

Taracsak et al. 2019). Mixing of plume-derived magmas with partial melts derived from the depleted lithospheric mantle was also described by Abratis et al. (2002) at El Hierro.

4.5.2 Connecting the Location of the Collected Samples with Their Origin at Depth: Pre-versus 2011 Eruptive Events

A key result is that the materials erupted from October 2011 until February 2012 can be related to their respective petrologic information at depth, i.e. to the rock's location on the seafloor within the Tagoro edifice. As a function of the location of the dredged samples and the geomorphological description of the Tagoro edifice (Somoza et al.2017), samples DA01a and DA08a correspond to the material forming the early volcanic edifice (i.e. October–December 2011), whereas the rest of the samples are from the last -and nowadays mostly exposed on the seafloor —edifice (January–February 2012), distributed from the summit area (DA05, DA06, DA07, DA09) up to the distal or apron parts of the lava flow (DA02 at middle of the lava flow body, and DA01b in the apron; Fig. 4.1, Table 4.1). Sample DA08m was ripped out from the pre-existing volcanic substrate (see also Stroncik et al. 2009) and collected within the same dredge of sample DA08a likely due to the dredge penetration in the thin layer of material from the first edifice at that location point (Fig. 4.1; Table 4.1).

This simple systematics between the location at surface of the 2011–2012 eruption material and their eruptive stage, as well as the common P–T estimates among all samples (i.e. same depth for the stagnated magma source) and the rest of geochemical similarities (including the isotopic data), demonstrate a fairly uninterrupted eruption episode emptying the entire plumbing system. This continuous magmatic pulse -mostly from the stagnation level- is also consistent with the seismic and deformation data supporting a partial emptying out of the deeper source as well (e.g. Martí et al. 2013a,b). Finally, the estimation of the plumbing system's volume at depth, by comparing the 2011–2012 pre- and post- eruption bathymetries, constrains a minimum volume value for the plumbing system emptying of c. 65×10^6 m^3 (Navarro 2016). This volume is higher than the previously estimated by petrological constraints (Martí et al. 2013a, b) and bathymetric maps (Rivera et al. 2013) of lower resolution than the here presented.

4.5.3 Syn-Eruption Magmatic Evolution: Eruptive Styles

Summarising the present results, a magma mixing mechanism at the stagnation level (c. 10–14 km depth) followed by a continuous eruptive event is the most plausible scenario explaining the similarities in composition of the studied samples along the whole 2011–2012 eruption period. However, three key petrographic and isotopic aspects help to constrain in more detail the syn-eruptive event. For instance, samples DA08a, belonging to the beginning of the eruption period (first edifice), is the only one revealing magma interaction with the sedimentary layers on the pre-2011 seafloor, which were heated up and partially melted by the arriving magma. This is evidenced by: (i) locally different glass microdomains in thin-section (Fig. 4.2i, j); (ii) slightly heavier bulk D/H values (Fig. 4.7); and (iii) a more evolved and fractionated geohemical composition than the rest of the samples (Fig. 4.3). The sedimentary layer was most probably fully consumed during the first edifice's building, and therefore could no longer interact with the following ascending magmas (corresponding to the rest of the studied samples) to be melted and partly assimilated. Sample DA01a, likely extracted from the first edifice as well, and compositionally identical to DA08a, shows none of the mentioned interaction features with the sediments, which may suggest that the sedimentary material had already been consumed by the time the magma corresponding to DA01a was erupted some days after 10th October 2011. The stable isotopes data reveal no interactions at the chamber or along the magma ascent to the surface, which supports a relatively fast eruption with no isotopic re-equilibrium in the magma at high T (e.g. O'Neill 1986; Bindeman 2008; Álvarez-Valero et al. 2016, 2020), as well as an eruption mainly controlled by deep-seated processes, with little influence from shallow crustal levels (see also Longprè et al. 2014). The higher MgO content in the bulk-rock of the last erupted samples in February 2012 (DA01b, DA02, DA09) supports the removal first of the most evolved magmas (DA01a, DA08a) at the start of the eruption in October 2011 (Fig. 4.4; Table 4.2).

Given the minor variation of the MgO content in the glasses with respect to the whole-rock, we suggest a rapid emptying of the plumbing system for the 2011–2012 eruption after at least two mixing events occurred ca. six and three months prior to eruption (see also Longprè et al. 2014; Albert et al. 2016). This may be related to a deeper chamber ascent from 20–24 to 10–14 km, where a slight differentiation started because the material escape to the surface was significantly fast (c. 45 days; up to three months according to Martí et al. 2013b) (Fig. 4.8). The arrival of the deeper and more mafic pulse most likely continued until the end of the eruptive event in February 2012. This was common for the pre- 2011 eruptions as evidenced by sample DA08m and previous studies (e.g. Stroncik et al. 2009). The Fe–Mg diffusion profile in the DA07 olivine of c. a month of mafic input before the ascent (c. 15 days) and open-system eruption, strongly supports this argument (see also Longprè et al. 2014).

4.6 Conclusions

The integration of the presented new results with previous literature builds and relates the effects of the magmatic process at depth with the observations at the surface of the Tagoro area. In view of all this we conclude that the 2011–2012 eruption forming the Tagoro edifice may have followed this sequence of magmatic-volcanic stages:

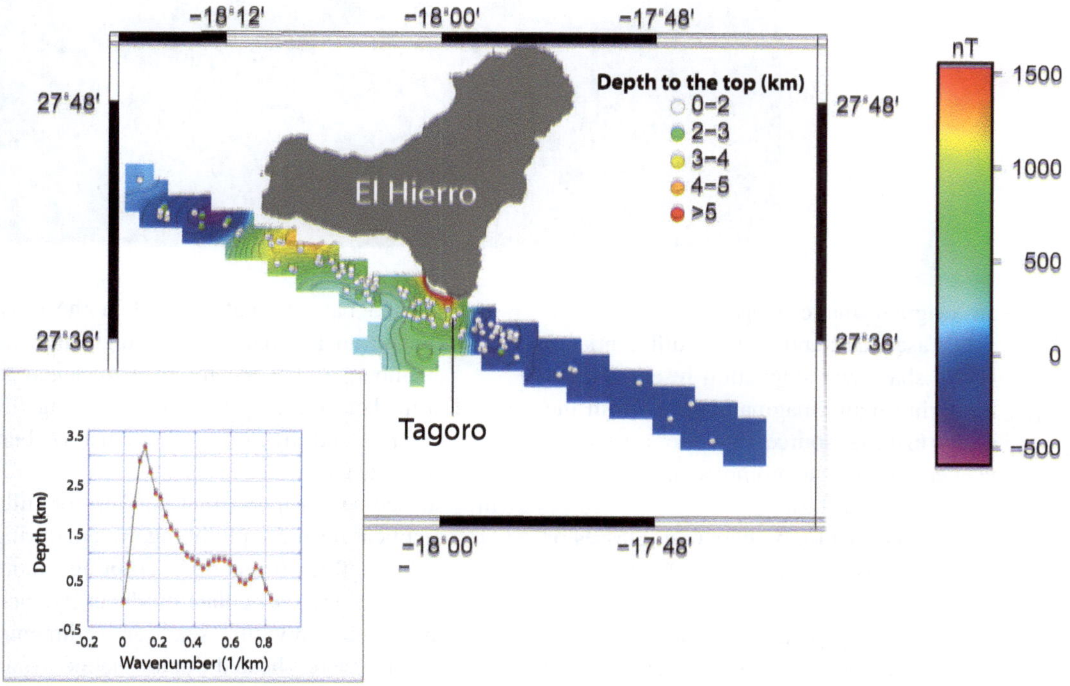

Fig. 4.8 Magnetometry map-profile and anomaly values (left-bottom corner). See the text for details

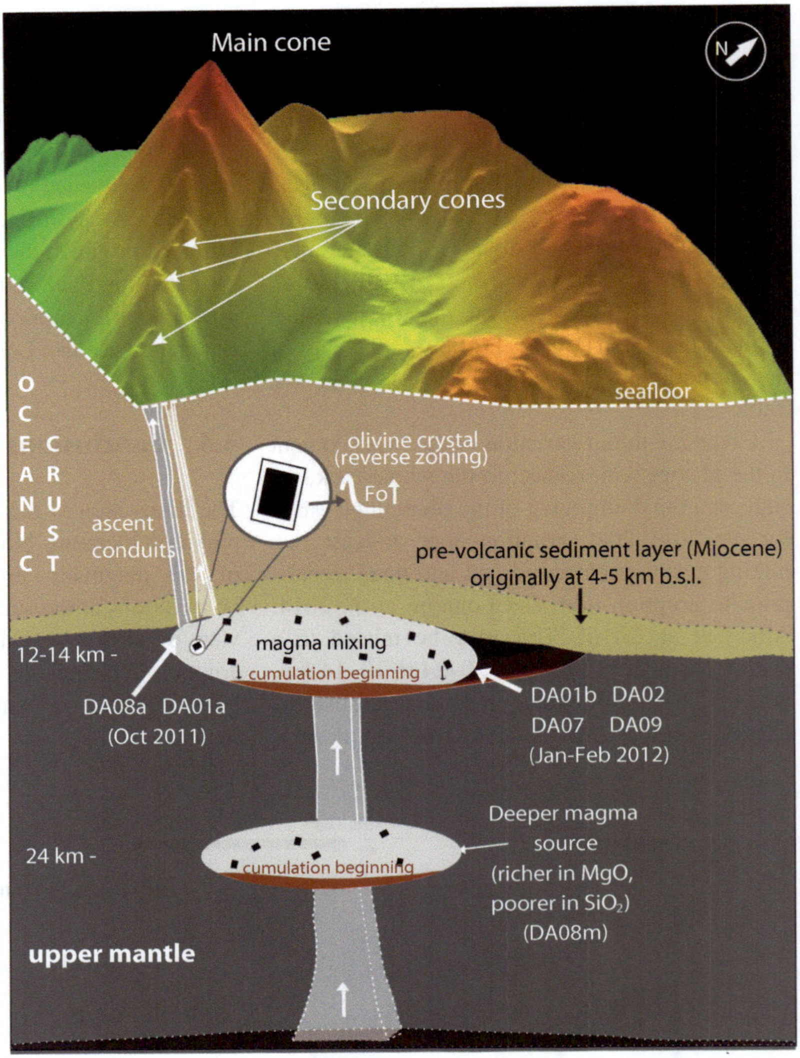

Fig. 4.9 3D summary scheme (not to scale) of the connection between the magmatic origin of the studied samples and their exposition at surface after five months of a continuous eruptive episode

(i) an original mafic magma at 20–24 km depth ascended and slightly differentiated into a shallower stagnation reservoir thus favouring mafic magma mixing (from the same magma source) by the arrival of deeper pulses that promoted instability in the latter (Fig. 4.9);

(ii) slight fractionation process (e.g. glasses of samples DA01b, DA02, DA09; Fig. 4.3) at a shallow reservoir (10–14 km depth) likely resulting in a crystal accumulation at the chamber bottom (slight higher crystallinity of sample DA09; Fig. 4.2). We are aware that the existence of small magma batches that stacked in the crust can be an alternative to a single magma chamber, which in any case managed to erupt because they mixed, increasing the volume and fracturing the crust (Albert et al. 2016);

(iii) subsequent eruption that partially or fully emptied the entire plumbing system during 4–5 months. It started erupting more massive (less vesicular) products interacting with previous seafloor sediments, which were heated and melted (and explosively emitted as part of the restingolitic material) at the beginning of the submarine eruption in October 2011 (e.g. Carracedo et al. 2012; Pérez-Torrado et al. 2012; Troll et al. 2012; Martí et al. 2013a).

(iv) the eruption of solid material continued until February 2012 building up the last and current seafloor edifice (Somoza et al. 2017) of more vesicular material (e.g. DA01b, DA02, DA06) but nearly identical in composition along the entire eruption.

(v) the magnetometry anomaly under Tagoro retrieved in 2016 is consistent with geophysical and oceanographic evidence of post-eruptive unrest periods (Benito-Saz et al. 2019) and current activity (e.g. Fraile-Nuez et al. 2016) in the Tagoro area. This also matches the location of the Moho and shallower chamber (Carracedo et al. 2012; Martí et al. 2013a, 2013b) nowadays in the region and the entire archipelago at c. 15 km depth).

4.7 Future Work Over the Next Decade

This book opens new potential research lines to enhance the knowledge of Tagoro volcano, and volcanological processes in general. From our perspective, this chapter releases interesting new questions that can be considered as interdisciplinary action points. For instance, we are envisaging the following promising aspects to deep through in the next years: (i) exhaustive microtextural study in detail of the different glass populations combined to advanced geochemical dataset. This will allow a better understanding of the magma mixing mechanisms at depth and sediments contribution at surface; (ii) a more detailed study on diffusion chronometry in olivine, clinopyroxene and plagioclase of the pre- and 2011–2012 eruptions. As mentioned, the two olivine profiles presented here, represent just a preliminary approach to be tested with the Longprè et al. (2014) data (that does not account for the EBSD correction). Our results encourage a detailed study of the phenocryst populations in Tagoro as they show different compositional profiles (i.e. different evolution) but similar ages. This will allow us to improve our knowledge on the processes (and their timescales) preceding the monogenetic eruptions in El Hierro; (iii) a novel and timely integration of advanced geophysical techniques, such as interferometric analysis and magnetometry, with petrologic-geochemical data attempting to constrain morphologies of magma sources at depth.

All this in turn should redound in direct implications for hazard assessment and eruption forecasting in the island and at any other submarine volcanic scenario worldwide, as complementary information to the classical techniques of combining seismics, ground deformation and free-gas measurements.

Acknowledgements This research was possible thanks to: (i) the funds provided by the MINECO and FEDER through the project VULCANO I (CTM2012-36317) and the project VULCANA (Vulcana-IEO-2015-2017) funded by the Spanish Institute of Oceanography, both granted to E.F.-N., that provided the oceanographic instruments, cruise and samples; (ii) the funds provided by the MINECO and MEC grants (Spanish Government) to A. M.A.-V. through the programs EXPLORA-CIENCIA (CGL2014—61775-EXP; MINECO), José Castillejo (CAS14-00189; MEC), and Programa Propio mod. 1B—2019 (USAL). J.M. is supported by EC Grant EVE (DG ECHO H2020 826292). A.P S is grateful for his PhD grant "Programa Propio III Universidad de Salamanca, cofounded by Banco de Santander". H.A. is a Serra Húnter Lecturer Professor at the University of Barcelona. A-V specially thanks all the colleagues and friends who made possible this research starting from the ones designing the oceanographic campaigns, to those sharing their work on the vessel's deck and acoustic lab, or carrying out geochemical analysis, as well as the undergraduate and graduate students involved in the research (some developing their master and graduate theses at USAL, e.g. I.N. and N.G., respectively). We also thank the careful and in-depth reviews by D. González García, an anonymous reviewer and editor. This research is also part of the PTI VOLCAN research initiatives.

References

Abratis M, Schmincke HU, Hansteen T (2002) Composition and evolution of submarine volcanic rocks from the central and western Canary Islands. Int J Earth Sci 91:562–582

Albarede F (1992) How deep do common basaltic magmas form and differentiate? J Geophys Res 97:10997–11009

Albert H, Costa F, Martí J (2015) Timing of magmatic processes and unrest associated with mafic historical monogenetic eruptions in Tenerife Island. J Petrol 56:1945–1966

Albert H, Costa F, Martí J (2016) Years to weeks of seismic unrest and magmatic intrusions precede monogenetic eruptions. Geology 44:211–214

Albert H, Costa F, Di Muro A, Herrin J, Métrich N, Deloule E (2019) Magma interactions, crystal mush formation, timescales, and unrest during caldera collapse and lateral eruption at ocean island basaltic volcanoes (Piton de la Fournaise, La Réunion). Earth Planet Sci Lett 515:187–199

Álvarez-Valero AM, Okumura S, Arzilli F, Borrajo J, Recio CM, Ban M, Gonzalo JC, Benítez JM, Douglas M, Sasaki O, Franco P, Gómez-Barreiro J, Carnicero A (2016) Tracking bubble evolution inside a silicic dike. Lithos 262:668–676

Álvarez-Valero AM, Burgess R, Recio C, de Matos V, Sánchez-Guillamón O, Gómez-Ballesteros M, Recio G, Fraile-Nuez E, Sumino H, Flores JA, Ban M, Geyer A, Bárcena MA, Borrajo J, Compaña JM (2018) Noble gas signals in corals predict submarine volcanic eruptions. Chem Geol 480:28–34

Álvarez-Valero AM, Gisbert G, Aulinas M, Geyer A, Kereszturi G, Polo-Sánchez A, Núñez-Guerrero E, Sumino H, Borrajo J (2020) δD and $\delta^{18}O$ variations of the magmatic system beneath Deception Island volcano (Antarctica): implications for magma ascent and eruption forecasting. Chem Geol 542:119595

Asimow P, Ghiorso M (1998) Algorithmic modifications extending MELTS to calculate subsolidus phase relations. Am Mineral 83:1127–1132

Balcells R, Gomez JA (1997a) Memorias y mapas geológicos del Plan MAGNA a escala 1:25.000 de las Hojas correspondientes a la isla de El Hierro. Hoja de Frontera, Instituto Geológico y Minero de España, Spain

Balcells R, Gomez JA (1997b) Memorias y mapas geológicos del Plan MAGNA a escala 1:25.000 de las Hojas correspondientes a la isla de El Hierro. Hoja de Sabinosa, Instituto Geológico y Minero de España, Spain

Ban M, Sagawa H, Miura K, Hirotani S (2008) Evidence for a short-lived stratified magma chamber: petrology of the Z-To5 tephra layer (c. 5.8 ka) at Zao volcano, NE Japan. Geol Soc Lond Spec Publ 304(1):149–168

Becerril L, Galindo I, Martí J, Gudmundsson A (2015) Three-armed rifts or masked radial pattern of eruptive fissures? The intriguing case of El Hierro volcano (Canary Islands). Tectonophysics 647–648:33–47

Becerril L, Galve JP, Morales JM, Romero C, Sánchez N, Martí J, Galindo I (2016) Volcano-structure of El Hierro (Canary Islands). J Maps 12(1):43–52

Benito-Saz MA, Sigmundsson F, Charco M, Hooper A, Parks M (2019) Magma flow rates and temporal evolution of the 2012–2014 post-eruptive intrusions at El Hierro, Canary Islands. J Geophys Res Solid Earth 124:12576–12592

Bigeleisen J, Perlman ML, Prosser HC (1952) Conversion of hydrogenic materials to hydrogen for isotopic analysis. Anal Chem 24:1536–1537

Bindeman I (2008) Oxygen isotopes in mantle and crustal magmas as revealed by single crystal analysis. Rev Min Geochem 69:445–478

Blakely RJ (1988) Curie temperature isotherm analysis and tectonic implications of aero-magnetic data from Nevada. J Geophys Res Solid Earth 93(B10):11817–11832

Blakely RJ (1995) Potential theory in gravity and magnetic applications. Cambridge University Press, New York, p 441

Blanco-Montenegro I, Torta JM, García A, Araña V (2003) Analysis and modelling of the aeromagnetic anomalies Gran Canaria (Canary Islands). Earth Planet Sci Lett 206:601–616

Borthwick J, Harmon RS (1982) A note regarding ClF_3 as an alternative to Br $_F5$ for oxygen isotope analysis. Geochim Cosmochim Acta 46:1665–1668

Carracedo JC, Badiola ER, Guillou H, de La Nuez J, Pérez Torrado FJ (2001) Geology and volcanology of the western Canaries: La Palma and El Hierro. Estudios Geol 57:175–273

Carracedo JC, Pérez-Torrado F, Rodríguez A, Soler V, Fernández JL, Troll V, Wiesmaier S (2012) The 2011 submarine volcanic eruption in El Hierro (Canary Islands). Geol Today 28(2):53–58

Carracedo JC, Troll VR, Zaczek K, Rodríguez-González A, Soler V, Deegan FM (2015) The 2011–2012 submarine eruption off El Hierro, Canary Islands: new lessons in oceanic island growth and volcanic crisis management. Earth Sci Rev 150:168–200

Carracedo JC (2011) Geología de Canarias I (Origen, evolución, edad y vulcanismo). Ed. Rueda. Madrid

Clayton RN, Mayeda TK (1963) The use of bromine pentafluoride in the extraction of oxygen from oxides and silicates for isotopic analysis. Geochim Cosmochim Acta 27:43–52

Domínguez I, Del Fresno C, Gomis A (2014) Seismicity patterns prior to the 2011 El Hierro Eruption. Bull Seismol Soc Am 104(1):1–9

Dóniz J, Romero C, Coello E, Guillén C, Sánchez N, García-Cacho L, García A (2008) Morphological and statistical characterisation of recent mafic volcanism on Tenerife (Canary Islands, Spain). J Volcanol Geotherm Res 173:185–195

Dowty E (1987) SHAPE copyright 1994, shape software 521 Hidden Valley Road, Kingsport, TN 37663 USA. http://www.shapesoftware.com/

Fitton GJ, Godard M (2004) Origin and evolution of magmas on the Ontong Java Plateau. Geol Soc Lond Spec Publ 229:151–178

Fraile-Nuez E, González-Dávila M, Santana-Casiano JM, Arístegui J, Alonso-González IJ, Hernánez-León S, Blanco MJ, Rodríguez-Santana A, Hernández-Guerra A, Gelado-Caballero MD, Eugenio F, Marcillo J, de Armas D, Domínguez-Yanes JF, Montero MF, Laetsch DR, Vélez-Belch P, Ramos A, Ariza AV,

Cómas-Rodríguez I, Benítez-Barrios VM (2012) The submarine volcano eruption at the island of El Hierro: physical-chemical perturbation and biological response. Sci Rep 2:486

Fraile-Nuez E, Santana-Casiano JM, Vázquez JT, Gómez-Ballesteros M, Sánchez-Guillamón O, Álvarez-Valero AM, Presas-Navarro C, González-Carballo M, Castro-Álamo A, Oosterbaan M, Domínguez-Yanes JF, Herrera I (2016) VULCANA 0316 project report. http://www.repositorio.ieo.es/e-ieo/handle/10508/10035

Fraile-Nuez E, et al (2022) Ten years of intense physical-chemical, biological and geological monitoring over Tagoro submarine volcano marine ecosystem: eruptive and degassing stages. In: Chapter 8: active volcanoes of the world. Springer, Berlin, Heidelberg

Fuster JM (1993) Geochronología de la Isla de El Hierro (Islas Canarias). Bol R Soc Esp Hist Nat 88(1–4): 85–97

Ghiorso M, Sack R (1995) Chemical mass transfer in magmatic processes IV. A revised and internally consistent thermodynamic model for the interpolation and extrapolation of liquid-solid equilibria in magmatic systems at elevated temperatures and pressures. Contr Mineral Petrol 119:197–212

Girona T, Costa F (2013) DIPRA: a user-friendly program to model multi-element diffusion in olivine with applications to timescales of magmatic processes. Geochem Geophys Geosyst 14(2):422–431

Godfrey JD (1962) The deuterium content of hydrous minerals from the east-central Sierra Nevada and Yosemite National Park. Geochim Cosmochim Acta 26:1215–1245

González PJ, Samsonov S, Pepe S, Tiampo KF, Tizzani P, Casu F, Fernández J, Camacho AG, Sansosti E (2013) Magma storage and migration associated with the 2011–2012 El Hierro eruption: Implications for shallow magmatic systems at oceanic island volcanoes. J Geophys Res Solid Earth 118:4361–4377

Guillou H, Carracedo JC, Pérez Torrado F, Rodríguez Badiola E (1996) K–Ar ages and magnetic stratigraphy of a hotspot-induced, fast grown oceanic island: El Hierro, Canary Islands. J Volcanol Geotherm Res 73:141–155

Hartman RR, Teskey DJ, Friedberg JL (1971) A system for rapid digital aeromagnetic interpretation. Geophysics 36:891–918

Irvine TN, Baragar WRA (1971) A guide to the chemical classification of the common volcanic rocks. Can J Earth Sci 8:523–548

Jenkin GRT (1988) Stable isotope studies in Caledonides of SW Connemara, Ireland. PhD thesis, U. Glasgow, UK

Ku CC, Sharp JA (1983) Werner deconvolution for automated magnetic interpretation and its refinement using Marquardt's inverse modeling. Geophysics 48:754–774

Le Maitre RW (1984) A proposal by the IUGS Subcommission on the systematics of igneous rocks for a chemical classification of volcanic rocks based on the total alkali silica (TAS) diagram: (on behalf of the IUGS subcommission on the systematics of igneous rocks). Austr J Earth Sci 31(2):243–255

Le Bas M, Le Maitre R, Streckeisen A, Zanettin B (1986) A chemical classification of volcanic rocks based on the total Alkali-Silica diagram. J Petrol 27(3):745–750

Longpré MA, Klügel A, Diehl A, Stix J (2014) Mixing in mantle magma reservoirs prior to and during the 2011–2012 eruption at El Hierro, Canary Islands. Geology 42(4):315–318

Longpré MA, Stix J, Klügel A, Shimizu N (2017) Mantle to surface degassing of carbon- and sulphur-rich alkaline magma at El Hierro, Canary Islands. Earth Planet Sci Lett 460:268–280

López C, Blanco MJ, Abella R, Brenes B, Cabrera VM, Casas B, Domínguez I, Felpeto A, Fernández de Villalta M, Del Fresno C, García O, García-Arias MJ, García-Cañada L, Gomis A, González-Alonso E, Guzmán K, Iribarren I, López-Díaz R, Luengo-Oroz N, Meletlidis S, Moreno M, Moure D, Pereda de Pablo J, Rodero C, Romero E, Sainz-Maza S, Sentre MA, Torres PA, Trigo P, Villasante-Marcos V (2012) Monitoring the volcanic unrest of El Hierro (Canary Islands) before the onset of the 2011–2012 submarine eruption. Geophys Res Lett 39(L13303): 1–7

Martí J, Pinel V, López C, Geyer A, Abella R, Tárraga M, Blanco MJ, Castro A, Rodróguez C (2013a) Causes and mechanisms of the 2011–2012 El Hierro (Canary Islands) submarine eruption. J Geophys Res Solid Earth 118:823–839

Martí J, Castro A, Rodríguez C, Costa F, Carrasquilla S, Pedreira R, Bolos X (2013b) Correlation of magma evolution and geophysical monitoring during the 2011–2012 El Hierro (Canary Islands) submarine eruption. J Petrol 54(7):1349–1373

Melián G, Hernández PA, Padrón E, Pérez N, Barrancos J, Padilla G, Dionis S, Rodríguez F, Calvo D, Nolasco D (2014) Spatial and temporal variations of diffuse CO_2 degassing at El Hierro volcanic system: relation to the 2011–2012 submarine eruption. J Geophys Res Solid Earth 119:6976–6991

Navarro I (2016) Petrología y geoquímica de la evolución subvolcánica de la erupción submarina de 2011 en El Hierro (Islas Canarias). Trabajo fin de máster - Universidad de Salamanca

Nimis P (1999) Clinopyroxene geobarometry of magmatic rocks. Part 2. Structural geobarometers for basic to acid, tholeiitic and mildly alkaline magmatic systems. Contr Mineral Petrol 135:62–74

Oglialoro E, Frezzotti ML, Ferrando S, Tiraboschi C, Principe C, Gropelli G, Villa IM (2017) Lithospheric magma dynamics beneath the El Hierro volcano, Canary Islands: insights from fluid inclusions. Bull Volcanol 79:70

O'Neill JR (1986) Theoretical and experimental aspects of isotopic fractionation. In: Valley JW, Taylor, HP, O'Neil JR (eds) Stable isotopes in high temperature geological processes, MSA, vol 16, pp 1–40

Padrón E, Pérez NM, Hernández PA, Sumino H, Melián GV, Barrancos J, Nolasco D, Padilla G, Dionis S,

Rodríguez F, Hernández Í (2013) Diffusive helium emissions as a precursory sign of volcanic unrest. Geology 41(5):539–542
Paonita A, Caracausi A, Martelli M, Rizzo AL (2016) Temporal variations of helium isotopes in volcanic gases quantify pre-eruptive refill and pressurization in magma reservoirs: the Mount Etna case. Geology 44 (7):499–502
Paonita A, Liuzzo M, Salerno G, Federico C, Bonfanti P, Caracausi A, Giuffrida G, La Spina A, Caltabiano T, Gurrieri S, Giudice G (2021) Intense over pressurization at basaltic open-conduit volcanoes as inferred by geochemical signals: the case of the Mt. Etna December 2018 eruption. Sci Adv 7(36):eabg6297
Pedrazzi D, Becerril L, Martí J, Meletlidis S, Galindo I (2014) Explosive felsic volcanism on El Hierro (Canary Islands). Bull Volcanol 76:863
Pellicer MJ (1975) Estudio vulcanológico, petrológico y geoquímico de la isla de El Hierro (Archipiélago Canario). Tesis Doctoral, Facultad de Ciencias Geológicas, Universidad Complutense de Madrid, p 179
Pellicer MJ (1977) Estudio volcanológico de la Isla de El Hierro, Islas Canarias. Estud Geol 33:181–197
Pérez-Torrado FJ, Carracedo JC, Rodríguez-González A, Soler V, Troll VR, Wiesmaier S (2012) La erupción submarina de La Restinga en la isla de El Hierro, Canarias: Octubre 2011-Marzo 2012. Estud Geol 68 (1):5–27
Putirka K (2005) Igneous thermometers and barometers based on plagioclase + liquid equilibria: tests of some existing models and new calibrations. Am Mineral 90:336–346
Putirka K (2008) Thermometers and barometers for volcanic systems. Rev Mineral Geochem 69:61–120
Putirka K, Johnson M, Kinzler R, Longhi J, Walker D (1996) Thermobarometry of mafic igneous rocks based on clinopyroxene-liquid equilibria, 0–30 kbar. Contr Mineral Petrol 123(1):92–108
Rivera J, Lastras G, Canals M, Acosta J, Arrese B, Hermida N, Micallef A, Tello O, Amblas D (2013) Construction of an oceanic island: insights from the El Hierro (Canary Islands) 2011–2012 submarine volcanic eruption. Geology 41:335–358
Roeder PL, Emslie R (1970) Olivine-liquid equilibrium. Contr Mineral Petrol 29(4):275–289
Romero C (1991) Las manifestaciones volcánicas históricas del archipiélago canario, Consejería de Política Territorial, Gobierno Autónomo de Canarias, 2 tomos, Santa Cruz de Tenerife
Roubault M, Fabries J, Touret J, Weisbrod A (1963) Determination des Mineraux des Roches au Microscope Polarisant. Lamarre-Poinat, Paris, p 365
Santana-Casiano JM, González-Dávila M, Fraile-Nuez E, de Armas D, González AG, Domínguez-Yanes JF, Escánez J (2013) The natural ocean acidification and fertilization event caused by the submarine eruption of El Hierro. Sci Rep 3:1140
Santana-Casiano JM, Fraile-Nuez E, González-Dávila M, Baker ET, Resing JA, Walker SL (2016) Significant discharge of CO_2 from hydrothermalism associated with the submarine volcano of El Hierro Island. Sci Rep 6:25686
Saunders S, Blundy J, Dohmen R, Cashman K (2012) Linking petrology and seismology at an active volcano. Science 336(6084):1023–1027
Scandone R, Malone SD (1985) Magma supply, magma discharge and readjustment of the feeding system of Mount St. Helens during 1980. J Volcanol Geotherm Res 23:239–262
Sharp ZD (1990) A laser-based microanalytical method for the in situ determination of oxygen isotope ratios of silicates and oxides. Geochim Cosmochim Acta 54:1353–1357
Shuey RT, Schellinger DK, Tripp AC, Alley LB (1977) Curie depth determination from aeromagnetic spectra. Geophys J Int 50(1):75–101
Somoza L, González FJ, Barker SJ, Madureira P, Medialdea T, de Ingancio C, Lourenço N, León R, Vázquez JT, Palomino D (2017) Evolution of submarine eruptive activity during the 2011–2012 El Hierro event as documented by hydroacoustic images and remotely operated vehicle observations. Geochem Geophys Geosyst 18:3109–3137
Spector A, Grant FS (1970) Statistical model for interpretating aeromagnetic data. Geophysics 14:293–302
Stroncik NA, Klügel A, Hansteen TH (2009) The magmatic plumbing system beneath El Hierro (Canary Islands): constraints from phenocrysts and naturally quenched basaltic glasses in submarine rocks. Contr Mineral Petrol 157:593–607
Taracsák Z, Hartley ME, Burgess R, Edmonds M, Iddon F, Longprè M-A (2019) High fluxes of deep volatiles from ocean island volcanoes: Insights from El Hierro, Canary Islands. Geochim Cosmochim Acta 258:19–36
Tárraga M, Martí J, Abella R, Carniel R, López C (2014) Volcanic tremors: good indicators of change in plumbing systems during volcanic eruptions. J Volcanol Geotherm Res 273:33–40
Taylor HP (1967) Oxygen isotope studies of hydrothermalmineral deposits. In: Barnes HL (ed) Geochemistry of hydrothermal ore deposits. Holt, Rinehart and Winston, p 670
Thébault E, Finlay CC, Beggan D, Alken P, Aubert J, Barrois O, Bertrand F, Bondar T, Boness A, Brocco L, Canet E, Chambodut A, Chulliat A, Coisson P, Civet F, Du A, Fournier A, Fratter I, Gillet N, Hamilton B, Hamoudi M, Hulot G, Jager T, Korte M, Kuang W, Lalanne X, Langlais B, Léger J-M, Lesur V, Lowes FJ et al (2015) International geomagnetic reference field: the 12th generation. Earth Planets Space 57:67–79
Troll VR, Klügel A, Longpré MA, Burchardt S, Deegan FM, Carracedo JC, Wiesmaier S, Kueppers U, Dahren B, Blythe LS, Hansteen TH, Freda C, Budd DA, Jolis EM, Jonsson E, Meade FC, Harris C, Berg SE, Mancini L, Polacci M, Pedroza K (2012) Floating stones off El Hierro, Canary Islands:

xenoliths of pre-island sedimentary origin in the early products of the October 2011 eruption. Solid Earth 3:97–110

Vázquez JT, et al (2016) Preliminary geomorphological analysis of the Tagoro Volcano underwater eruption (submarine slope of El Hierro Island), Abstract Book V Simposium Internacional de Ciencias del Mar, Universidad de Alicante, Alicante, Spain

Vázquez JT, et al (2022) Geomorphological analysis of Tagoro volcanic edification along eruption and degassing phase. Chapter 7: Active Volcanoes of the World. Springer, Berlin, Heidelberg

Yamada Y, Kohno H, Murata M (1995) A low dilution fusion method for major and trace element analysis of geological samples. Adv X-Ray Anal 26:33–44

Magma Storage and Migration in El Hierro During the Period 2011–2014

5

I. Domínguez Cerdeña, M. Charco, E. González-Alonso, C. del Fresno, M. A. Benito-Saz, and L. García-Cañada

Abstract

The 2011–2014 volcanic episode of El Hierro has become an excellent laboratory for studying the magma plumbing system beneath the island, with tens of works in the literature focusing on the geophysical, geodetic and/or petrological perspective. This chapter mainly focuses on the results obtained through the analysis of the seismicity and ground deformation during the different periods of the activity, including the three months of pre-eruptive unrest, the 5-month-long submarine volcanic eruption, and the magmatic intrusions during the following 2 years after the end of the eruption. All these results, together with those from other petrological and gravimetric studies, have allowed us to obtain a joint description of magma transport and storage at El Hierro Island. Most interpretations point to the presence of a main deep magma storage system below the center of the island feeding both the eruption and the later sill-like magmatic intrusions. The stagnation of the post-eruptive magma accumulations could indicate the presence of stress barriers inhibiting the ascent of magma. This structure may have an important influence on the growth of the island, with similar a contribution from magma storage under the crust as from the eruption itself. Despite all the accumulated knowledge, there are still open questions to be addressed.

Keywords

Geophysics · Volcano seismicity · Geodetic monitoring · Magmatic systems · Ocean island volcanoes

5.1 Introduction

Magma supply, storage and transport are among the critical processes that govern volcanic activity (e.g., Poland et al. 2014). There are several well-known magma plumbing systems on intraplate volcanic islands. In Hawai'i there is both horizontal transport of magma for 20 km at 30–35 km depth and near-vertical migration over ~ 30 km to a shallow magma chamber (Wright and Klein 2006), while, in La Réunion magma rises toward Piton de La Fournaise through complex, inclined pathways that are influenced by the fabric of the oceanic crust under the island (Michon et al. 2015). However,

I. Domínguez Cerdeña (✉)
Centro Geofísico de Canarias, Instituto Geográfico Nacional, Tenerife, Spain
e-mail: ifdominguez@mitma.es

M. Charco
Instituto de Geociencias, CSIC-UCM, Madrid, Spain

E. González-Alonso · C. del Fresno · M. A. Benito-Saz · L. García-Cañada
Instituto Geográfico Nacional, Madrid, Spain

P. J. González (ed.), *El Hierro Island*, Active Volcanoes of the World,
https://doi.org/10.1007/978-3-031-35135-8_5

little is known about the magma plumbing systems in long-dormant volcanic islands like the Canary Islands. Moreover, there have not been many recent monogenetic basaltic eruptions in volcanic islands, and fewer of them fully monitored (Albert et al. 2016). There were no direct geophysical or geodetic observations of the magma plumbing system before the 2011 eruption in El Hierro Island, and the only information about this structure came from petrological studies (e.g., Stroncik et al. 2009).

In 2011, a magmatic unrest episode in El Hierro culminated in a 5 months long submarine volcanic eruption (López et al. 2012) followed by 6 short-duration magmatic intrusions between 2012 and 2014 (Benito-Saz et al. 2017). This was the first fully monitored eruption in the Canary Islands and has produced tens of scientific studies providing an outstanding opportunity to explore magma storage and transport through geodetic and seismic methods.

In this contribution we review published geodetic and seismic studies about El Hierro volcanic episode, together with geological results, in order to provide a common interpretation about the magma storage and transport. Firstly, the main available data sets used in the literature are briefly reviewed, followed by a description of the observed geodetical and seismological signals during the different pre-eruptive, syn-eruptive and post-eruptive periods. Finally, the different results in the literature are discussed to achieve a joint interpretation of the magma plumbing system and a list of open questions related to the volcanic process is provided.

5.2 Volcano Monitoring Network in El Hierro

At the beginning of the seismic crisis on 19 July 2011, the volcano monitoring system on the island of El Hierro was scarce. It consisted of a couple of seismic stations on the island which, together with 3 other stations on the neighboring islands of La Palma and La Gomera, allowed for the location of only the larger events. There was also a GNSS station (FRON) belonging to Grafcan (www.grafcan.es). Since the beginning of the unrest the IGN (López et al. 2012) and other institutions (Pérez et al. 2012; Prates et al. 2013) deployed a monitoring network on the island, including seismic, geodetic, gravimetric, and geochemical stations. For the study of magma migration below the island presented in this chapter we have relied on information from scientific publications on the subject, most of which are based on data obtained by the IGN seismic and geodetic networks as well as radar images.

Figure 5.1 shows the evolution of the seismic and geodetic networks deployed on El Hierro Island and the periods of operation of each station. It should be noted that on 21 July there were already 6 seismic stations on the island, 4 of which transmitted real-time data. By October 2011, the IGN network in the island was composed of 9 real-time seismic stations, mostly with short period sensors, and 10 GNSS stations, one of them shared with the University of Cádiz (UCA). This joint network was in operation from the beginning of the volcanic eruption and during all the post-eruptive intrusions and it has already been described in the literature, both the seismic (López et al. 2012; Domínguez Cerdeña et al. 2014) and the geodetic networks (López et al. 2012; Domínguez Cerdeña et al. 2018).

In addition, the monitoring network on the surrounding areas served as support and used, in the case of the GNSS network, to process GNSS data in a regional network (Benito-Saz et al. 2017). In the case of the seismic observations, data from other islands stations was not used for location as only the larger events were recorded and farther stations observations ($\Delta \geq 80$ km) considerably increased the time residuals due to the inadequate velocity model used (Domínguez Cerdeña et al. 2014). However, 8 broad band seismic stations on the neighboring islands of La Palma, La Gomera and Tenerife were used to help constrain the magnitude and focal mechanism of the largest earthquakes.

The deployment of the network had some difficulties to properly cover all the geography, and not all the stations in El Hierro Island had

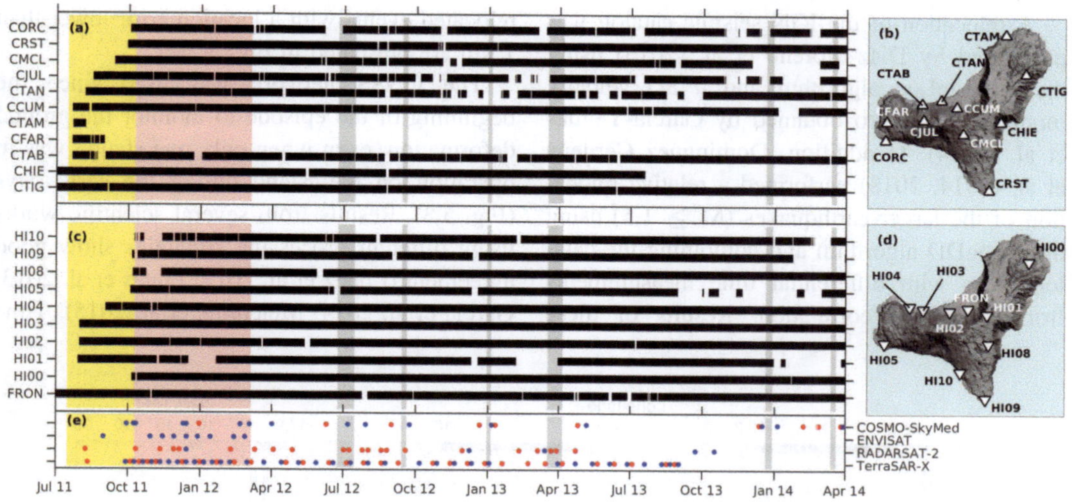

Fig. 5.1 Evolution of the IGN seismic and geodetic volcano monitoring networks in El Hierro during the period 2011–2014. **a** Seismic data available from each seismic station. **b** Location of the IGN seismic stations in El Hierro. **c** GNSS data available from each station at El Hierro. **d** Location of the IGN/Grafcan/UCA GNSS stations in El Hierro. **e** Radar data from COSMO-SkyMed, ENVISAT, RADARSAT-2 and TerraSAR-X satellites in ascending (red) and descending (blue) orbits used for the different publications (López et al. 2012; González et al. 2013; Cong et al. 2015; Benito-Saz et al. 2017). Vertical shadows in **a**, **c**, **e** indicate the pre-eruptive period (yellow), the eruptive period (pink) and the post-eruptive intrusions (gray)

adequate communications for real time transmission. In these cases, the emergency communications system of the Canary Islands Government was occasionally used for data transmission. Some of the data outages (Fig. 5.1) were caused by this lack of communications, as well as other problems including vandalism, power supply problems—due to the low insolation of the stations in the north of the island or the deterioration of batteries and solar panels of stations close to the sea due to salinity and corrosion.

When the volcanic activity was detected in El Hierro in July 2011, no radar satellites were acquiring images on a regular basis over the island, so only a short number of acquisitions were available during this period. It was not until September 2011 that some satellites such as COSMO-SkyMed, RADARSAT-2 or TerraSAR-X were programmed to capture images with a permanent frequency over the area. Those satellites kept monitoring the island during the eruption and throughout the 2012–2014 unrest periods, except for the Envisat satellite, which was at the end of its operational life and stopped acquiring images in 2012 (Fig. 5.1).

5.3 Seismic and Geodetic Results

Seismic and geodetic data allowed scientists to infer the processes that occurred below El Hierro Island between 2011 and 2014. The IGN manually reviewed seismic catalog contains approximately 25,000 earthquakes between 2011 and 2014, including the pre-eruptive crisis, the eruption and the successive magmatic intrusions that took place after the end of the eruption. However, due to the high rate of the seismicity, it was impossible to manually locate all the events, and the number of earthquakes that actually occurred was much larger. Only during the three months before the eruption, the IGN seismic catalog registered 10,000 earthquakes (López et al. 2012), however, through new automatic algorithms we know now that at least 25,000 earthquakes took place during this period (Díaz Suárez et al. 2019).

A relocation of the IGN seismic catalog was performed by Díaz-Moreno et al. (2015) using the NonLinLoc algorithm and a 3D velocity model of El Hierro obtained by García-Yeguas et al. (2014). In addition, Domínguez Cerdeña et al. (2014, 2018) performed a relative relocation of the larger earthquakes (M ≥ 1.5) using the HypoDD algorithm and combining the catalog picks with differential time measurements from waveform correlation. Results of these relocated events with a location error lower than 1 km are presented in Fig. 5.2a.

The GNSS network was used since the beginning of the episode to monitor the ground deformation, even when only one station was in operation on the island during the early days (Fig. 5.3). Results from several scientific works using different processing strategies show good agreement (López et al. 2012; Prates et al. 2013; García et al. 2014; Meletlidis et al. 2015). Time

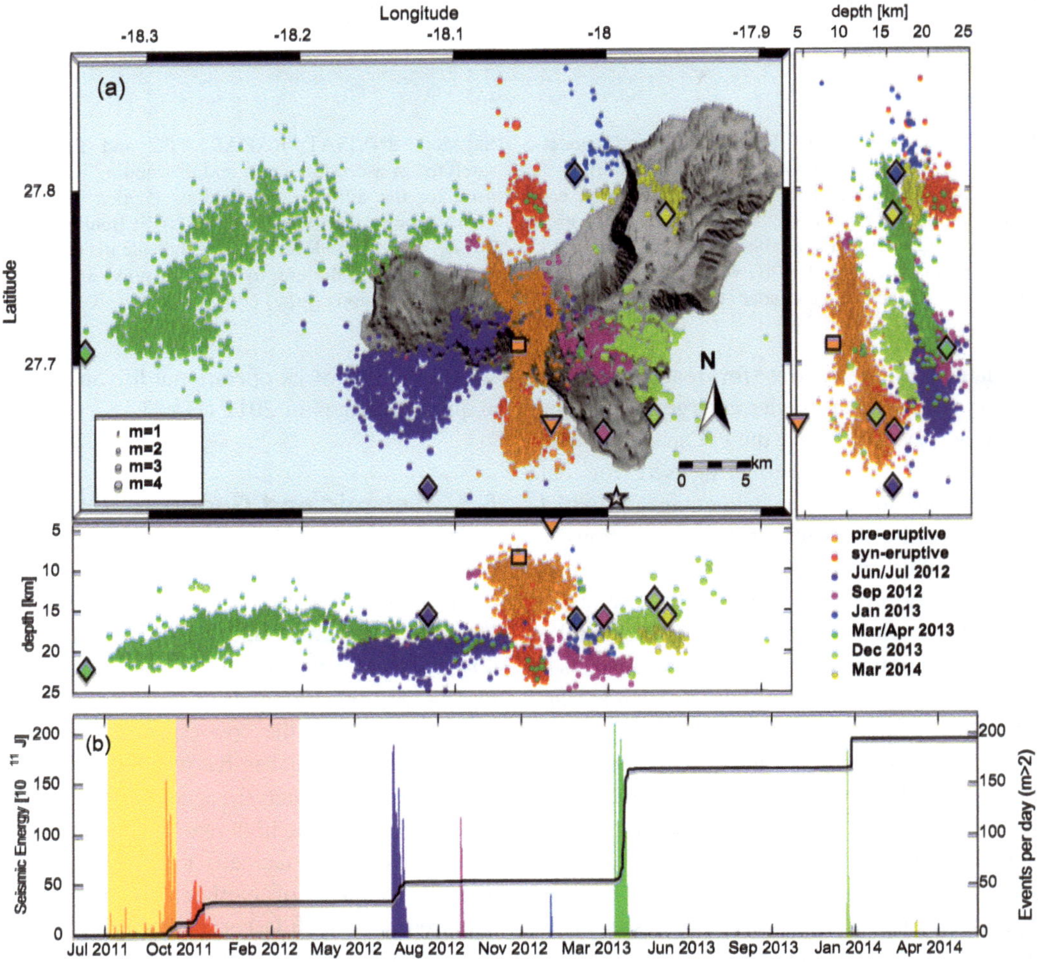

Fig. 5.2 Seismicity recorded at El Hierro between 2011 and 2014 including the pre-eruptive, and the eruptive period and the six post-eruptive intrusions. **a** HypoDD relocations of the seismicity (from Domínguez Cerdeña et al. 2014, 2018). White star indicates the main vent of the Tagoro eruption. Geodetic models of sill like magmatic intrusions for two pre-eruptive periods 5 May–8 Aug 2011 (orange square) and 31 Aug–30 Oct 2011 (orange triangle) from González et al. (2013) and for the six posteruptive magmatic intrusions (colored diamonds) from Benito-Saz et al. (2019). **b** Histogram of earthquakes with magnitude > 2 (right) and cumulative seismic energy release (left)

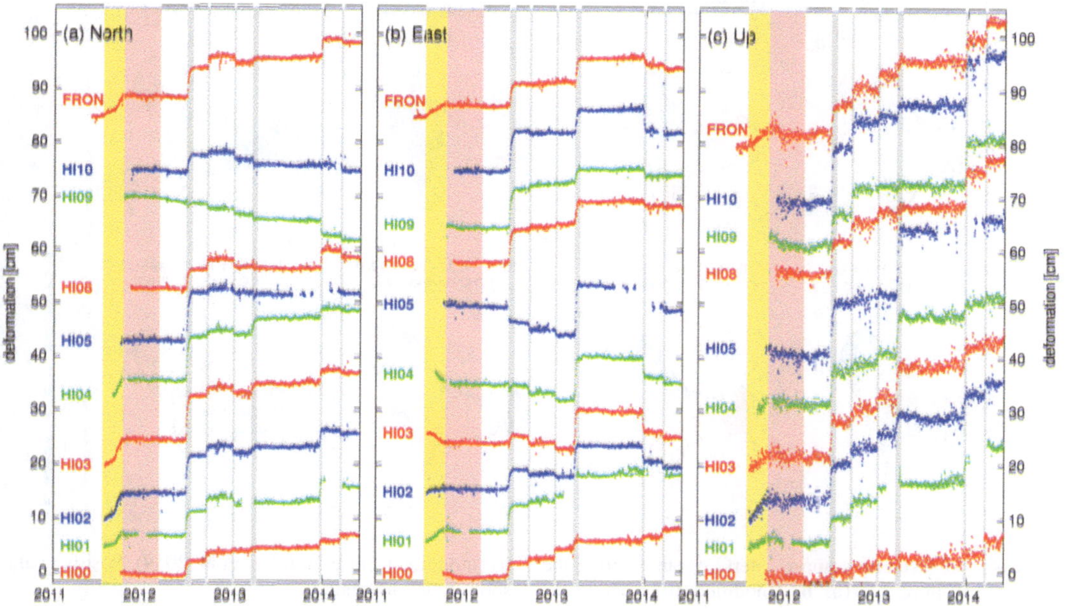

Fig. 5.3 Daily ground surface deformation at El Hierro Island from 2011 to 2014 in **a** north, **b** east, and **c** vertical (up) directions recorded at the ten GNSS stations installed on the island. Yellow shadow shows the pre-eruptive interval, red shadow the eruptive period and gray shadows the six post-eruptive intrusive events. Modified from Domínguez Cerdeña et al. (2018)

series of GNSS coordinates are presented in Fig. 5.3. InSAR results also provide valuable information about the surface deformation pattern (López et al. 2012; González et al. 2013; Cong et al. 2015; Benito-Saz et al. 2017, 2019). Representative interferograms for pre-eruptive and post-eruptive periods are shown in Fig. 5.4. Furthermore, both GNSS and InSAR data were inverted, together or independently, to estimate the position and volume of the deformation source for each of the different periods (González et al. 2013; Meletlidis et al. 2015; López et al. 2017a; Benito-Saz et al. 2017, 2019). A summary of modelling results is showed in Table 5.1.

5.3.1 Pre-eruptive Magma Emplacement (Jul.–Oct. 2011)

Though the volcanic unrest at El Hierro begun in mid-July 2011, and the low rate of background seismicity on the island (with about 50 earthquakes per year during the previous 15 years), López et al. (2017b) found several early precursor signals of possible magma emplacement since 2003, some years before the onset of the unrest. These included fractal dimensions changes in the seismic data of the only station available in El Hierro since 1997 (CHIE) and regional deformation patterns observed by GNSS stations in the Canary Islands but also affected the south of the Iberian Peninsula and northwest Africa (López et al. 2022). These signals, mostly found in deformation and seismic noise fractal analysis, were consistent with the idea of magma storage beneath the crust for years before eruption forming an ephemeral magma accumulation zone, rather than a magma chamber because geobarometric data do not indicate long-term storage of the magma at crustal depth (Klügel et al. 2015 and references there in).

According to López et al. (2012) clear signs of deformation were measurable from the 7 July in

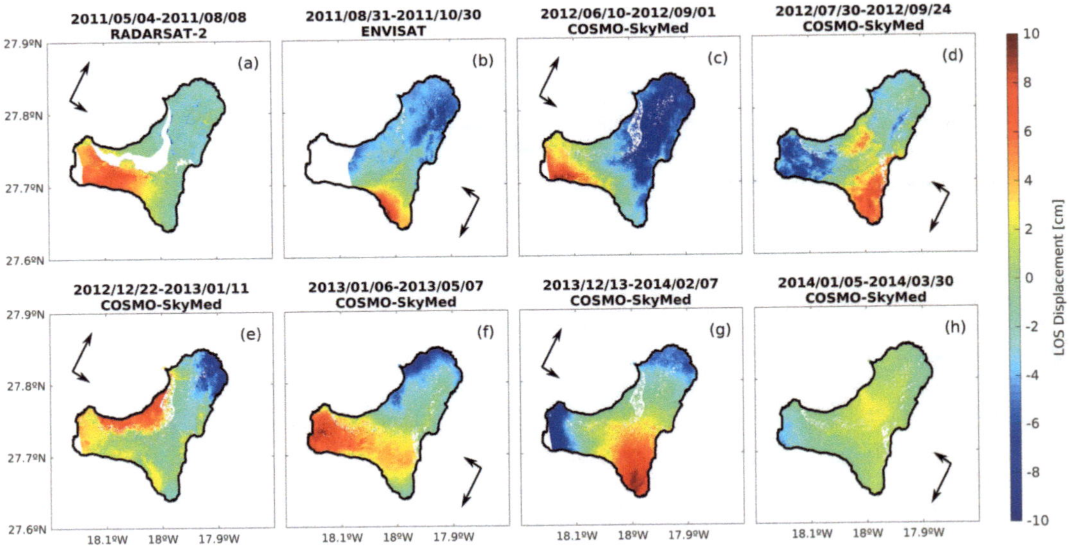

Fig. 5.4 Observed unwrapped interferograms during the pre-eruptive intrusion (**a**, **b**) (modified from González et al. 2013) and during the post-eruptive intrusions (**c**–**h**) (modified from Benito-Saz et al. 2019). Positive values show LOS shortening

the only GNSS station available on the island at that time (FRON, Fig. 5.3) and the anomalous seismicity started on 19 July. However, the seismic unrest begun on 16 July with the occurrence of a few tens of events. From these dates and until the eruption onset, the spatio-temporal evolution of the seismicity and deformation was similar indicating a magmatic intrusion that evolved with time. In general, the source center of deformation followed a similar evolution to the migration of the earthquake epicenters (Prates et al. 2013; García et al. 2014) although they were always located at shallower depths and further south than the seismic sources (González et al. 2013; García et al. 2014; Meletlidis et al. 2015; López et al. 2017a).

During the first week of the unrest, the seismicity was located in the center of the island and migrated northward (Meletlidis et al. 2015). A north-eastward displacement at FRON station started on 7 July (López et al. 2012) and the volcanic origin of the deformation was confirmed when more stations were installed. Due to the lack of deformation data during these dates, not possible to obtain the source position. From that moment on, a magmatic sill began to intrude in the northern part of the island at a depth of $\sim$ 10 km, which gradually opened the crust as it was shown by the seismicity epicentral distribution (Domínguez Cerdeña et al. 2014). This seismicity was characterized by a Gutenberg Richter *b-value* greater than 2 (Ibáñez et al. 2012; Roberts et al. 2016), possibly indicating the presence of fluids in the area of rupture. In addition, ground deformation showed a radial pattern around Tanganasoga volcano, located in the centre of the island, reaching a maximum of 10 cm of shortening in LOS (Line Of Sight) direction. Deformation results indicated a magma source at the same depth for both spherical point and sill-like sources using InSAR data (González et al. 2013). Shallower depths were obtained when modeling only GNSS data with spherical point sources (Table 5.1). The joint mechanism of the main earthquake families during this first stage of the unrest has shown that most of the energy was released with thrust fault mechanism earthquakes, with a vertical T axis (del Fresno 2016), which is consistent with the inflation observed at GNSS stations and InSAR data.

From mid-September onwards, seismicity began to migrate southward and deeper (Domínguez Cerdeña et al. 2014). It was during this period that the deepest earthquakes occurred

Table 5.1 Geodetic models of the magmatic intrusions from different publications

Source	Period	GNSS/InSAR	Mogi depth (km)	Sill depth (km)	Seismic depth (km)
Pre-eruptive					
González et al. (2013)	5 May–8 Aug 2011	InSAR	8.4 ± 4.1	9.7 ± 2.5	10.2 ± 1.5
	Stack R2S6[a]	InSAR	10.2 ± 4.7	11.2 ± 1.8	–
	31 Aug–30 Oct 2011	InSAR	4.2 ± 2.4	4.9 ± 3.8	12.7 ± 1.9
Meletlidis et al. (2015)	3–15 Aug 2011	GNSS	4.9	–	10.2 ± 1.3
	15 Aug–3 Sep 2011	GNSS	4.9	–	10.3 ± 0.6
	3 Sep–1 Oct 2011	GNSS	3.8	–	12.7 ± 1.7
	1–7 Oct 2011	GNSS	5.0	–	13.4 ± 1.7
López et al. (2017a)	31 Jul–8 Sep 2011	GNSS	$4.9_{2.3}^{14.0}$	–	10.3 ± 0.9
	8–21 Sep 2011	GNSS	$5.4_{4.1}^{21.6}$	–	11.2 ± 0.7
	21 Sep–9 Oct 2011	GNSS	$5.1_{4.2}^{11.6}$	–	13.5 ± 1.5
	3 Oct–9 Oct 2011	GNSS	$12.6_{3.0}^{22.9}$	–	13.1 ± 1.6
Post-eruptive					
Benito-Saz et al. (2019)	Jun–Jul 2012	Both	11.6 ± 0.4	15.0 ± 1.5	20.0 ± 1.2
	Sep 2012	Both	12.1 ± 1.4	16.4 ± 1.2	20.9 ± 2.2
	Dec 2012	Both	10.6 ± 1.2	14.6 ± 1.4	17.2 ± 3.1
	Mar–Apr 2013	Both	17.9 ± 3.3 10.8 ± 0.9	26.8 ± 2.5 21.3 ± 2.4	17.4 ± 1.9
	Dec 2013	Both	9.8 ± 0.7	13.6 ± 1.1	16.2 ± 1.3
	Mar 2014	Both	11.8 ± 1.9	15.4 ± 1.4	18.0 ± 0.5

The table shows the period modelled, the data source used (InSAR, GNSS or both), depths of the Mogi model or sill model (when available) and average depth of the seismicity during the same period (from relocated earthquakes of Fig. 5.2)
[a] A stack of four RADARSAT-2 ascending interferograms with the 8 August 2011 as common slave image

(13 km). Evolution of the deformation trend was stable until early September when accelerated and changed direction to the north indicating a causative source located offshore. González et al. (2013) registered a maximum motion onshore of 10–12 cm towards the satellite whereas GNSS stations detected a northward movement with maximum accumulated horizontal displacements of 3.5–4 cm (Meletlidis et al. 2015) and an uplift of $\sim$ 1 cm (FRON station). Modelling results for this period locate a point source at a depth of $\sim$ 5 km (González et al. 2013; García et al. 2014; Meletlidis et al. 2015). According to del Fresno (2016), results of the moment tensor inversion of the largest earthquakes located during this period indicate thrust or vertical mechanisms when considering a double couple model and significant isotropic and CLVD (Compensated Linear Vector Dipole) components when considering a full MT model, these earthquakes, with dominant vertical T axes ($\theta \leq 45°$), can be interpreted as volume increase in the focal region due to the magma intrusion.

The 3 October represents an important change in the activity. On the one hand, the focal mechanism solutions showed a different pattern: larger double couple components which corresponded to strike slip or normal fault mechanisms (del Fresno 2016). On the other hand, a strong change in the surface deformation rate (López et al. 2012; Prates et al. 2013) as well as in strain and the velocities of the medium (López et al. 2017a). Modelling results for this period located point pressure sources at 12.6 km, deeper than in previous phases at only between 4 and 5 km deep (Table 5.1). In this last stage, seismicity occurred slightly shallower ($\sim$ 12 km) including the largest event of the series, which took place on 8 October, with a magnitude Mw = 4.0 and a focal mechanism showing no volume change (del Fresno et al. 2015). This earthquake coincides with the maximum release of seismic energy, maximum strain rate (López et al. 2017a) and changes in the gravity field (Sainz-Maza et al. 2014). Vertical GNSS components of all stations had showed uplift during the whole pre-eruptive period reaching a maximum of 5 cm; on 8 October the deformation stopped. From this moment on, there is an aseismic ascent of the magma to the surface with just a small swarm of shallow earthquakes (4 km to 1 km depth) in the last 30 h before the eruption onset (Domínguez Cerdeña et al. 2014).

5.3.2 Eruptive Phase (Oct. 2011–Mar. 2012)

There is some controversy on the exact date of the beginning of the eruption, with some authors pointing to the 10 October and the beginning of the volcanic tremor signal (López et al. 2012; Martí et al. 2013a) while others consider the 12 October, corresponding to the "visual confirmation" on the sea surface (Ibáñez et al. 2012) and changes on the tremor signal (González et al. 2013). Numerous publications have described the evolution of the eruption of Tagoro volcano in terms of the observable effects on the sea surface, the results of bathymetric studies, and the petrology of the emitted products (e.g., Carracedo et al. 2012; Martí et al. 2013a, b; Rivera et al. 2013; Longpré et al. 2014).

One piece of evidence for the Tagoro eruption was the detection of a non-harmonic continuous volcanic tremor (López et al. 2012; Tárraga et al. 2014). The tremor showed a strong variation on amplitude, with a maximum during the first days of eruption. Analysis of the fractal dimensions of the tremor signal suggests a change in the geometry of the vent from a fracture plane to a smaller dimension, presumably in the form of a conduit as was established on 12 October (López et al. 2014). Tárraga et al. (2014) analyzed the tremor signal and interpreted their results as indicating an interconnection of the whole plumbing system and a dependency of eruption dynamics on its behavior. They related variations in tremors to changes in the stress conditions of the plumbing system, dimensions of the conduit and vent, intensity of the explosive episodes, and rheological changes in the erupting magma.

Throughout the eruptive period, the volcanic tremor signal was dominant in the seismic records and hindered the analysis of the seismicity. Despite this, the seismic catalog includes more than 2000 earthquakes in this period. During the first week, just a few earthquakes were detected in the region of the last pre-eruptive seismicity. A more active period started on the 17 October with a new cluster of seismicity at 20–25 km depth in the North part of the island, and only a week later another shallower swarm began at 10–15 km depth (Martí et al. 2013b; Meletlidis et al. 2015). This eruptive seismicity shows a progressive increment of the magnitude and energy. The hypocentral relocation reveals two clearly separated clusters (Fig. 5.5) in a smaller area than the IGN seismic catalog. The *b-value* obtained by Ibáñez et al. (2012) was close to 1 for this period, however, separate analysis of the two clusters shows a higher *b-value* at the beginning of their activity (Fig. 5.5c).

During the first two months of eruption, there was a deflation measured by GNSS stations which mainly affected the vertical components, being remarkable only for some stations (Fig. 5.5a; Meletlidis et al. 2015). Correlations between strong seismicity during the eruption and sub daily GNSS positions were found for some stations in El Golfo, mostly in their height components (Prates et al. 2013). However, no significant deflation was found for the whole island which might point to a recharge of the system rather than to a deep depressurization (Prates et al. 2013; García et al. 2014). Since December 2011 until the end of the eruption deformation results did not show significant displacements.

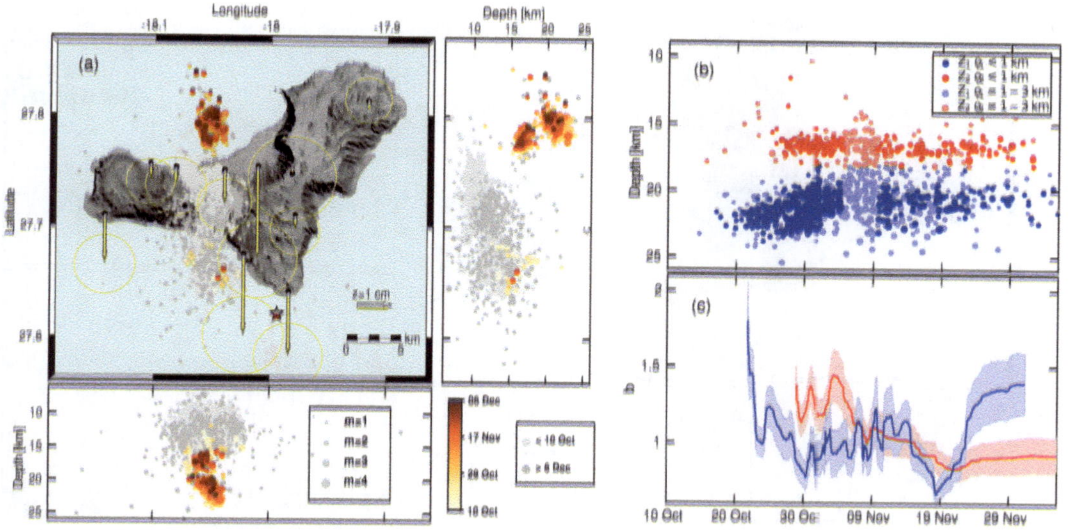

Fig. 5.5 Relocated seismicity and surface deformation during the El Hierro eruption. **a** Main seismic swarm (colored dots) during the eruption (10 Oct–6 Dec), the pre-eruptive seismicity (light gray dots) and the rest of the syn-eruptive seismicity (7 Dec–15 Mar, dark gray dots). Vertical deformation during the period 10 Oct–31 Dec for each GNSS station (yellow arrows). **b** Depth evolution of the seismicity for the deeper seismic swarm (blue) and shallower swarm (red). Light and full colored distinguish between high and low uncertainty locations. **c** Evolution of the b-value for the deeper seismic swarm (blue) and shallower swarm (red). Light colored shadows show error bars

Through the eruption was not officially reported as having ended until 5 March, there is evidence suggesting that it could have ended by 15 February, as suggested by the cessation of tremor and a change in waveform similarity (Sánchez-Pastor et al. 2018).

5.3.3 Post-eruptive Magmatic Intrusions (Jun. 2012–Mar. 2014)

After the 2011–2012 submarine eruption, six intrusive events were detected on the island from June 2012 to March 2014, as revealed by seismic swarms and ground deformation (García et al. 2014; Díaz-Moreno et al. 2015; Klügel et al. 2015; Telesca et al. 2016; Lamolda et al. 2017; Benito-Saz et al. 2017, 2019; Domínguez Cerdeña et al. 2018).

Each intrusion lasted between a few days and three weeks and was characterized by the near-simultaneous initiation of seismicity and ground deformation (Lamolda et al. 2017). However, seismicity was located ~ 3–8 km deeper than inferred geodetic deformation source models, with seismic swarms occurring at 15–23 km depths and optimal geodetic models located in the ~ 13–16 km depth range (Table 5.1).

These intrusions exhibited initial flow rates of ~ 300 m^3/s, decaying exponentially with time and were inferred to be sill-like. During the two largest intrusions that occurred in June–July 2012 and March–April 2013, magma migrated laterally for > 15 km in a period of ~ 20 days with a volume increase of > 120 × 10^6 m^3 beneath the island. The shortest events, < 1 week-long, intruded between ~ 24 and 44 × 10^6 m^3 of magma beneath the volcano (Benito-Saz et al. 2019). Despite the contrast in magma volume between the six intrusions, they have a similar mean magma volume rate, with values between 64 and 99 m^3/s (Domínguez Cerdeña et al. 2018) due to the different duration of the processes.

We have found a new feature in the seismic and geodetic data in the periods between some of the post-eruptive magmatic intrusions. There are two periods of seismic activity located in the

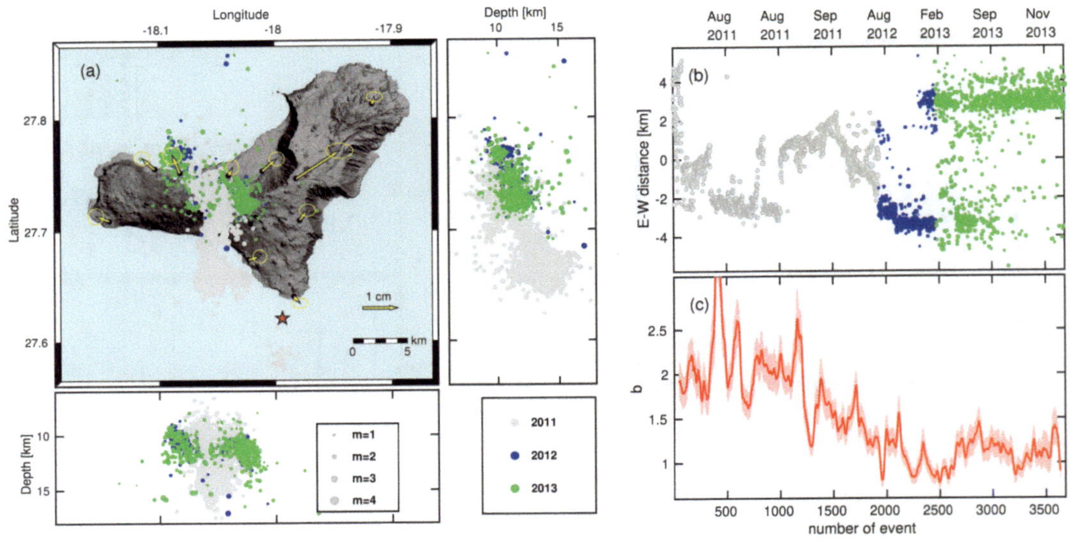

Fig. 5.6 Seismicity and deformation produced by the northern magmatic sill. **a** Joint relocation of the seismicity during the pre-eruptive unrest (gray dots), Aug–Sep 2012 (blue dots) and Jul–Dec 2013 (green dots). Horizontal deformation during the period July–Dec 2013 (yellow arrows). **b** Evolution in E-W direction of the relocated seismicity during the pre-eruptive unrest (gray dots), 2012 (blue dots) and 2013 (green dots). **c** Evolution of the b value of the whole process. X-axis for **b** and **c** is in number of events

central part of the island before the second and fifth intrusions (September 2012 and December 2013) (Fig. 5.2). The activity in August–September 2012 and July–December 2013 was located at around 10 km depth in the north part of the island, where the pre-eruptive seismicity began in July 2011. The relative relocation of this seismicity (Fig. 5.6a) shows its relation with the magmatic sill observed in July–August 2011 (Domínguez Cerdeña et al. 2014). There are two clusters of alternating activity with increasing separation with time (Fig. 5.6b). Moreover, a slight deformation of ~ 1 cm is detected in various GNSS stations with the source located in between both clusters (Fig. 5.6a). The *b-value* is around 1 suggesting the absence of fluids (e.g., Zobin 2012) at the location of these earthquakes (Fig. 5.6c) however, contrary to the *b-value* obtained for pre-eruptive seismicity.

In total, from June 2012 to March 2014 > 20 cm of uplift and > 10 cm of horizontal ground deformation was detected in the central-western part of the island, corresponding to approximately 0.32–0.38 km^3 of magma intruded beneath different parts of the volcano (Benito-Saz et al. 2017, 2019). Cong et al. (2015) used TerraSAR-X images from August 2011 to June 2013 to calculate averaged velocities for this period. Results show maximum deformation velocities between 8 and 20 cm/year.

Since March 2014, no significant ground deformation has been detected on the island and remnant seismic activity has occurred at a rate of ~ 15 earthquakes per month, decaying through the successive years.

5.4 Discussion

5.4.1 Comparison of the Unrest Episodes

Synthesis of the seismic and geodetic results allows a robust comparison of unrest episodes at El Hierro. The 2011–2014 unrest episodes registered at El Hierro signify the emplacement of multiple magmatic intrusions at depth beneath the island. The 2011 intrusion culminated in a volcanic eruption, whereas 2012–2014 intrusions did not, which constitutes a clear demonstration that magmatic intrusions do not often culminate in eruption, as previously discussed by e.g. Biggs et al. (2014). All magmatic intrusions started in the center of the island, where quasi-periodic pulses of magma from depth stagnate at 10–16 km b.s.l. The temporal pattern of the magmatic intrusions occurrence does not suggest any correlation with known periodic physical phenomena. Recently, Miguelsanz et al. (2021) showed that ocean and body tides modulated the seismic activity at El Hierro, establishing a correlation between earthquake occurrence times and tidal stress values and rates. However, the pre-eruptive intrusion developed over a period of three months while the post-eruptive intrusions lasted only between 3 and 20 days. Spatially, magma associated with pre-eruptive intrusions and the longest of the post-eruptive intrusions migrates laterally from the center of the island. The pre-eruptive intrusion propagated laterally for > 15 km to the south whereas the June–July 2012 intrusion propagated laterally for ~ 14 km to the southwest and the March–April 2013 intrusion migrated for ~ 28 km to the west (Benito-Saz et al. 2019). A conspicuous characteristic of this lateral migration is their spatial occurrence in a successive anticlockwise rotation through the three rifts of the island (Domínguez Cerdeña et al. 2018). This may indicate the existence of a complex structure below El Hierro that responds to stress changes: the alterations provoked by a magmatic intrusion in one of the rifts triggered the opening of a new magma finger in another of the rifts in a succession. Moreover, after the largest post-eruptive unrest periods (June 2012, March 2013) there were some reactivations of the central sill. Figure 5.6 shows the seismicity and deformation related to such reactivations (August–September 2012 and July–December 2013). Such a series of events suggest a causality, since the likelihood of a random distribution leading to this sequence is very low. Though the rifts do not appear to have a deep origin (Sainz-Maza et al. 2017) and despite the non-existence of a three-armed construction of the island (Becerril et al. 2015) there seems to be a structure beneath the island related with the rifts as suggested by some authors (Carracedo 1996; Münn et al. 2006). Furthermore, Domínguez Cerdeña et al. (2018) found differences in the ratio of the seismic and geodetic moment of the intrusions related with the capacity of the caprock to transform stress energy accumulated by the deformation in seismic events (Pedersen et al. 2007). Such contrast in the moment ratio indicates differences in the rheology of the rock beneath each of the rifts on El Hierro.

The crustal response to deformation produced by magma intrusions did not differ significantly from one intrusive episode to another (Telesca et al. 2016). During the six unrest periods (2012–2014), the displacement detected by GNSS and InSAR (Benito-Saz et al. 2019) or using only GNSS results (García et al. 2014; Meletlidis et al. 2015; Domínguez Cerdeña et al. 2018) were of the same magnitude or even larger than during the pre-eruptive phase (Figs. 5.3 and 5.4). They show a peak of 12 cm in March/April 2013 (HI05 station) when a velocity of up to 2 cm per day was measured (Domínguez Cerdeña et al. 2018). Domínguez Cerdeña et al. (2018) grouped the six post-eruptive intrusions in three pairs: June 2012 and March 2013 intrusions located in the western rift of the island, September 2012 and December 2013 swarms located in the southern rift, and December 2012 and March 2014 unrests located in the northeast rift. Apart from these three groups, now we have highlighted a new shallow magmatic activity in the North in two periods, August–September 2012 and July–December 2013 (Fig. 5.6) that can be considered a new group related to the reactivation of the central sill. There are some similarities between the three different groups mentioned, including the mean magma volume rates (Domínguez Cerdeña et al. 2018) and the linear relationship between the seismic energy and the intrusion volume (Lamolda et al. 2017). Geodetic data from the June–July 2012 and March–April 2013 post-eruptive intrusions suggest a volume > 120 × 10^6 m^3 (Benito-Saz et al. 2019) whereas the volume of the shorter intrusions such as the December 2013 was ~ 40 × 10^6 m^3, similar to the volume estimated for the pre-eruptive intrusion (González et al. 2013; López et al. 2017a, b).

Deep intrusions during the pre-eruptive and post-eruptive phases caused seismic swarms. The magnitude of the earthquakes registered during post-eruptive seismic swarms was similar or even larger to those recorded during the pre-eruptive period in the days prior to the eruption. However, there are several important differences between the pre-eruptive and post-eruptive intrusions. Seismicity during the 2011 pre-eruptive intrusion occurred at 9–11 km depth, whereas in the post-eruptive intrusions it occurred deeper, at ~ 13–22 km. Furthermore, seismic energy released is much larger for western rift intrusions (Benito-Saz et al. 2019) although no clear pattern was observed indicating that intrusions were becoming less energetic or that they were going to stop occurring.

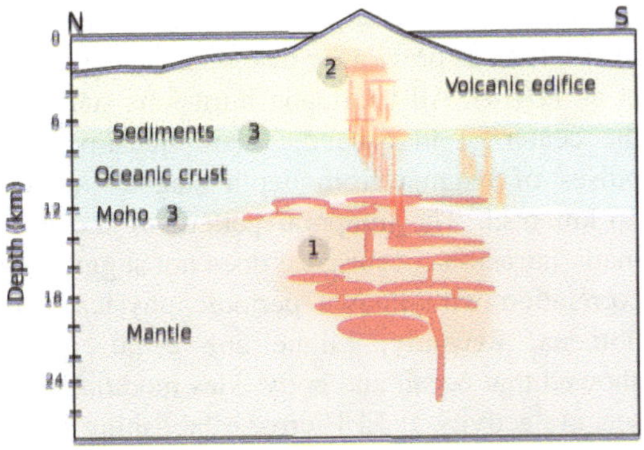

Fig. 5.7 Schematic interpretation of magma storage. (1) Main magma feeding reservoir constrained from petrologic and seismic data. Geodesy provides an indirect constraint; (2) High seismic velocity anomaly (low density material) under Tanganasoga volcano that favors magma stalling and lateral migration; (3) areas of ephemeral accumulation of magma. Magma transport involves sill intrusions and lateral migration within lower crust. At shallower levels, ephemeral magma accumulation can occur at pre-volcanic sediments layer

5.4.2 Inferences on the Magma Plumbing System

Volcanoes' magmatic evolution is mainly controlled by their crustal and mantle plumbing systems. The unrest episodes of El Hierro provide many novel findings on the sub-surface structure and plumbing system of the island. Whereas the analyses of erupted materials through experimental petrology constitutes the main tool to advance our understanding of the long-term evolution of magmatic systems, geodetic and geophysical data are in valuable to constrain magma location and volume. In particular, at basaltic intraplate oceanic islands with low eruption rates such as El Hierro, both seismicity and ground deformation allow us to constrain the magma plumbing system.

Before Tagoro eruption, Stroncik et al. (2009) proposed a general model for the El Hierro plumbing system that consisted of a volcanic edifice underlain by a plexus of partly interconnected magma pockets, sills and dikes within the uppermost mantle between ~ 14 and 30 km depth, and most likely extending to deeper levels. Joint interpretation of all available published geodetic and seismic results, building on the foundation of petrological studies, allows us to constrain the location and geometry of magma accumulation zones. Figure 5.7 displays a schematic view of our interpretation of the El Hierro plumbing system. It contains several elements: (1) a main magma reservoir beneath the center of the island, located at the uppermost mantle; (2) structural/rheological discontinuities and features that may inhibit upward magma flow; (3) ephemeral areas of magma accumulation, mainly located at these discontinuities, where magma has been accumulated before lateral migration. Curiously, the location of the main area of magma storage corresponds quite well with the inferred feeding system of the only polygenetic volcano of the island, Tanganasoga volcano (Carracedo et al. 2001; Manconi et al. 2009).

Most of the seismic tomographic models suggest the existence of a magma accumulation zone in the upper mantle of uncertain origin at a depth of 20–25 km, from where magma aseismically rises until an unrest begins. Martí et al. (2017) showed a low P-wave velocity zone, probably imaging a reservoir-type structure in the lithospheric mantle, at 12–24 km depth. Similarly, Gorbatikov et al. (2013) found rather low shear-wave velocities in the area for depths 20–25 km, suggesting the presence of magmatic melt material in the area. García-Yeguas et al.

(2014) also observed a low-velocity anomaly beneath a shallower high velocity anomaly, interpreted as a batch of magma coming from the mantle. Clinopyroxene-melt barometry performed on phenocryst rims (Longpré et al. 2014), the reverse zoning pattern in olivine erupted at the first stage of the eruption episode, and the subsequent magma residence times from this compositional zoning indicate that the main magma reservoir from where the bulk of magma was delivered for the eruption is located in the uppermost mantle (~ 20–25 km).

Strong and deep seismicity (15–25 km) registered in the north of the island during the first two months of the eruption seems to be correlated with the presence of this main magma storage region located in the upper mantle. A combination of petrological and geophysical analysis by Martí et al. (2013a) pointed out a scenario with seismicity produced by shear fracturing in response to gravitational and tectonic readjustments of the plumbing system. Furthermore, they found a clear change in the composition of the petrological samples taken after December 2011, coinciding with a change in the seismicity pattern, northern seismicity at 15–25 km depth disappears (Fig. 5.5) as seismicity at shallower levels increase (Meletlidis et al. 2015). González et al. (2013) proposed that seismicity and its correlation with volume changes estimated in an ephemeral magma reservoir located in the lower crust could be caused by the collapse of the deeper part of the magma plumbing system due to magma withdrawal and/or chamber walls relaxation. The location of the earthquakes during this period (Fig. 5.5) indicates a seismicity tracing the north part of the main feeding system and the presence of increasing seismicity suggest an increasing pressurization (González et al. 2013; Rivera et al. 2013) instead of just readjustments taking into account the decrease in magma rate at the end of this seismic swarm on December 2011.

Furthermore, magma intrusions responsible for the post-eruptive unrest would have been supplied with magma from the overpressurized area in the uppermost mantle discussed above. Benito-Saz et al. (2019) showed a clear exponential decay of magma volume change time-series at least at the long-lasting post-eruptive events. Few physical models explain surface deformation with this temporal pattern (e.g., Lengliné et al. 2008; Dzurisin et al. 2009; Anderson and Segall 2011; Reverso et al. 2014; Walwer et al. 2019; Rodríguez-Molina et al. 2021). At El Hierro, we suggest a hydraulic pressurization of the crust due to a continuous magma supply from a deeper source located at the central area of the island in concordance with seismic and petrological results. The quantitative analyses of the physics and dynamic of magma accumulation zones often requires simple geometrical representations of these areas. However, the presence of melt in lower crust and upper mantle is likely to be distributed in a complex system of dykes, sills and conduits that may not coalesce into a large clearly defined form (e.g., Marsh 2015). At El Hierro, the non-uniform and limited spatial coverage of the seismic network due to the size of the island restricts the resolution of tomographic images at lower crust/upper mantle levels (e.g., Lees 2007). Furthermore, the geodetic signature due to the uppermost mantle accumulation area is not directly identified due to the high ratio of the depth to the dimensions of the island. Therefore, we cannot rule out the idea of Stroncik et al. (2009), of a magma plexus within this area which results in the reservoir having an oblate shape. Magma pressurization and/or physico-chemical conditions of melt, buoyancy forces, stresses of the host rocks and structural/geomorphological/rheological features of the media would strongly influence magma pathways, the preferred location for new vents to open as well as eruption possibilities.

Geodetic signals during pre-eruptive and post-eruptive unrest episodes show that magma from the uppermost mantle intruded and accumulated into the lower oceanic crust directly beneath the central part of the island (Table 5.1). Magma storage at this depth is consistent with an intrusion stalled at the base of the oceanic crust where stress barriers due to thermal and mechanical properties of the host rock could inhibit magma flow (e.g., Martel 2000; Burov et al. 2003). P-wave tomographic images by Martí et al. (2017)

map a high velocity body beneath the central part of the island reaching about 8–10 km depth. This body could inhibit magma ascent directly above the mantle reservoir. It has a similar geometry to the high-density body mapped at 6–10 km depths by Montesinos et al. (2006) on the basis of gravimetric modelling. García-Yeguas et al. (2014) also mapped a high-velocity anomaly at 10–12 km. Furthermore, Moho depths around El Hierro have been estimated at about 12–15 km using gravity and seismic reflection data (Watts 1994; Ranero et al. 1995) and recent seismic tomographic images (Martí et al. 2017).

Depth estimates of this accumulation area at the pre-eruptive stage could be biased. The static image of the magmatic system provided by InSAR constrained a spherical reservoir (although a sill-like reservoir could not be disregarded) located at ~ 9.5 km (González et al. 2013). Interpretations based on continuous GNSS stations constrained a shallower point pressure source depth (~ 5 km; Meletlidis et al. 2015; López et al. 2017a). Such discrepancies could be due to the accuracy of both estimates that is limited by the temporal and spatial coverage of the available geodetic data. González et al. (2013) used a single acquisition geometry whereas the GNSS network was not fully operational until September 2011. However, the depth of this pre- and co-eruption reservoir constrained by InSAR is similar to the intrusion depths given by joint interpretation of GNSS and InSAR data with better spatiotemporal coverage and frequency of post-eruptive events (Table 5.1). Furthermore, fluid inclusion geobarometry (Longpré et al. 2014) and the equilibration of olivine crystals and diffusion modelling (Martí et al. 2013a) constrain short-term stagnation levels during multistage magma ascent and the depth where magma started to differentiate (~ 12–15 km) by fractional crystallization in this newly formed magma reservoir at the central part of the island, respectively. Both petrological and geodetic inferred the ephemeral character of this reservoir at the lower oceanic crust. In particular, volume time series obtained during pre-, syn- and post-eruptive deformation episodes (Fig. 6 of González et al. 2013 and Figs. 5–8 of Benito-Saz et al. 2019) suggested that magma stagnation at this level would persist from weeks to months.

Intrusions may be not confined to the lower oceanic crust. Geobarometric and InSAR data for El Hierro indicate that additional ephemeral magma accumulation may have occured at shallower depths (González et al. 2013; Hansteen et al. 1998), e.g., within a pre-volcanic sedimentary layer located between the top of the oceanic crust and the volcanic edifice at 4–5 km depth (Ranero et al. 1995). Although seismic images were not able to resolve this layer (e.g., Martí et al. 2017) given its small thickness (~ 1 km), this interface would generate a sufficient density/rheology contrast to trap ascending magmas. As well as the central reservoir, this shallower reservoir which is located slightly offshore at Mar de las Calmas coastline at 4.5 km depth (Fig. 5.2) was obtained from one single acquisition geometry interferogram by González et al. (2013). However, the presence of some basanites contaminated with silicic material (xenopumices) in earlier products of the eruption requires magma stagnation at shallow depths (e.g., Meletlidis et al. 2012; Troll et al. 2012; Sigmarsson et al. 2013). Volume time series constrain magma stagnation at least from days to weeks at this shallow level at the final pre-eruptive stage. Moreover, this magma accumulation may be related with the shallow seismicity that occurred during the last 30 h before the beginning of the eruption with depths starting at 4–5 km to finally reach 1–2 km (Domínguez Cerdeña et al. 2014).

5.4.3 Lateral Migration of Magma

Dikes and sills are perhaps the most common means of magma transport, as suggested by their abundance in eroded volcanoes (e.g., Gudmundsson 1983; Walker 1987), as well as monitoring data from active volcanoes (e.g., Pollard et al. 1983; Woods et al. 2019). Transport of magma within El Hierro occurs through multiple deep sill-like intrusions at lower oceanic crust that move laterally beyond volcanic edifice.

Elastic modelling of ground deformation during pre-eruptive and post-eruptive unrest episodes is compatible with both spherical and sill-like sources (Table 5.1). However, Benito-Saz et al. (2019) determined that the opening area of post-eruptive intrusions favors sill-like intrusions which are thin compared to their lateral extent.

The spatial distribution of the seismicity at the beginning of the pre-eruptive and post-eruptive unrest episodes, as well as the hydraulic connection between a deep magma source and the intrusions suggested by deformation time-series, indicate a long-term magma ascent pathway beneath the center of the island, from where magma pulses would be deflected into sills once they reach ~ 13–16 km depth. We note that there are other examples of basaltic systems dominated by sill intrusion over dykes (e.g., Bagnardi et al. 2013). Contrary to volcanic regions in extensional tectonic settings such as Iceland, in an intraplate volcanic region such as the Canary Islands, the regional stress field is expected to be compressive. At El Hierro, intrusions stalled at the level of the crust/mantle discontinuity, where magma encounters a weak horizontal interface. The intrusive events started beneath the highest part of the island and evolved outwards through lateral pathways following a radial pattern, as dictated by topographic stress changes. Furthermore, at the center of the island, there is a body with high density contrast that intrudes upwards from the lower crust. All of these scenarios predict sill formation as previously discussed by several authors (e.g., Menand 2008; Gudmundsson 2011; Menand et al. 2010, 2011; Maccaferri et al. 2011).

One of the main puzzles that has not yet been unequivocally resolved is the apparent inclination of the seismicity mapped during several of the swarms towards the South (Fig. 5.2), and, if truly exits, its causative origin. Although Domínguez Cerdeña et al. (2014, 2018) and Díaz-Moreno et al. (2015) show evidence for the real existence of this tilt there are still some doubts. The poor geometrical coverage of the seismic stations, and the depth of the seismicity which was in many cases outside the seismic network —make doubtful any structure found. Moreover, magma source evolution inferred from geodetic data show decreasing depth of volume changes with time (González et al. 2013; Meletlidis et al. 2015; Benito-Saz et al. 2019) whereas seismicity locations goes deeper. If this feature were true, its origin would be uncertain since there is not a 3D model of the island that shows structures with this shape (García-Yeguas et al. 2014; Montesinos et al. 2006; Sainz-Maza et al. 2017). Tomography performed by Martí et al. (2017) showed a sort of inclined structure. Nevertheless, it is not yet clear whether the obtained inclination is the result of accumulation of uncertainties due to the azimuthal gap of seismic network, which affects the southern seismicity most strongly.

Underplating, or magma storage under the crust, have become an important contribution on the growth of some volcanic islands (Klügel et al. 2005 and references there-in). The total volume of magma storage beneath the island during intrusive events (2011–2014) could be estimated at 0.34–0.42 km^3 (González et al. 2013; Prates et al. 2013; Meletlidis et al. 2015; López et al. 2017a, b; Benito-Saz et al. 2019) while the total volume emitted during the eruption was 0.25–0.40 km^3 (Martí et al. 2013b; Rivera et al. 2013; Somoza et al. 2017). Therefore, in rough numbers, each source (magma storage or erupted) provides a similar addition of ~ 0.006–0.007% to the total volume of the island, estimated to be ~ 5500 km^3 (Schmincke and Sumita 2010). However, whereas the eruptive volume mainly contributes to the growth of the nearest area of the new submarine volcanic cone located off-shore at the southern end of the island, El Hierro experienced an important homogeneous and cumulative uplift > 20 cm during the whole process (2011–2014). Interestingly, during each post-eruptive intrusion, magma intruded from the center of the island towards areas not yet affected by previous intrusions during the pre-, syn- and post-eruption periods. Magma pathways would have been conditioned by stress changes caused by previous magma emplacements, showing a radial pattern where most of intruded volume would remain in the central part of the island. Thus, the pattern of magma pathways and storage contributes to a relatively homogeneous uplifting of the entire volcanic edifice. A magmatic underplating process could explain the observed rates of uplift at the current shield building stage in the island as seen at e.g., La Palma (Staudigel and Schmincke 1984; Hildenbrand et al. 2003). However, it has been discussed whether this is strictly underplating or not (Klügel et al. 2015; Martí et al. 2017). In any case, this process has clearly been an important source of growth for a volcanic island, observed in real time using geodetic techniques.

5.5 Summary and Open Questions

This chapter presents the first attempt to understand the whole magmatic process which took place at El Hierro between 2011 and 2014, by collecting all known pieces of the puzzle to build a picture of its magma plumbing system. Most of the geodetical, geophysical and petrological data point to a magma plumbing system fed by a main magma reservoir beneath Tanganasoga volcano, located within the uppermost mantle. The presence of sill-like magma batches growing outward beneath the island, and the apparent relation between them, indicate a common origin of the whole process at the center of the island.

While the initial accumulation of magma in an ephemeral 10 km deep reservoir ended with the Tagoro eruption, the post-eruptive intrusions stalled 5–10 km deeper, indicating the existence of stress barriers inhibiting magma flow upwards to produce new eruptions. The possibility of even shallower (4–5 km) short term magma reservoirs prior to the eruption has also been proposed.

There is a clear relation between all magmatic intrusions and the main deep magma reservoir. Post-eruptive intrusions seem to follow a successive anticlockwise rotation of the position of the magma fingers, possibly caused by modification of the stress field by previous activity. Two noteworthy features have been observed during these intrusions. First, the eventual seismic and geodetic signals observed in the location of the initial magmatic sill seem to indicate a relationship between the pre-eruptive and post-eruptive phenomena. Second, the position of these magmatic intrusions concurs with the three rifts of the island. However, it is unclear if that fact suggests the presence of a deep three-armed structure of the island, since geological and gravimetric studies point to the contrary.

The growth of El Hierro Island may be influenced by the magma plumbing system. This volcanic episode involved a similar volume increase contribution from the magma storage beneath the island as from the eruption itself. This fact could indicate the way volcanic islands grow, with a significant role played by the magma storage process, as has been found in the neighboring island of La Palma.

Despite all the knowledge produced by ten years of scientific research on the 2011–14 El Hierro volcanic episode, there still remain several open questions. First, further work remains in distinguishing why the 2011 intrusion resulted in an eruption, while post-eruptive intrusions did not, an issue which is especially important during volcanic crises. Just before eruption, geodetic data registered by GNSS showed an interesting strong decay superimposed on the overall inflationary trend that merits further investigation. In addition, the conspicuous inclined shape of the seismicity -not matching the structure of the magmatic deformation—should be further analyzed.

Even though this activity was well monitored, there is a lack of data in the early stages of the unrest. This fact, in addition to the small size of the island compared with the depth of the intrusions, produced a poor spatial coverage for the geodetic and seismicity, increasing the uncertainties in the results obtained. Nevertheless, Geodesy and Seismology have proven to be essential in the reconstruction of the magma plumbing system in volcanic areas, and powerful tools for monitoring monogenetic volcanic fields where the location of possible eruptions is an additional unknown.

References

Albert H, Costa F, Martí J (2016) Years to weeks of seismic unrest and magmatic intrusions precede monogenetic eruptions. Geology 44(3):211–214. https://doi.org/10.1130/G37239.1

Anderson K, Segall P (2011) Physics-based models of ground deformation and extrusion rate at effusively erupting volcanoes. J Geophys Res 116:B07204. https://doi.org/10.1029/2010JB007939

Bagnardi M, Amelung F, Poland MP (2013) A new model for the growth of basaltic shields based on deformation of Fernandina volcano, Galápagos Islands. Earth Planet Sci Lett 377–378:358–366. https://doi.org/10.1016/j.epsl.2013.07.016

Becerril L, Galindo I, Martí J, Gudmundsson A (2015) Three-armed rifts or masked radial pattern of eruptive fissures? The intriguing case of El Hierro volcano (Canary Islands) Tectonophysics 647:33–47. https://doi.org/10.1016/j.tecto.2015.02.006

Benito-Saz MA, Parks MM, Sigmundsson F, Hooper A, García-Cañada L (2017) Repeated magmatic intrusions at El Hierro Island following the 2011–2012 submarine eruption. J Volcanol Geoth Res 344:79–91. https://doi.org/10.1016/j.volgeores.2017.01.020volgeores.2017.01.020

Benito-Saz MA, Sigmundsson F, Charco M, Hooper A, Parks M (2019) Magma flow rates and temporal evolution of the 2012–2014 posteruptive intrusions at El Hierro, Canary Islands. J Geophys Res Solid Earth 124:12576–12592

Biggs J, Ebmeier SK, Aspinall WP, Lu Z, Pritchard ME, Sparks RSJ, Mather TA (2014) Global link between deformation and volcanic eruption quantified by satellite imagery. Nat Commun 5. https://doi.org/10.1038/ncomms4471

Burov E, Jaupart C, Guillou-Frottier L (2003) Ascent and emplacement of buoyant magma bodies in brittle-ductile upper crust. J Geophys Res 108:2177, 20 p. http://doi.org/10.1029/2002JB001904

Carracedo JC (1996) Morphological and structural evolution of the western Canary Islands: hotspotinduced three-armed rifts or regional tectonic trends? J Volcanol Geoth Res 72:151–162. https://doi.org/10.1016/0377-0273(95)00080-1

Carracedo JC, Badiola ER, Guillou H, De la Nuez J, Pérez-torrado FJ (2001) Geology and volcanology of La Palma and El Hierro, Western Canaries. Estud Geol 57:175–273. http://doi.org/10.3989/egeol.01575-6

Carracedo JC, Pérez-Torrado FJ, Rodríguez-González A (2012) The ongoing volcanic eruption of El Hierro, Canary Islands. Eos Trans AGU 93(9):89–90. https://doi.org/10.1029/2012EO090002

Cong X, Eineder M, Fritz T (2015) Advanced SAR interferometric techniques for monitoring nonlinear volcanic deformation—test site El Hierro Island. In: Proceedings of fringe 2015. http://seom.esa.int/fringe2015/files/Contribution282.pdf

Del Fresno C (2016) Determinación de la Fuente Sísmica a distancias regionales: Aplicación a la Serie de El Hierro 2011. Ph.D. thesis, Universidad Complutense de Madrid

Del Fresno C, Domínguez Cerdeña I, Cesca S, Buforn E (2015) The 8 October 2011 earthquake at El Hierro (Mw 4.0): focal mechanisms of the mainshock and its foreshocks. Bull Seismol Soc Am 105:330–340

Díaz Suárez EA, Domínguez Cerdeña I, del Fresno C, Galván Fraile J, Medina Hernández J (2019) Automatic detection system tested in the 2011 pre-eruptive unrest of El Hierro eruption. AGU meeting 2019

Díaz-Moreno A, Ibanez JM, De Angelis S, Garcia-Yeguas A, Prudencio J, Morales J, Tuvè T, Garcia L (2015) Seismic hydraulic fracture migration originated by successive deep magma pulses: the 2011–2013 seismic series associated to the volcanic activity of El Hierro Island. J Geophys Res Solid Earth 120:7749–7770

Domínguez Cerdeña I, del Fresno C, Gomis Moreno A (2014) Seismicity patterns prior to the 2011 El Hierro eruption. Bull Seismol Soc Am 104(1):567–575

Domínguez Cerdeña I, García-Cañada L, Benito-Saz MA, del Fresno C, Lamolda H, Pereda de Pablo J, Sánchez Sanz C (2018) On the relation between ground surface deformation and seismicity during the 2012–2014 successive magmatic intrusions at El Hierro Island. Tectonophysics 744:422–437

Dzurisin D, Lisowski M, Wicks CW (2009) Continuing inflation at Three Sisters Volcanic Center, Central Oregon Cascade Range, United States, from GNSS, leveling, and InSAR observations. Bull Volcanol 71:1091–1110. https://doi.org/10.1007/s00445-009-0296-4

García A, Fernández-Ros A, Berrocoso M, Marrero JM, Prates G, De la Cruz-Reyna S, Ortiz R (2014) Magma displacements under insular volcanic fields, applications to eruption forecasting: El Hierro, Canary Islands, 2011, 2013. Geophys J Int 179:322–334. https://doi.org/10.1093/gji/ggt505

García-Yeguas A, Ibáñez JM, Koulakov I, Jakovlev A, Romero-Ruiz MC, Prudencio J (2014) Seismic tomography model reveals mantle magma sources of recent volcanic activity at El Hierro Island (Canary Islands, Spain). Geophys J Int 199(3):1739–1750. https://doi.org/10.1093/gji/ggu339

González PJ, Samsonov SV, Pepe S, Tiampo KF, Tizzani P, Casu F et al (2013) Magma storage and migration associated with the 2011–2012 El Hierro eruption: implications for crustal magmatic systems at oceanic island volcanoes. J Geophys Res Solid Earth 118:4361–4377

Gorbatikov AV, Montesinos FG, Arnoso J, Stepanova MY, Benavent M, Tsukanov AA (2013) New features in the subsurface structure model of El Hierro Island (Canaries) from low-frequency microseismic sounding: an insight into the 2011 Seismo-volcanic crisis. Surv Geophys 34:463–489. https://doi.org/10.1007/s10712-013-9240-4

Gudmundsson A (1983) Form and dimensions of dykes in eastern Iceland. Tectonophysics 95(3–4):295–307. https://doi.org/10.1016/0040-1951(83)90074-4

Gudmundsson A (2011) Deflection of dykes into sills at discontinuities and magma-chamber formation. Tectonophysics 500:50–64

Hansteen H, Klügel A, Schmincke HU (1998) Multi-stage magma ascent beneath Canary Islands: evidence from fluid inclusions. Contrib Miner Petrol 132:48–64

Hildenbrand A, Gillot P, Soler V, Lahitte P (2003) Evidence for a persistent uplifting of La Palma (Canary Islands), inferred from morphological and radiometric data. Earth Planet Sci Lett 210:277–289

Ibáñez JM, De Angelis S, Díaz-Moreno A, Hernández P, Alguacil G, Posadas A, Pérez N (2012) Insights into the 2011–2012 submarine eruption off the coast of El Hierro (Canary Islands, Spain) from statistical analyses of earthquake activity. Geophys J Int 191:659–670

Klügel A, Hansteen TH, Galipp K (2005) Magma storage and underplating beneath Cumbre Vieja volcano, La Palma (Canary Islands). Earth Planet Sci Lett 236:211–226

Klügel A, Longpré MA, García-Cañada L, Stix J (2015) Deep intrusions, lateral magma transport and related uplift at ocean island volcanoes. Earth Planet Sci Lett 431:140–149

Lamolda H, Felpeto A, Bethencourt A (2017) Time lag between deformation and seismicity along monogenetic volcanic unrest periods: the case of El Hierro Island (Canary Islands). Geophys Res Lett 44:6771–6777

Lees JM (2007) Seimic tomography of magmatic systems. J Volcanol Geotherm Res 167:37–56

Lengliné O, Marsan D, Got J-L, Pinel V, Ferrazzini V, Okubo PG (2008) Seismicity and deformation induced by magma accumulation at three basaltic volcanoes. J Geophys Res Solid Earth 113:B12305. http://doi.org/10.1029/2008JB005937

Longpré MA, Klügel A, Diehl A, Stix J (2014) Mixing in mantle magma reservoirs prior to and during the 2011–2012 eruption at El Hierro, Canary Islands. Geology 42:315–318

López C, Blanco MJ, Abella R, Brenes B, Cabrera Rodríguez V, Casas B, Domínguez Cerdeña I, Felpeto A, Fernández de Villalta M, del Fresno C et al (2012) Monitoring the volcanic unrest of El Hierro (Canary Islands) before the onset of the 2011–2012 submarine eruption. Geophys Res Lett 39: L13303

López C, Martí J, Abella R, Tárraga M (2014) Applying fractal dimensions and energy-budget analysis to characterize fracturing processes during magma migration and eruption: 2011–2012 El Hierro (Canary Islands) submarine eruption. Surv Geophys 35:1023–1044. https://doi.org/10.1007/s10712-014-9290-2

López C, Benito-Saz MA, Martí J, del-Fresno C, García-Canada L, Albert H, Lamolda H (2017a) Driving magma to the surface: the 2011–2012 El Hierro volcanic eruption. Geochem Geophys Geosyst 18:3165–3184

López C, García-Canada L, Martí J, Domínguez Cerdeña I (2017b) Early signs of geodynamic activity before the 2011–2012 El Hierro eruption. J Geodyn 104:1–14. https://doi.org/10.1016/j.jog.2016.12.00

López C, Luengo Oroz N, Felpeto A, Torres-González PA, Meletlidis S, García-Cañada L, Sainz-Maza S, del Fresno C, Benito-Saz MA, Blanco MJ (2022) Geophysical, geodetic and geochemical evidence for precursory activity: the 2011–2012 Tagoro submarine eruption (in this volume)

Maccaferri F, Bonafede M, Rivalta E (2011) A quantitative study of the mechanisms governing dike propagation, dike arrest and sill formation. J Volcanol Geotherm Res 208:39–50

Manconi A, Longpre MA, Walter TR, Troll VR, Hansteen TH (2009) The effects of flank collapses on volcano plumbing systems. Geology 37(12):1099–1102. http://doi.org/10.1130/G30104A.1

Marsh BD (2015) Magma chambers. In: Sigurdsson H et al (eds) The encyclopedia of volcanoes, 2nd edn. Academic Press, Amsterdam, pp 185–201. https://doi.org/10.1016/B978-0-12-385938-09.00008-0

Martel SJ (2000) Modeling elastic stresses in long ridges with the displacement discontinuity method. Pure Appl Geophys 157(6–8):1039–1057. https://doi.org/10.1007/s000240050016

Martí J, Castro A, Rodríguez C, Costa F, Carrasquilla S, Pedreira R, Bolos X (2013a) Correlation of magma evolution and geophysical monitoring during the 2011–2012 El Hierro (Canary Islands) submarine eruption. J Petrol 54:1349–1373

Martí J, Pinel V, López C, Geyer A, Abella R, Tárraga M, Blanco MJ, Castro A, Rodríguez C (2013b) Causes and mechanisms of the 2011–2012 El Hierro (Canary Islands) submarine eruption. J Geophys Res 118:1–17

Martí J, Villaseñor A, Geyer A, López C, Tryggvason A (2017) Stress barriers controlling lateral migration of magma revealed by seismic tomography. Sci Rep 7:40757. https://doi.org/10.1038/srep40757

Meletlidis S, Di Roberto A, Pompilio M, Bertagnini A, Iribarren I, Felpeto A, Torres PA, D'Oriano C (2012) Xenopumices from the 2011–2012 submarine eruption of El Hierro (Canary Islands, Spain): constraints on the plumbing system and magma ascent. Geophys Res Lett 39:L17302. https://doi.org/10.1029/2012GL052675

Meletlidis S, Di Roberto A, Domínguez Cerdeña I, Pompilio M, García-Cañada L, Bertagnini A, Benito-Saz M, Del Carlo P, Sainz-Maza Aparicio S (2015) New insight into the 2011–2012 unrest and eruption of El Hierro Island (Canary Islands) based on integrated geophysical, geodetical, and petrological data. Ann Geophys 58:S0546

Menand T (2008) The mechanics and dynamics of sills in layered elastic rocks and their implications for the growth of laccoliths and other igneous complexes. Earth Planet Sci Lett 267:93–99

Menand T, Daniels KA, Benghiat P (2010) Dyke propagation and sill formation in a compressive tectonic environment. J Geophys Res 115(Article No B08201). http://doi.org/10.1029/2009JB006791

Menand T, de Saint-Blanquat M, Annen C (eds) (2011) Emplacement of magma pulses and growth of magma bodies. Tectonophysics 500:1–112

Michon L, Ferrazzini V, Di Muro A, Villeneuve N, Famin V (2015) Rift zones and magma plumbing system of Piton de la Fournaise volcano: how do they differ from Hawaii and Etna? J Volcanol Geoth Res 303:112–129. https://doi.org/10.1016/j.jvolgeores.2015.07.031

Miguelsanz L, González PJ, Tiampo KF, Fernández J (2021) Tidal influence on seismic activity during the 2011–2013 El Hierro volcanic unrest. Tectonics 40:e2020TC006201. http://doi.org/10.1029/2020TC006201

Montesinos FG, Arnoso J, Benavent M, Vieira R (2006) The crustal structure of El Hierro (Canary Islands) from 3-D gravity inversion. J Volcanol Geotherm Res 150:283–299. https://doi.org/10.1016/j.jvolgeores.2005.07.018

Münn S, Walter T, Klügel A (2006) Gravitational spreading controls rift zones and flank instability on El Hierro, Canary Islands. Geol Mag 143(3):257–268. https://doi.org/10.1017/S0016756806002019

Pedersen R, Sigmundsson F, Einarson P (2007) Controlling factors on earthquake swarms associated with magmatic intrusions; constraints from Iceland. J Volcanol Geotherm Res 162:73–80. https://doi.org/10.1016/j.jvolgeores.2006.12.010

Pérez NM, Padilla GD, Padrón E, Hernández PA, Melián G, Barrancos J, Dionis S, Nolasco D, Rodríguez F, Calvo D, Hernández I (2012) Precursory diffuse CO_2 and H_2S emission signatures of the 2011–2012 El Hierro submarine eruption, Canary Islands. Geophys Res Lett 39 (16):L16311. https://doi.org/10.1029/2012GL052410

Poland MP, Miklius A, Montgomery-Brown EK (2014) Magma supply, storage, and transport at shield-stage Hawaiian volcanoes. In: Poland MP, Takahashi TJ, Landowski C (eds) Characteristics of Hawaiian volcanoes. USGS professional paper 1801

Pollard DD, Delaney PT, Duffield WA, Endo ET, Okamura AT (1983) Surface deformation in volcanic rift zones. Tectonophysics 94(1–4):541–584. http://doi.org/10.1016/0040-1951(83)90034-3

Prates G, García A, Fernández-Ros A, Marrero JM, Ortiz R, Berrocoso M (2013) Enhancement of sub-daily positioning solutions for surface deformation surveillance at El Hierro volcano (Canary Islands, Spain). Bull Volcanol 75(6):1–9

Ranero CR, Torné M, Banda E (1995) Gravity and multichannel seismic reflections constraints on the lithospheric structure of the Canary Swell. Mar Geophys Res 17:519–534. https://doi.org/10.1007/BF01204342

Reverso T, Vandemeulebrouck J, Jouanne F, Pinel V, Villemin T, Sturkell E et al (2014) A two-magma chamber model as a source of deformation at Grímsvötn Volcano, Iceland. J Geophys Res Solid Earth 119:4666–4683. http://doi.org/10.1002/2013JB010569

Rivera J, Lastras G, Canals M, Acosta J, Arrese B, Hermida N, Micallef A, Tello O, Amblas D (2013) Construction of an oceanic island: insights from the El Hierro (Canary Islands) 2011–2012 submarine volcanic eruption. Geology 41(3):355–358. http://doi.org/10.1130/G33863.1

Roberts NS, Bell AF, Main IG (2016) Mode switching in volcanic seismicity: El Hierro 2011–2013. Geophys Res Lett 43. https://doi.org/10.1002/2016GL068809

Rodríguez-Molina S, González PJ, Charco M, Negredo AM, Schmidt DA (2021) Time-scales of inter-eruptive volcano uplift signals: Three Sisters Volcanic Center, Oregon (United States). Front Earth Sci. https://doi.org/10.3389/feart.2020.577588

Sainz-Maza S, Arnoso J, González Montesinos F, Martí J (2014) Volcanic signatures in time gravity variations during the volcanic unrest on El Hierro (Canary Islands). J Geophys Res Solid Earth 119:5033–5051. https://doi.org/10.1002/2013JB010795

Sainz-Maza S, Montesinos FG, Marti J, Arnoso J, Calvo M, Borreguero A (2017) Structural interpretation of El Hierro (Canary Islands) rifts system from gravity inversion modelling. Tectonophysics 712:72–81. https://doi.org/10.1016/j.tecto.2017.05.010

Sánchez-Pastor P, Obermann A, Schimmel M (2018) Detecting and locating precursory signals during the 2011 El Hierro, Canary Islands, submarine eruption. Geophys Res Lett 45:10288–10297. https://doi.org/10.1029/2018GL079550

Schmincke HU, Sumita M (2010) Geological evolution of the Canary Islands. Görres-Verlag GMBH, Koblenz, 188 p

Sigmarsson O, Laporte D, Carpentier M, Devouard B, Devidal JL, Marti J (2013) Formation of U-depleted rhyolite from a basanite at El Hierro, Canary Islands. Contrib Miner Petrol 165(3):601–622. https://doi.org/10.1007/s00410-012-0826-5

Somoza L, González FJ, Barker SJ, Madureira P, Medialdea T, de Ignacio C, Lourenço N, León R, Vázquez JT, Palomino D (2017) Evolution of submarine eruptive activity during the 2011–2012 El Hierro event as documented by hydroacoustic images and remotely operated vehicle observations. Geochem Geophys Geosyst 18:3109–3137. https://doi.org/10.1002/2016GC006733

Staudigel H, Schmincke HU (1984) The Pliocene seamount series of La Palma/Canary Islands. J Geophys Res 89:11195–11215

Stroncik NA, Klügel A, Hansteen TH (2009) The magmatic plumbing system beneath El Hierro (Canary Islands): constraints from phenocrysts and naturally quenched basaltic glasses in submarine rocks. Contrib Miner Petrol 157:593–607. https://doi.org/10.1007/s00410_008-0354-5

Tárraga M, Martí J, Abella R, Carniel R, López C (2014) Volcanic tremors: good indicators of change in plumbing systems during volcanic eruptions. J Volcanol Geoth Res 273:33–40. https://doi.org/10.1016/j.jvolgeores.2014.01.003

Telesca L, Lovallo M, López C, Martì J (2016) Multi-parametric statistical investigation of seismicity occurred at El Hierro (Canary Islands) from 2011 to 2014. Tectonophysics. https://doi.org/10.1016/j.tecto.2016.01.045

Troll V, Klügel A, Longpré M-A, Burchardt S, Deegan FM, Carracedo JC, Wiesmaier S, Kueppers U, Dahren B, Blythe LS, Hansteen TH, Freda C, Budd DA, Jolis EM, Jonsson E, Meade FC, Harris C, Berg SE, Mancini L, Polacci M, Pedroza K (2012) Floating stones off El Hierro, Canary Islands: Xenoliths of pr-island sedimentary origin in early products of the October 2011 eruption. Solid Earth 3:97–110. https://doi.org/10.5194/se-3-97-2012

Walker GPL (1987) In: Decker RW, Wright TL, Stauffer PH (eds) The dike complex of Koolau volcano, Oahu; internal structure of a Hawaiian rift zone, chap 41, vol 2. U.S. Geological Survey professional paper 1350, Volcanism in Hawaii, pp 961–993. Also available at http://pubs.usgs.gov/pp/1987/1350/

Walwer D, Ghil M, Calais E (2019) Oscillatory nature of the Okmok volcano's deformation. Earth Planet Sci Lett 506:76–86. https://doi.org/10.1016/j.epsl.2018.10.033

Watts AB (1994) Crustal structure, gravity anomalies and flexure of the lithosphere in the vicinity of the Canary Islands. Geophys J Int 119:648–666

Woods J, Winder T, White RS, Brandsdóttir B (2019) Evolution of a lateral dike intrusion revealed by relatively-relocated dike-induced earthquakes: the 2014–15 Bárðarbunga-Holuhraun rifting event, Iceland. Earth Planet Sci Lett 506:53–63. https://doi.org/10.1016/j.epsl.2018.10.032

Wright TL, Klein FW (2006) Deep magma transport at Kilauea volcano Hawaii Lithos 87(1–2) 50–79 https://doi.org/10.1016/j.lithos.2005.05.004

Zobin V (2012) Introduction to volcanic seismology, 2nd edn. Elsevier, Amsterdam

Geophysical, Geodetic and Geochemical Evidence for Precursory Activity: The 2011–2012 Tagoro Submarine Eruption

6

C. López, N. Luengo-Oroz, A. Felpeto, P. A. Torres-González, S. Meletlidis, L. García-Cañada, S. Sainz-Maza, C. Del Fresno, M. A. Benito-Saz, and M. J. Blanco

Abstract

Observational data recorded during the reawakening of volcanic activity in El Hierro, which ended with 2011–2012 Tagoro eruption, allow us to study the precursory activity associated with this event. This volcanic activity occurred between 1971 Teneguía eruption and the 2021 eruption in Cumbre Vieja (both in La Palma island) and represents the first such event in the Canary Islands that has been fully recorded from beginning to end by a multi-parameter monitoring network. The lack of previous instrumental data makes these data unique and a case study for a broader understanding of monogenetic volcanism. In this chapter we review previous works to collect the precursors of the historical eruptions in the Canaries, discussing which were most relevant to forecast the evolution of the eruptive processes. We find solid signs of the undergoing magmatic process in El Hierro in the evolution of the seismicity, the surface deformations and gas emissions, the changes in media properties and the appearing of gravity anomalies. Results can contribute to a better interpretation of the observational data and the correct volcanic risk assessment of future events.

Keywords

Eruption forecasting · Geophysical monitoring · Geodetic monitoring · Geochemical monitoring · Canary Islands

6.1 Introduction

Since the beginning of humanity, predicting natural phenomena is a constant challenge, linked to a community's prosperity, decline or disappearance. Volcanoes have always attracted humans; their fertile fields and the available natural resources have made them a cornerstone for development and progress (Baxter 2005). Thus, forecasting their eruptions is crucial. The knowledge of their destructive potential (Vesuvius 79 AD, Krakatoa 1883, Mt Pelee and St Vincent 1902) promoted the advance of science, especially in the last 40 years after the 1980 St Helens eruption (Driedger et al. 2020). Despite this progress, the correct forecast of the next eruption is still a pressing issue in almost all volcano observatories. The main difficulty is the identification and correct interpretation of those

C. López (✉) · A. Felpeto · L. García-Cañada · S. Sainz-Maza · C. Del Fresno · M. A. Benito-Saz
Instituto Geográfico Nacional, Madrid, Spain
e-mail: clmoreno@mitma.es

N. Luengo-Oroz · P. A. Torres-González · S. Meletlidis · M. J. Blanco
Instituto Geográfico Nacional, Tenerife, Spain

P. J. González (ed.), *El Hierro Island*, Active Volcanoes of the World,
https://doi.org/10.1007/978-3-031-35135-8_6

anomalous signals, that are precursors of volcanic activity. And an additional problem is that each volcano can have a different anomalous behaviour, depending on its geodynamic context, the magma involved, and its evolution. The real challenge is the proper interpretation of these precursors in real time, while they are occurring; this is crucial for the advance assessment of whether or not the process will culminate in an eruption. We say that a volcanic signal is called a precursor when it indicates the approach of a meaningful change in the state of a volcano. A precursor has to be able to reflect the variations in the underlying processes occurring in the subsurface that have an influence on what it could be expected in the future. These precursors, that include seismic activity, ground deformation, chemical and physical changes in gas emissions, etc., could take place days, weeks, months or even years before an eruption. Unfortunately, eruptions at monogenetic volcanoes are difficult to be observed or studied. Although this type of volcanism is the most widespread on Earth and occurs in every known tectonic setting, very few monitoring networks exist compared with those deployed in composite stratovolcanoes or calderas (Kereszturi and Németh 2012). Only a few eruptions of this type have been accurately described, and even fewer have had qualitative and quantitative data recorded (de la Cruz-Reyna and Yokoyama 2011). Some of the cases of monogenetic volcanoes eruptions were Jorullo 1759 (Guilbaud et al. 2011) and Paricutin 1943 (Foshag and Gonzalez 1956) in Mexico, Teneguía 1971 (Barker et al. 2015) and the recent eruption in Cumbre Vieja 2021 (Longpré 2021) in Spain, East-Izu 1930–1989 (Yamamoto et al. 1991) in Japan, Heimaey 1973 (Mattsson and Hoskuldsson 2003) and Gjálp 1996 (Gudmundsson 2000) in Iceland. Studies of monogenetic volcanism, especially small-volume basaltic systems, revealed that there are similarities and common features not only in the type of the products and deposits, but also in the pre-eruptive behaviour, concerning specific precursors. Seismicity is still the most reliable precursor of a volcanic unrest whereas the uncertainty of the localization of the future eruptive vent or fissure is significant. Ground deformation is also a reliable precursor, however in the first stages of unrest remains difficult due to the low-volume magma batches involved and the depths from where they rise. The heat and gas, released from the rising magma, have also been used as a precursor for this type of volcanism (Hincks et al. 2014). Currently, with the technological developments, other monitoring techniques such as gravimetry, InSAR analysis, infrasound networks or satellite observations can be used to effectively monitor monogenetic volcanism (National Academies of Sciences, Engineering, and Medicine 2017). In this work, we compile and analyse available geophysical and geochemical signals to assess which precursory parameters were most relevant to forecast the evolution of El Hierro volcanic activity that preceded the submarine eruption of Tagoro. This eruption that started on 10 October 2011 at 2 km south from El Hierro Island, provides valuable data for the study of the monogenetic volcanism in the Canary Islands.

6.2 Precursory Activity in Historical Eruptions in the Canaries

Since 1585, there have been at least 14 volcanic eruptions in the Canary Islands, seven of them in La Palma Island, four in Tenerife, two in Lanzarote and one in El Hierro (Fig. 6.1). This activity results in an average of three eruptions per century. These eruptions have had quite different durations, from 10 days (Chinyero, 1909) to 2055 days (Timanfaya, 1730–36).

Of all of them, only the more recent ones, Teneguía (La Palma, 1971), Tagoro (El Hierro, 2011–12) and the last one on Cumbre Vieja (La Palma, 2021) have had instrumental monitoring before the onset of the eruption, and only the last two have had multi-technique monitoring. Therefore, the identification of precursors in most of the historical eruptions in the Canary Islands is based on a collection of multiple historical documentary sources, but not on instrumental data. The primary work on the compilation of documentary sources of historical eruptions in the Canaries is that of Romero Ruiz

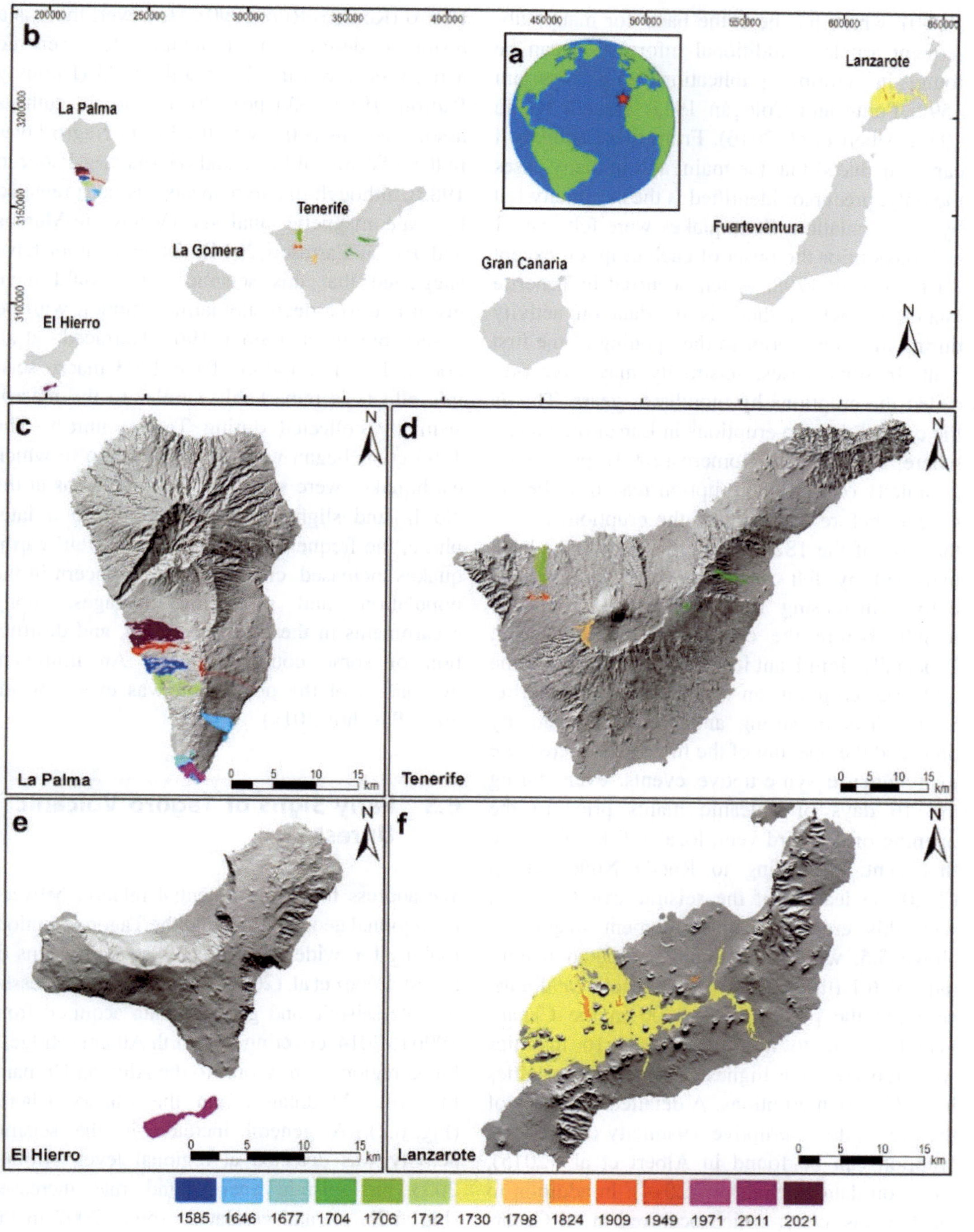

Fig. 6.1 **a** Location of the Canary Archipelago (red star) in the globe. **b** Overview map of the Canary Archipelago (UTM coordinates, zone 28N) **c** La Palma, **d** Tenerife, **e** El Hierro and **f** Lanzarote showing well documented historical lava flows. Colour scale at the bottom shows the year of the beginning of each eruption. Modified from Longpré and Felpeto (2021)

(1990), which has been the basis for many subsequent articles. Additional information can be found in multiple publications (e.g., Torriani 1592; Ponte and Cologán 1911; Bonelli Rubio 1950; Albert et al. 2016). From these sources it can be deduced that the main, and in many cases the only, precursor identified is the seismicity felt by the population. Earthquakes were felt from 1 to 7 days prior the onset of each eruption except for the one in 1798, which occurred in Tenerife Island, for which there is no data on activity during the weeks prior to the opening of the first vent. In some cases, seismicity may have preceded the eruptions by months or years. This is the case of the two eruptions in Lanzarote Island, where, according to Romero (2003), precursory seismicity of 1730–36 eruption may have begun 4 years before the onset of the eruption; and in the case of the 1824 eruption, it may have been preceded by felt seismicity starting 11 years before, increasing in intensity and frequency months before the opening of the first vent. Especially significant in terms of seismicity is the 1704–05 eruption on the island of Tenerife. Seven days of strong and frequent seismicity preceded the opening of the first vent. There were many intense syn-eruptive events, even during the 16 days of volcanic hiatus prior to the opening of the third vent, located 9 km from the first vent. According to Rueda Núñez et al. (2020), at least 8 of the seismic events related with this eruption had a moment magnitude above 5.5, with a maximum computed magnitude of 6.1 (the highest of the whole catalogue, covering the period 1341–2000 in the Canary Islands). Due to house collapses 16 fatalities were reported, the highest number of casualties in all Canarian eruptions. A detailed summary of the pre- and syn-eruptive seismicity of 1704–05 eruption can be found in Albert et al. (2015), based on data by Sánchez (2014). In addition to the felt seismicity, which occurred in all Canarian historical eruptions, a few other precursors have been reported prior to some of the eruptions. These include gaseous emissions, ground cracking and subterranean rumble. Prior to the Tagoro eruption, there is no reliable documentation of any eruption in historical times on El Hierro (Romero Ruiz 1990). However, there is a historical description of notable 'felt' seismic activity occurred in March–July 1793 (Darias y Padrón 1980; Sánchez 2014). Some authors associated this activity to the Lomo Negro lavas in the NW area of the island (Hernández-Pacheco 1982), although this hypothesis has been rejected by paleomagnetic analysis (Villasante-Marcos and Pavón-Carrasco 2014). Other authors have suggested that this seismic crisis could have given rise to a deep submarine eruption with no observable effects (Bravo 1968; Carracedo et al. 2001). The description of the 1793 macro seismic effects is remarkably similar to the macrosismicity collected during Tagoro unrest. The 1793 crisis began with an initial phase in which earthquakes were strongly felt in the areas at the North, and slightly felt at the NE. In a later phase, the frequency and number of 'felt' earthquakes increased, creating a great concern in the population, and producing damages, along escarpments in the N of the island, and destruction of some country houses. An imminent evacuation of the population was even considered (Sánchez 2014).

6.3 Early Signs of Tagoro Volcanic Unrest

We address firstly the potential relation between the regional geodynamics and the Tagoro eruption, looking for evidence of precursory early signs of unrest. López et al. (2017a) analysed and processed reliable seismic and geodetic data acquired from 1996 to 2014, covering the North Atlantic Ridge, a broad region from Azores to the Alboran Domain, the Atlas Mountains, and the Canary Islands (Fig. 6.2). A general increase in the seismic activity was detected at regional level. Around 2003 the seismic energy and rate increased (Fig. 6.2b, c), and escalated around 2007 in the Canaries and in the Atlas (Fig. 6.2b, c). In El Hierro, seismic activity increased in 2003, maintaining a higher rate from 2007 onwards (Fig. 6.2c, d). These increases are real, the methodology used accounted for the improvements in network capabilities or instrumental effects.

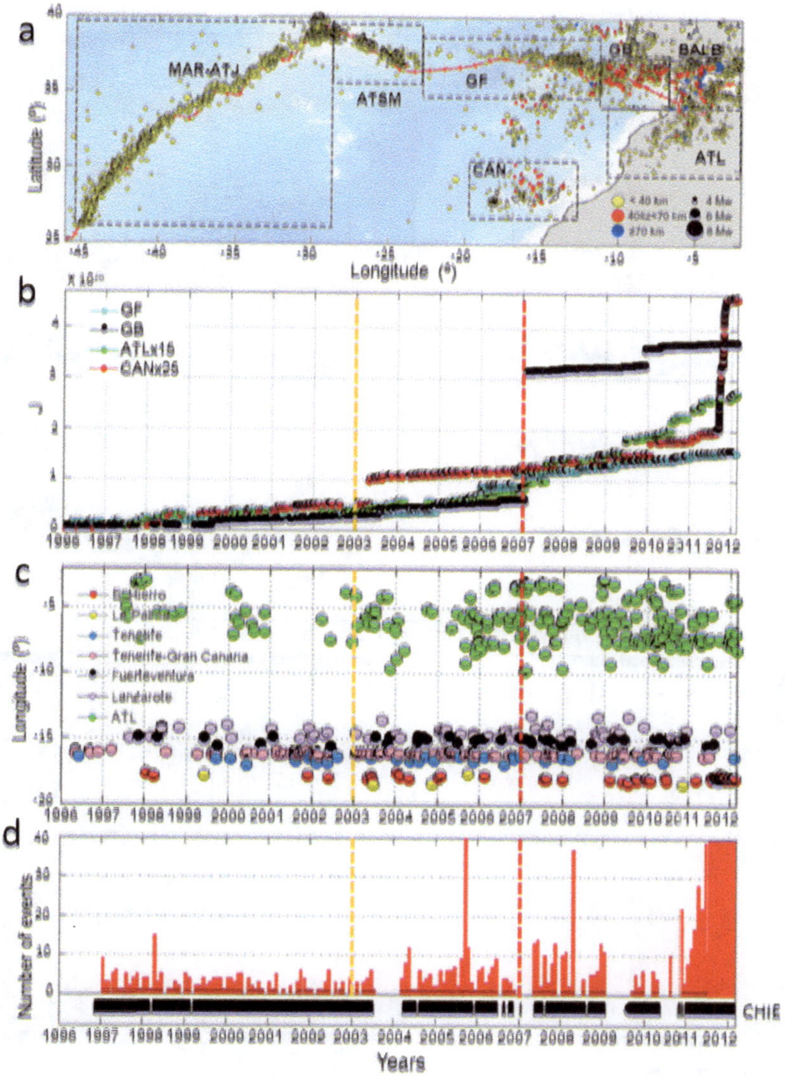

Fig. 6.2 1996–2012 comparative seismic analysis. **a** $M_W > 3.0$ events along the North Atlantic Ridge, including the Azores Triple Junction (MAR-ATSM), the Azores (ATSM), the Gloria Fault (GF), the Gorringe Bank (GB), Alboran Domain (BALB), the Atlas Mountains (ATL) and the Canary Islands (CAN). **b** Accumulated seismic energy for $Mw \geq Mc$ events in GF, GB, ATL and CAN. **c** Longitudinal evolution of the seismic activity in CAN and ATL. **d** Monthly event histogram (red bars) registered at the CHIE seismic station on El Hierro, black bars indicate data availability. Vertical lines indicate the increase on seismic activity in 2003 (yellow) and 2007 (red). Modified from López et al. (2017a)

There are also early signs of anomalous regional surface displacements. Figure 6.3 shows time-series coordinates (Fig. 6.3a) and horizontal and vertical velocities (Fig. 6.3b, c) calculated at GNSS stations in the Azores, the Iberian Peninsula, Morocco and the Canary Islands.

From October 2002 to October 2005, the horizontal surface coordinates showed a counter-clockwise rotation in LPAL (drift = 53° SW); MAS1, (drift = 54° SW) and RABT (drift = 25° SW) (Fig. 6.3b), coinciding with the abrupt westwards jump recorded at SFER station, which

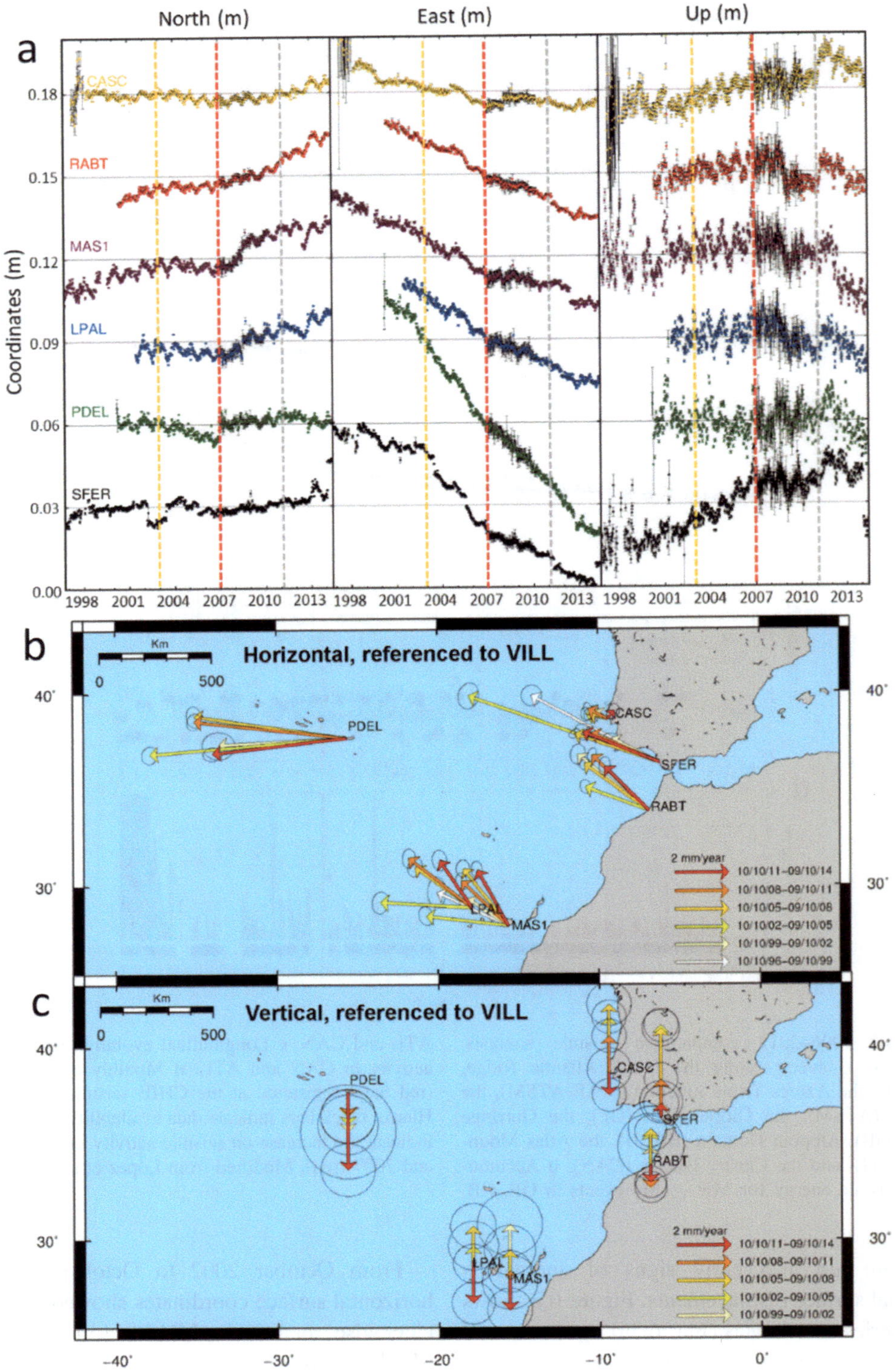

Fig. 6.3 1996–2014 GNSS surface deformation and velocities at GNSS stations in the Azores (PDEL, San Miguel, Portugal), the Iberian Peninsula (CASC, Cascais, Portugal; VILL, Villafranca and SFER, San Fernando, Spain), Morocco (RABT, Rabat) and the Canary Islands (LPAL, La Palma and MAS1, Gran Canaria). **a** Coordinates series. Vertical lines indicate main trend changes in 2003 (yellow), 2007 (red) and 2011 (grey), **b** horizontal and **c** vertical vectors in relation to VILL station (European Plate). Modified from López et al. (2017a)

registered several changes in the magnitude of the horizontal components but no rotation. The stress field gradually reverted to steady compressive NW–SE behaviour while the vertical deformation component continues to increase, firstly at RABT station and subsequently at LPAL and MAS1 stations, until the onset of the eruption (Fig. 6.3c).

Early signs of local micro seismicity and deformation are also found in the Canaries region (López et al. 2017a). Time varying fractal dimension *(FD)* analysis (it is a statistical index that measures the complexity, geometrical dimension and pattern of a source field) performed on the seismic data was used to identify the different stages of the seismic activity and to infer the geometry and path of the seismicity associated with the process of migration of magma and associated fluids to reach the Earth's surface. Results showed, in both CCAN (Tenerife) and CHIE (El Hierro) seismic stations, a decrease in *DF* from 2003 onwards (Fig. 6.4a), an additional decrease at the beginning of 2011 and a further decrease in CHIE during the ongoing eruption (October 2011–March 2012). The relative deformation between the two GNSS stations LPAL-MAS1 showed a tendency to shorten northwards and lengthen westwards, thus confirming an extension between the stations from 2003 (Fig. 6.4b). Starting in January 2011, LPAL-MAS1 trend changed, with a positive northward deformation episode that was more clearly observed from July 2011 onwards, coinciding with the beginning of the pre-eruptive unrest. With the onset of the submarine eruption in October 2011, the northward deformation was quickly reversed and at the end of the eruption the relative deformation remained stable (Fig. 6.4b).

In addition to the local seismicity and surface displacements, changes on some geochemical parameters were also detected before the seismic unrest in July 2011. Padrón et al. (2008) reported significant increases in the diffuse emission of

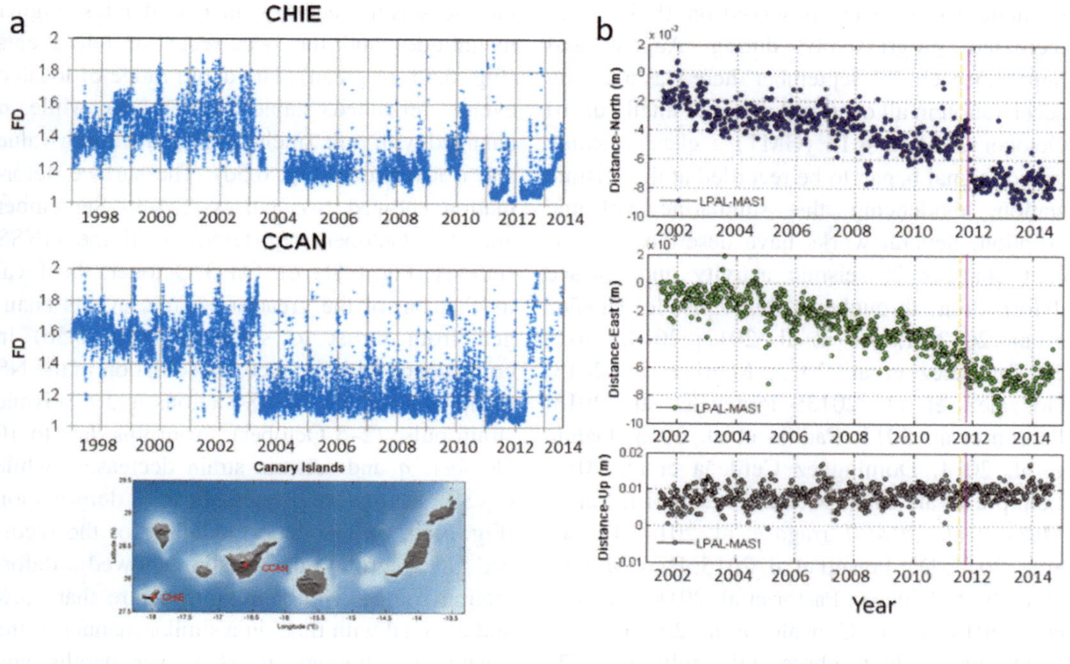

Fig. 6.4 Local seismic and geodetic analysis. **a** Time varying fractal dimensions FD at CHIE and CCAN seismic stations. **b** Time evolution of the three-component (north, east, upward) GNSS LPAL-MAS1 relative deformation. Vertical lines indicate unrest start (July 2011, in yellow) and the eruption onset (October 2011, in red). Modified from López et al. (2017a)

carbon dioxide several days before the main four seismic events that took place around El Hierro Island in 2004 in March, April, and October. An automatic geochemical station (HIE01) reported anomalous data (hourly basis) in the centre of the island (Fig. 6.6). In addition, analysis of samples of deep-water corals growing in the surroundings of the submarine eruption suggested that magmatic ^{3}He emission took place some weeks before the beginning of the seismic unrest (Álvarez-Valero et al. 2017).

6.4 July–October 2011 Tagoro Volcanic Unrest

Here, we focus on the precursory signals recorded during the volcanic unrest (19 July to 10 October 2011). During these months, the rapid changes in seismic and geodetic observations, well above the base level of activity in previous years, pointed to a volcanic reactivation. It was the first time that activity was considered as 'precursory'. The first anomalous seismic and geodetic activity were observed on 19 July and increased progressively during August and September. On 27 September there was a drastic acceleration in all observed data and finally on 10 October 2011 at 4:15 (GMT) a clear volcanic tremor signal began to be recorded at the seismic station, evidencing the submarine volcanic eruption. Several works have described significant changes in seismic activity and surface deformation throughout this episode (Ibáñez et al. 2012; López et al. 2012, 2014, 2016, 2017b; Pérez et al. 2012; Martí et al. 2013; González et al. 2013; Prates et al. 2013; Padrón et al. 2013; Padilla et al. 2013; García et al. 2014; Domínguez Cerdeña et al. 2014; Longpré et al. 2014; Melián et al. 2014; Sainz-Maza et al. 2014; Tárraga et al. 2014; Telesca et al. 2014; Del Fresno et al. 2015; Díaz-Moreno et al. 2015; Sánchez-Pastor et al. 2018; Bartolini et al. 2018; Torres-González et al. 2019).

During a first phase (19 July to ~ 22 August), the seismic activity was mostly located to the north of the island and consisted of clustered low magnitude volcano-tectonic events occurring in seismic swarms, at a stable depth range of 8–12 km (Fig. 6.5a). During this first period, the contribution of the tensile micro-seismicity was notable (López et al. 2017b), with high *b-values* (it measures the frequency of smaller earthquakes relative to larger ones) and low values of the Poisson ratio (it measures the ratio of transverse contractional strain to the longitudinal extensional strain) (Fig. 6.5d, e). Besides the occurrence of seismic events, a small horizontal deformation was recorded in the GNSS stations (Fig. 6.5b, c). At the end of August, activity intensified showing a clear uni-directional migration of the seismicity below the Moho discontinuity, towards the south, that became more intense in the period 9–26 September. In this period, seismicity increased in magnitude while *b-value* decreased (Fig. 6.5d) and deformation accelerated (Fig. 6.5b, c). Maximum depth of the seismicity increased to > 15 km, with the consequent σ ratio increasing (Fig. 6.5d).

Beginning on 27 September, a new phase of the underlying process was identified, with intense seismic activity, increased rates, higher magnitudes and the occurrence of felt events (Fig. 6.5d, in green). The depth range of located events remained stable (12–18 km) while σ changed with time (including its maximum value and a reversion) (Fig. 6.5d). The surface deformation showed two pulses (27–30 September and 2–5 October) registered in all the GNSS stations (Fig. 6.5b, c). On 3 October, the focal mechanism of the greatest seismic events changed from thrust to strike-slip (Fig. 6.5b) in coincidence with the second oscillation in the NS component of the GNSS stations and a seismic strain pulse (2–5 October). From this date to 10 October, σ and seismic strain decreased, while GNSS stations registered further deformation (Fig. 6.4b, c). Geodetic modelling of the recorded GNSS ground deformation showed a deformation source with a pressure centre that grew and evolved with time, in a similar manner as the seismicity, although in shallower depths and slightly further to the south.

On 8 October, the highest magnitude seismic event of the unrest occurred, with 4.0 Mw. Its

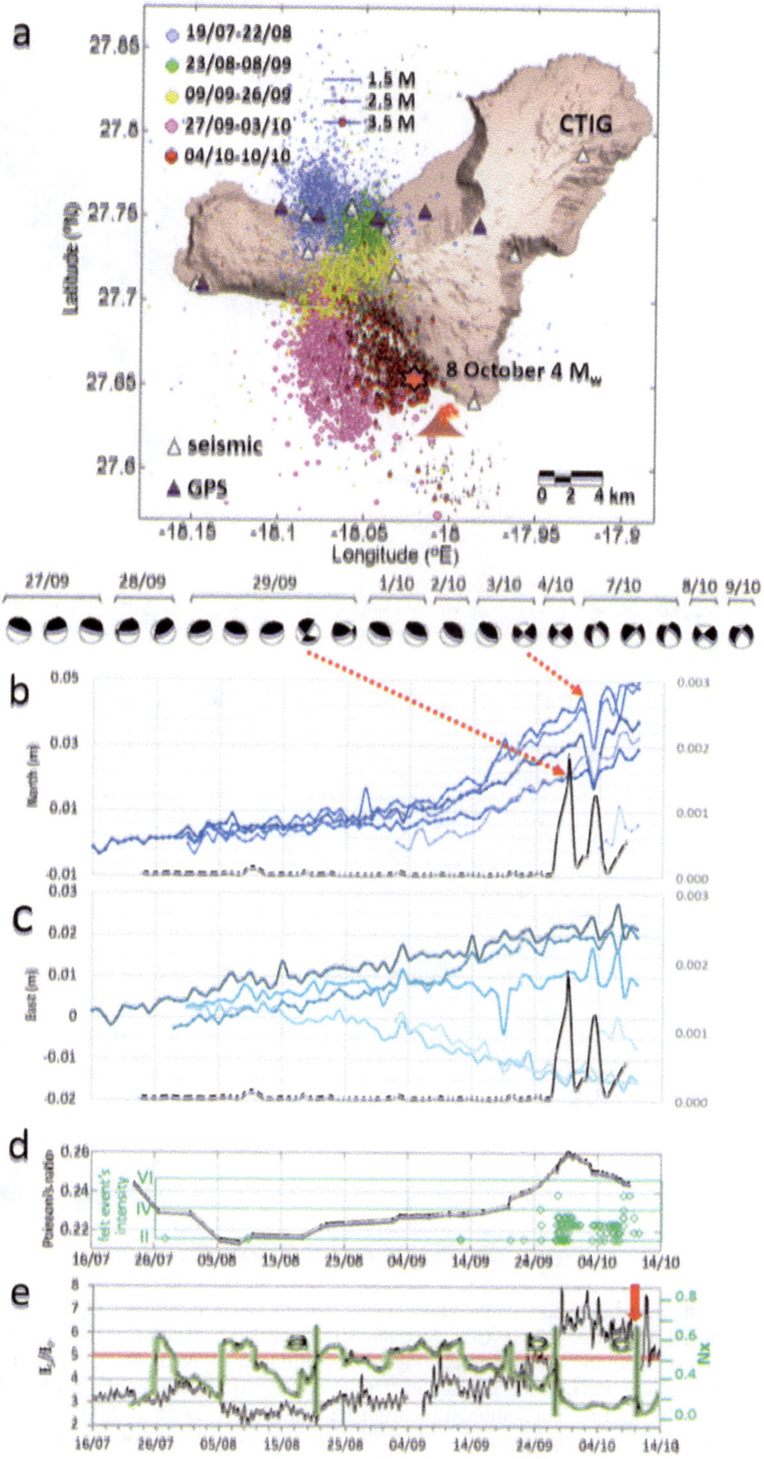

Fig. 6.5 **a** Evolution of seismicity during unrest. **b** North GNSS components and seismic strain with focal mechanism solutions for the greatest seismic events located in the period 27 September–9 October. **c** East GNSS components and seismic strain. **d** Poisson's ratio (black), Gutenberg-Richter b-value (red) and felt event's intensity included in IGN catalogue (green) evolution. **e** Tensile/shear type of fracturing (measured by the rate of the energy measured on seismic P and S phases, Es/Ep) time evolution (black) and Shannon entropy power Nx (green). Modified from Telesca et al. (2014), López et al. (2017b). The thick red arrow points to the eruption onset

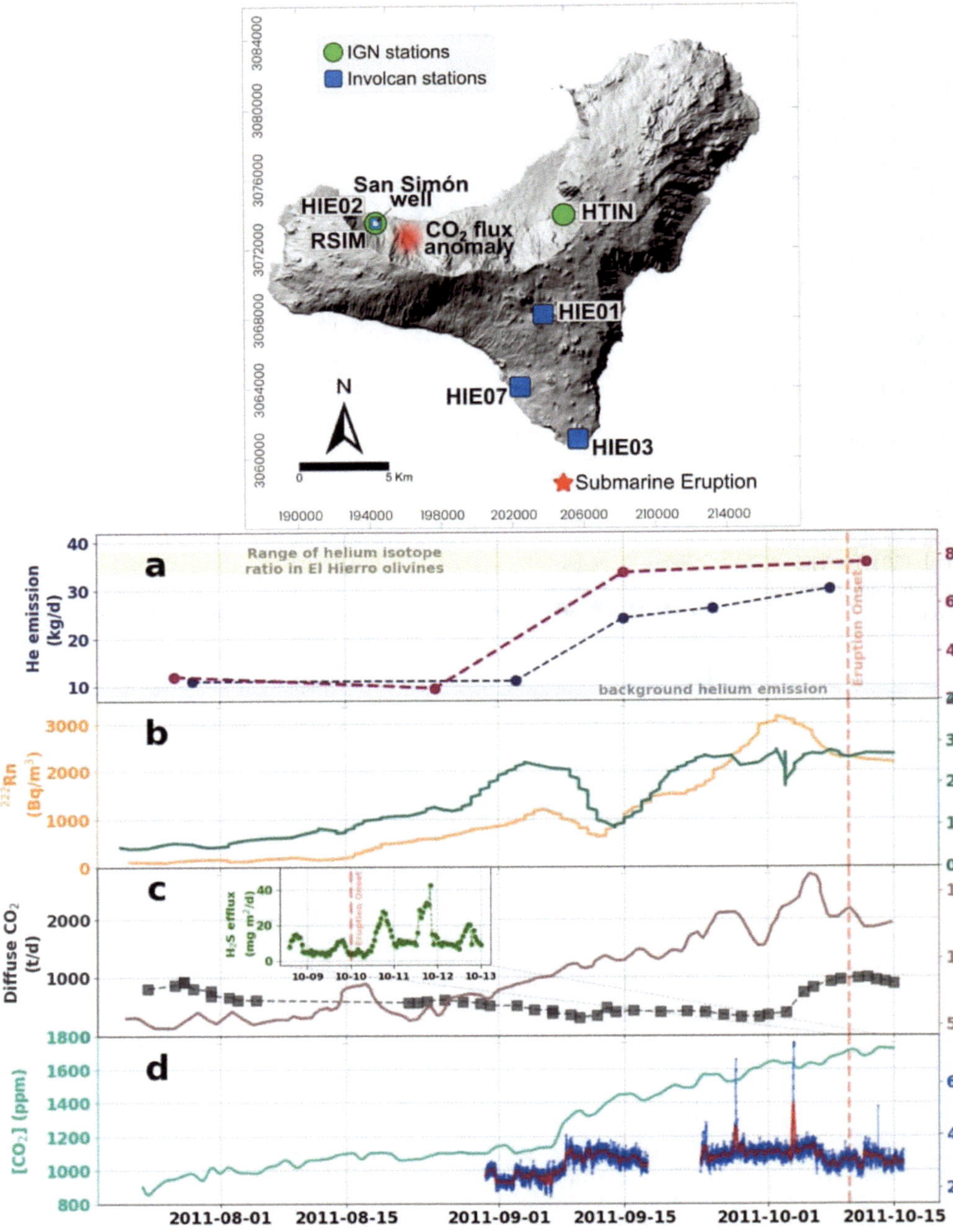

Fig. 6.6 Situation map with the geochemical stations and the location of the diffuse CO_2 flux anomaly and the submarine eruption. Geochemical signals: **a** diffusive helium emission across the island and $^3He/^4He$ corrected ratio ($(R/R_a)_{corr}$) in San Simón well, from Padrón et al. (2013), **b** soil ^{222}Rn activity and $^{222}Rn/^{220}Rn$ ratio (168 h moving average) in HIE02 from Padilla et al. (2013), **c** diffuse CO_2 emission in surveys across the island from Melián et al. (2014), soil CO_2 efflux (24 h moving average) in HIE01 and soil H_2S emission in HI07 from Pérez et al. (2012), **d** low pass filtered CO_2 concentration in air in HTIN and ^{222}Rn concentration in air in RSIM from Torres-González et al. (2019)

focal mechanism was modelled by Del Fresno et al. (2015), obtaining a pure double-couple mechanism with null isotropic (dilatational) component, which discards any volume changes in the source due to a magma intrusion at that date. The modelled Coulomb stress changes associated with the modelled point pressure source on the 4.0 Mw plane solution showed a positive 0.5 MPa stress transfer with an excellent fit with the 3–8 October seismic locations (in dark red in Fig. 6.5a), suggesting a process of stress transfer triggering seismicity on pre-existent faults (López et al. 2017b).

Telesca et al. (2014) applied the Fisher Information Measure (FIM) that quantifies the amount of organization or order of a system, and the Shannon entropy power (*Nx*), that quantifies the degree of unpredictability, to the RSAM signal (continuous seismic ground motion amplitude) measured at CHIE station during the unrest. The temporal variation of FIM values has a few high values (Fig. 6.5e) at the beginning of the unrest process that authors interpreted as corresponding to the deformation necessary to accommodate the overpressure arising after the first injection of magma under the north of the island. From end August to end September, FIM shows constantly low values (and persistent high *Nx* values) associated with massive hydraulic fracturing of the host rock as the magma pressure progressively increased. From end September to 10 October, minimum *Nx* values suggest that an irreversible process was occurring (Fig. 6.4e). Ibáñez et al. (2012) and Domínguez Cerdeña et al. (2014) studied likewise the evolution of the seismic *b-value* showing similar results, high *b-values* during the first phases of unrest and a lowering to *b-value* = 1 before the onset of the eruption.

Besides geophysical signs, several geochemical parameters being monitored showed distinctive features related to the volcanic unrest process. Continuous diffuse emissions of CO_2 and H_2S (hourly basis) were measured by Pérez et al. (2012) at two geochemical stations; one (HIE01) located in the centre of the island, in the conjunction of the three volcanic rifts, and the second one in the south (HIE07) (Fig. 6.6). At HIE07, which was installed on 7 October H_2S increased to a maximum value of 42 mg/(m^2 d) on 12 October (Fig. 6.6c). Also, an increase in CO_2 flux in HIE01 was measured from 23 August to 5 October, when the maximum value [18.9 g/(m^2 d)] was reached (Fig. 6.6c). Several authors made discrete measurements of diffuse CO_2 flux in different areas of the island. López et al. (2012) detected an abnormal CO_2 flux emission, reaching 621 g/(m^2 d), in a small area near Sabinosa village (Fig. 6.6), two months before the eruption started. Melián et al. (2014) measured increases in the emission of diffuse CO_2 flux across the whole island ten days before the occurrence of the 4.0 Mw seismic event and two weeks before the onset of the submarine eruption (Fig. 6.6c). Padrón et al. (2013), detected changes in diffusive helium emissions from the soil one month before the beginning of the eruption, measuring in 473 sampling points across the island (Fig. 6.6c). They also measured $^3He/^4He$ ratio in dissolved gases in groundwater in San Simón well (Fig. 6.6), finding an important increase one month prior to the eruption, with a maximum value of 8.2 R_A (R_A = $^3He/^4He$ ratio in air) (Fig. 6.6a). In addition, soil ^{222}Rn activity and $^{222}Rn/^{220}Rn$ ratio, monitored by Padilla et al. (2013) at two geochemical stations (HIE02 and HIE03, Fig. 6.6), showed important increases before the onset eruption (Fig. 6.6b). Finally, by measuring CO_2 and ^{222}Rn in air in separate sites, located in the north of the island, Torres-González et al. (2019) detected geochemical anomalies associated to the volcanic process. On the one hand, from mid-July to 7 September, a clear increase in CO_2 concentration in the air (Fig. 6.6d) in a sub-horizontal water mining tunnel (HTIN) located in El Golfo embayment (Fig. 6.6), was observed (from 825 $\pm$ 19 to 2305 $\pm$ 35 ppm), and causally related to ground deformation. On the other hand, from 7 September to the onset of the submarine eruption, an abrupt increase in CO_2 and ^{222}Rn concentrations in the same tunnel and in a well located in the NW (RSIM) (Fig. 6.6) respectively were registered. The ^{222}Rn concentration in air measured in the well was associated to both ground deformation and seismic energy release.

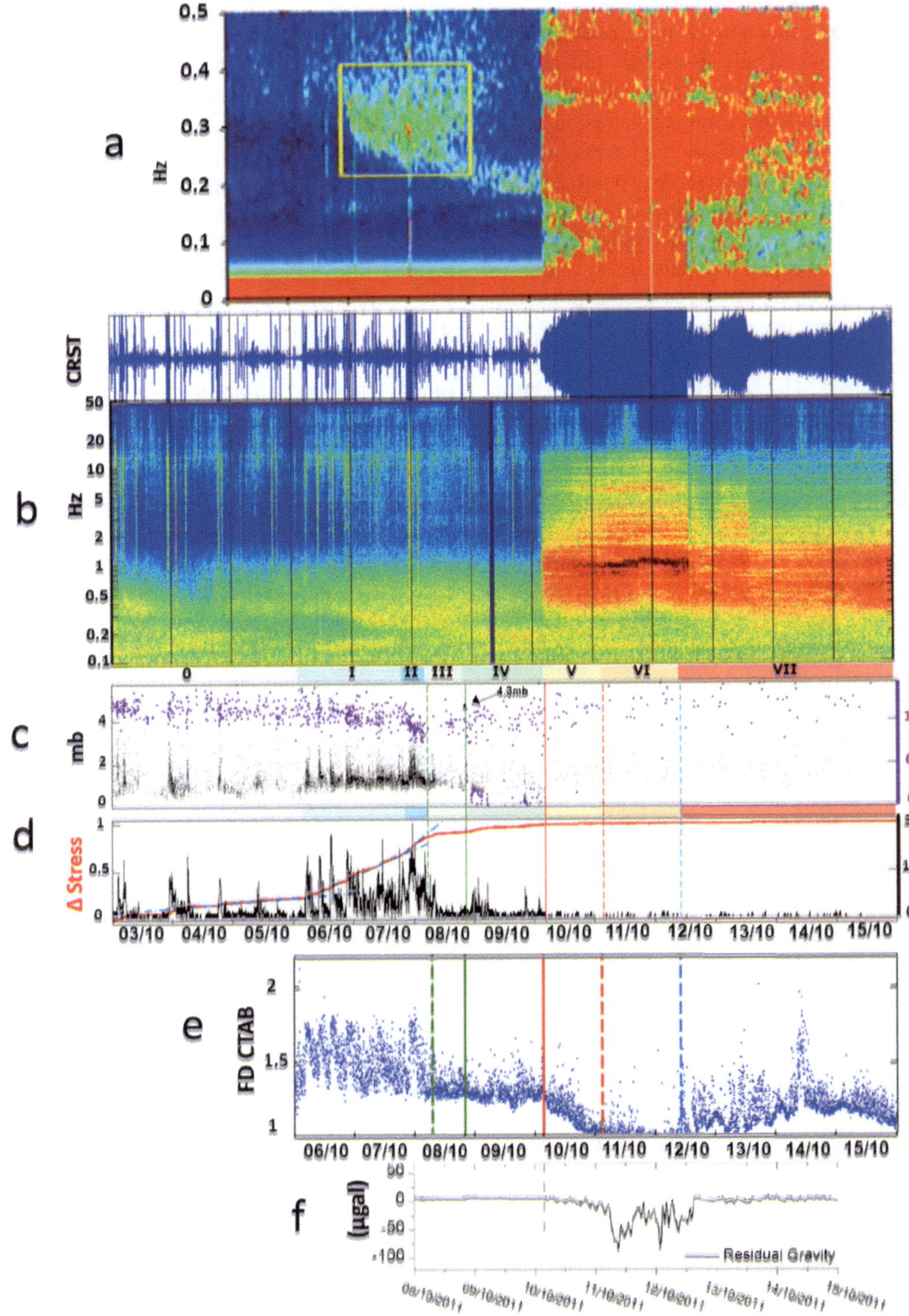
a
Hz
b
CRST
Hz
0
I
II
III
IV
V
VI
VII
c
mb
4.3mb
Depth (km)
d
Δ Stress
(Nº det)/dt
03/10 04/10 05/10 06/10 07/10 08/10 09/10 10/10 11/10 12/10 13/10 14/10 15/10
e
FD CTAB
06/10 07/10 08/10 09/10 10/10 11/10 12/10 13/10 14/10 15/10
f
(µgal)
Residual Gravity
08/10/2011 09/10/2011 10/10/2011 11/10/2011 12/10/2011 13/10/2011 14/10/2011 15/10/2011

◀ **Fig. 6.7** Very short-term signs. **a** Spectrum of high-frequency gravity signal recorded by the gPhone-054 gravimeter at AU, **b** ground-motion spectrogram of the vertical component of the closet to the eruption site station, CRST, and its trace, **c** seismic magnitude (in black) and location-depth time evolution (in purple) of events, **d** seismic rate (in black) and its accumulated value (in red). Colour bands labelled 0-VII indicate distinct phases defined in López et al. (2014), **e** temporal variations of seismic fractal dimension, FD, at CTAB seismic station, **f** residual gravity signal calculated from the gPhone-054 gravimeter record at site AU. The 4.0 Mw event and the onset of the seismic tremor are indicated on the graph with full green and red vertical lines. The dashed green vertical line indicates the beginning of the reduction in the seismic rate. The dashed red vertical line indicates the beginning of the fall in FD shorter values. The dashed blue vertical line indicates the onset of the appearance of light-green discoloured seawater stained by volcanic gases. Modified from Sainz-Maza et al. (2014), López et al. (2014)

6.5 Very Short-Term Signs of Imminent Eruption

We look for short-term precursors of the ultimate processes occurring at depth below El Hierro, that preceded the ascent of pressurized magma to the surface. We collect from López et al. (2014) and Sainz-Maza et al. (2014) signs observed during the last days of pre-eruptive unrest (from 6 October to the beginning of the eruption on 10 October) and the first days of volcanic tremor (10–15 October) (Fig. 6.7). During 6–8 October, the volcano-tectonic seismic activity strongly increased both in rate and in size, reflecting the seismic strain a permanent deformation accumulation, with a maximum rate of 1.4×10^{-6} strain h^{-1} (Fig. 6.7c, d), the highest throughout the entire process. On 8 October, after the occurrence of the 4.0 M_W earthquake, the seismic activity, the seismic strain, and the released seismic energy fell drastically. From this day, the seismic strain reflected a smaller but persistent permanent deformation with a maximum rate of 0.3×10^{-6} strain $h^{=1}$, which was maintained until 10 October (Fig. 6.7b–d). Up to 8 October, FD showed cyclic variations with a dominant value of FD > 1.4 during the most intense seismic swarms. On 10 October at 04:10 UTC, a clear emergent tremor signal was registered by all the seismic stations, and a decrease to a shorter FD dimension began (Fig. 6.7g), which reached minimum values of ~ 1 in all stations by the end of 11 October. From 10 October (06:00) to 12 October (14:30), the volcanic tremor reached its maximum energy on a broad frequency band rich in high frequencies (Fig. 6.7). On 12 October, FD ~ 1 suggested that the minimum dimension conduit geometry was stablished (Fig. 6.7g).

During this period, results provided by the analysis of the data obtained by the gravimeter gPhone#054, located in the site 'Aula de la Naturaleza' (AU) (Sainz-Maza et al. 2014) ties in with previous signals. Figure 6.7a shows the moving window spectrum of the gravity signal. Since 6 October until the eruption onset 0.2–0.4 Hz gravity constricted signals in frequency domain were recorded. A extensive discussion about this signal can be found in Sainz-Maza et al. (2014). The origin of these observed frequencies is usually linked to pressure pulses which propagate to the sea floor generating micro seismicity (Longuet-Higgins 1950).

Usually, atmospheric disturbances are the most plausible cause of these phenomena. Here, in light of the result, it would be quite probable that pressure pulses generated because of the continuous crustal fracturing. According to Sainz-Maza et al. (2014), in this case the source of the micro seismicity should be under the surface. These signals were interpreted assuming that frequencies between 0.2 and 0.4 Hz point to magma accumulation in some part of the Moho or in some heterogeneity inside the crust. After 4.0 Mw event, a more constricted 0.2 Hz frequency dominated the spectrogram till the beginning of the tremor. As per Sainz-Maza et al. (2014), it could represent magma ascent through a confined area.

Immediately after the submarine eruption onset on 10 October, there was a decrease in the gas pressure in the magma beneath the island which was reflected in: (i) a drop in the diffuse

H_2S and CO_2 at HIE07 and HIE01 stations respectively (Pérez et al. 2012) (Fig. 6.6c), (ii) a decrease in the diffusive helium emission between 20 and 26 October 2011 (Padrón et al. 2013) (Fig. 6.6a) and, (iii) a slowdown in CO_2 increasing trend and a decrease in background ^{222}Rn level at HTIN and RSIM stations respectively (Torres-González et al. 2019) (Fig. 6.6d).

6.6 Which Precursory Parameters Were Most Relevant for Eruption Forecasting in Tagoro 2011?

Here we summarize the sequence of undergoing processes and the associated precursory signals from the first signs of early unrest in 2003, to the eruption onset in October 2011 (Fig. 6.8). The precursors identified during the early signs of unrest (2003–2011) at regional scale include an increase of the seismic activity and an E-W drift and uplift deformations that affected the southern Iberian Peninsula, northwest Africa, and the Canary Islands in 2003 and 2007. The coherent and joint evolution of seismicity and deformation and its rapid evolution, supports a scenario of a magmatic unrest in the region. In the Canaries, we find a similar seismic evolution and a concurrent stress-shift episode measured by GNSS stations coherent patterns occurring locally from 2003 to the beginning of the unrest in July 2011. All these precursors have been interpreted as result of evidence for magma ascent towards the base of the crust below the Canaries. Before El Hierro unrest, notable geochemical anomalies were reported, including a significant increase in diffuse CO_2 flux from the soil before the occurrence of some seismic events in 2004 and magmatic ^3He emission some weeks before the beginning of the seismic unrest in July 2011. In El Hierro, during the first phase of unrest (July–August 2011) anomalous activity includes the appearance of clustered swarm seismic activity

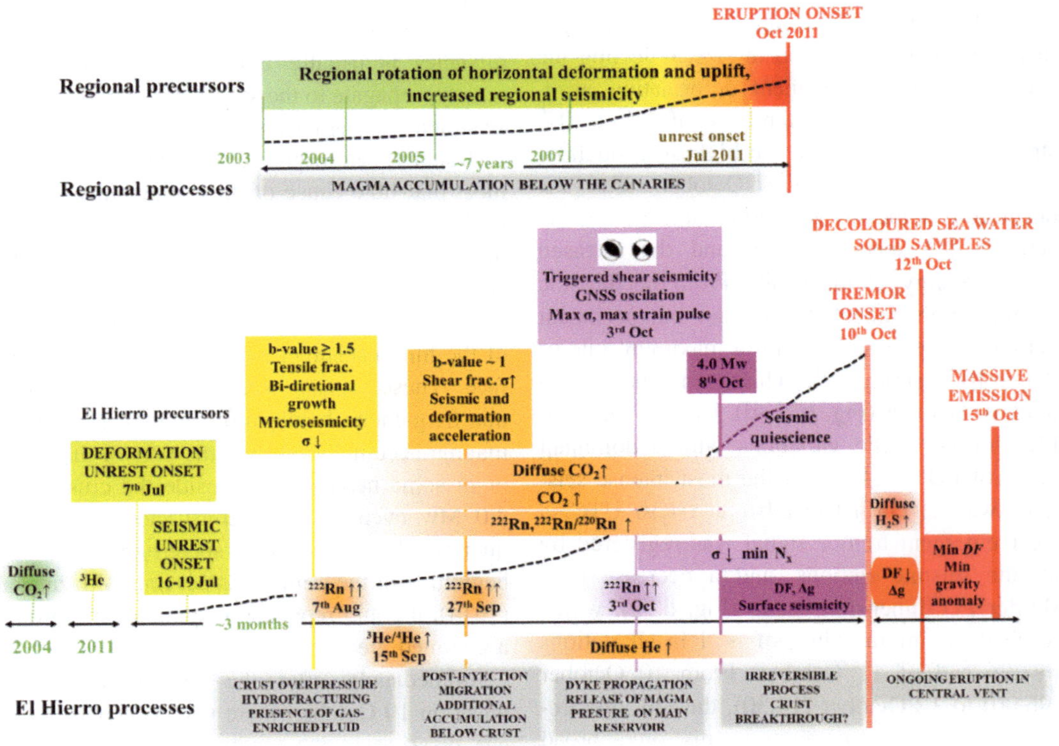

Fig. 6.8 Sketch including main processes and long-term regional and local short-term precursors of Tagoro eruption

of low magnitudes and *b-value* $\geq$ 1.5, tensile fracturing and media velocity changes. These features are interpreted as evidence for magma emplacement below El Hierro, promoting the over-pressurization and hydro-fracturing of the edifice. A fluid injection (gas-enriched) close to the crust-mantle discontinuity has been proposed to explain previous unrest features, and some other like bi-directional seismicity growth pattern observed during that period (López et al. 2014). This fluid intrusion could not ascent through the high velocity crust in that area (Martí et al. 2017), but instead, it over pressurized the entire island, promoting extension and tensile fracturing. The observation of a spatial positive CO_2 flux anomaly in a fractured area at the north of El Hierro between 22 July and 14 August (López et al. 2012) could be congruent with the described state. Longpré et al. (2014) proposed that this injection of magma was accommodated in a pre-existing mantle reservoir using petrological data.

A seismic migration was observed from 21 August to ~ 26 September, suggesting unidirectional growth of the fluid injection, driven by structural or regional stress gradients. In this sense, the exact coincidence between the unidirectional growth path and the lateral heterogeneity at 10–15 km below the Moho discontinuity is notable (Gorbatikov et al. 2013). Díaz-Moreno et al. (2015), in their analysis of the temporal and spatial distribution of the seismic activity occurring on El Hierro suggest the existence of stress diffusion and hydraulic fracturing during this period.

From 27 September onwards, we find signals that have been interpreted as evidence of a process that drive to a critical state after an increased magma pressurization, diking, the crossing of the probable mantle-crust boundary and the triggering of seismic events in pre-existing faults. The related signs include significant changes in several gaseous parameters (He, CO_2, $(R/R_a)_c$, $^{222}Rn/^{220}Rn$), and the acceleration of the seismic activity and deformations, with very intense swarms of VT events, alternation of *b-values* ~ 1, alternation of shear and tensile behaviour, maximum Poisson's ratio and the change from thrust to strike slip type of fracturing. Besides, we find higher modelled point-pressure source volumes moving to the south in congruence with the seismic activity migration, pulses and maximum deformation as reflected by the GNSS oscillation, gravimetric anomalies associated with fluid flow and minimum *Nx* values. During the reverse faulting period (27 September to 3 October), volcanic stress due to magmatic accumulation likely dominated, being the source of this fracturing clearly below the crustal base. The change in focal mechanisms from thrust to strike-slip, and the GPS oscillations registered on 3 October have been interpreted as evidence for magma lateral migration from its source and into the dyke, allowing the decreasing of the magma pressure at the base of the crust. The rapid variations in Poisson's ratio recorded in this period would reflect the abrupt changes in the stress and deformation states, which cause fracturing, and the active transport of fluids (melt), that caused extensive fracturing in a positive feedback system that ultimately leads to eruption (Koulakov et al. 2013).

We highlight the drop-off in the seismic activity, that started some hours before the occurrence of the 8 October 4.0 Mw event and lasted until the beginning of the tremor signal on 10 October, and the appearing of shallow seismicity occurring for the first time during the unrest (0–3 km of depth). The occurrence of 8 October 4.0 M_w event, as well as other low magnitude seismicity observed during the period of 3–8 October, has been interpreted as result of Coulomb stress transfer of magma overpressure in pre-existing faults, reflecting the escalating process of magma overpressure. In fact, as result of the 4.0 Mw 8 October event, the effective pressure, σ_f, reached a critical value of 5 MPa that would have surpassed the strain-rate threshold necessary to initiate such a catastrophic fracture and thus opened a path through the crust to the surface (López et al. 2014). During these days, minimum *Nx* values pointed to the occurrence of a predictable process. The gravimetric analysis performed by Sainz-Maza et al. (2014) identified evidence for magma ascent after the 4.0 Mw 8 October event and the beginning of the tremor signal. The 'deep' and 'shallow' seismicity registered from this event to

10 October, could represent the depths of the bottom and top edges (12–3 km) of such ascent. According to the observational data presented here, at this stage, the system reached a critical point due to magma accumulation and overpressure, and it is likely that magma was flowing towards the surface from 8 October onwards. This hypothesis is supported by the observation of elevated levels and a gradual increase in the H_2S flux recorded at a station located on the coastline of the Mar de Las Calmas (close to the seismic epicentral area) prior to the onset of the eruption (Pérez et al. 2012).

Finally, we review the precursors that can shed light on the timing of the eruption onset and its geometry. As the Tagoro eruption was submarine, it is not straightforward to know the exact onset of the emission of magma on the ocean floor, and there is an uncertainty of two days during which its initiation has been discussed. The tremor signal started on 10 October at 04:10 UTC, indicating fluid flow. Right from its beginning, while tremor amplitude increased, *FD* and the gravity anomaly started diminishing until 11 October at 06:00 UTC. During 10 October, the only indirect evidence of the ongoing process was the appearance on the sea surface of many dead deep-sea fish. Increasing tremor amplitudes have commonly been observed on other volcanoes and are thought to occur when ascending magma vesiculates due to decompression. On 12 October 14:30 h and thereafter, a sudden decrease in the tremor amplitude was recorded at all stations, with a shift of the frequency band to smaller values. The appearance of eruptive signals on the sea surface this day (stained water and small fragments of juvenile material) corroborates that volcanic activity was occurring at the submarine vent. The $FD \sim 1$ and the gravity anomaly reaching its minimum, suggest that the conduit to the surface was established and steady.

We find solid precursors of the Tagoro eruption in the appearing of changes in seismicity (seismic acceleration, seismic quiescence, changes in *b-value*, type of faulting, thrust to strike-slip focal mechanism, changes in *FD, FIM, Nx*), changes in the media properties (*Vp/Vs* or σ), changes in deformation (deformation acceleration, changes in coordinate series and velocities, deformation oscillations, increase in pressure source modelled), changes in gravity (Δg) and changes in emitted gases (CO_2, He, $^3He/^4He$, ^{222}Rn, H_2S). We believe that precursors cannot be selected nor evaluated independently, but rather as a sequence, the appearance of each one conditioning the subsequent observables and increasing the probability of eruption. Thus, we tentatively propose the following sequence of precursors for El Hierro eruption, which can serve as a reference for a future phenomenon of monogenetic volcanic activity in the Canary Islands.

At long term and regionally, starting from ~ 7 years before, we highlight the rotation of horizontal regional deformations, the uplift and the increase in regional seismicity. At short term and locally, the sequence: 19 July, (3 months before eruption) the appearing above background level of anomalous seismicity and deformation, 27 September (~ two weeks before), acceleration of observables and onset of notable felt seismicity, 3 October (1 week before) the change of focal mechanism from inverse to normal, induced seismicity in pre-existing faults, deformation oscillations, maximum anomalies of the media parameters, maximum stress and strain, 8 October (1–2 days before), maximum seismic event followed by relative quiescence and appearance of surface seismicity, gravimetric anomalies, maximum persistence of micro seismicity (Δg, *FD*) and signs of a predictability on the process (*Nx*).

Some of these changes were observed during the reactivation monitoring and others have been obtained later from an in-depth analysis of the acquired data. The study of these changes, their relationship with the variations of the subsurface processes and the influence of each of them in the subsequent development of the activity, has been of great importance to deepen the knowledge of the monogenetic volcanic phenomenon in general and specifically in the environment of the Canary Islands.

Finally, we highlight that this eruption was the first multi-technique monitored event at the

Canary Island and that despite the difficulties, we were able to observe many volcanic precursors. If this eruption had not had an instrumental network, the only evidence of unrest would have been the felt seismicity starting two weeks before the submarine eruption, which is consistent with the precursors observed in most of the historical eruptions in the Archipelago. The acquired experience is essential for monitoring upcoming volcanic activity on the island and in the Canaries, and it will contribute to a better near real-time interpretation of the observed signals and a correct volcanic risk assessment.

References

Albert H, Costa F, Martí J (2015) Timing of magmatic processes and unrest associated with mafic historical monogenetic eruptions in Tenerife Island. J Petrol 56:1945–1966. https://doi.org/10.1093/petrology/egv058

Albert H, Costa F, Martí J (2016) Years to weeks of seismic unrest and magmatic intrusions precede monogenetic eruptions. Geology 44(3):211–214. https://doi.org/10.1130/G37239.1

Álvarez-Valero AM, Burgess R, Recio C, de Matos V, Sánchez-Guillamón O, Gómez-Ballesteros M (2017) Noble gas signals in corals predict submarine volcanic eruptions. Chem Geol. http://doi.org/10.1016/j.chemgeo.2017.05.013

Barker A, Troll V, Carracedo J, Nicholls P (2015) The magma plumbing system for the 1971 Teneguía eruption on La Palma, Canary Islands. Contrib Miner Petrol 170. http://doi.org/10.1007/s00410-015-1207-7

Bartolini S, López C, Becerril L, Sobradelo R, Martí J (2018) A retrospective study of the pre-eruptive unrest on El Hierro (Canary Islands): implications of seismicity and deformation in the short-term volcanic hazard assessment. Nat Hazards Earth Syst Sci 18:1759–1770. https://doi.org/10.5194/nhess-18-1759-2018

Baxter P (2005) Human impacts of volcanoes. In: Marti J, Ernst G (eds) Volcanoes and the environment. Cambridge University Press, Cambridge, pp 273–303. http://doi.org/10.1017/CBO9780511614767.011

Bonelli Rubio JM (1950) Contribución al estudio de la erupción del Nambroque o San Juan (Isla de La Palma) 24 de Junio - 4 de Agosto de 1949, Talleres del Instituto Geográfico y Catastral. Instituto Geográfico y Catastral, Madrid

Bravo T (1968) Hidrogeología de la isla de El Hierro. Estudios Canarios 11–13:88–90

Carracedo JC, Badiola ER, Guillou H, de la Nuez J, Pérez-Torrado FJ (2001) Geology and volcanology of La Palma and El Hierro, Western Canaries. Estud Geol 57:175–273

Darias y Padrón DV (1980) Noticias generales históricas sobre la isla de El Hierro, una de las Canarias, 2nd edn. Goya Ediciones, Santa Cruz de Tenerife, 287 p

de la Cruz-Reyna S, Yokoyama I (2011) A geophysical characterization of monogenetic volcanism. Geofisica Internacional 50:465–484

Del Fresno C (2016) Determinación de la Fuente Sísmica a distancias regionales: Aplicación a la Serie de El Hierro 2011. Ph.D. thesis, Universidad Complutense de Madrid

Del Fresno C, Domínguez Cerdeña I, Cesca S, Buforn E (2015) The 8 October 2011 earthquake at El Hierro (Mw 4.0): focal mechanisms of the mainshock and its foreshocks. Bull Seismol Soc Am 105:330–340

Díaz-Moreno A, Ibáñez JM, De Angelis S, García-Yeguas A, Prudencio J, Morales J, Tuvè T, García L (2015) Seismic hydraulic fracture migration originated by successive deep magma pulses: the 2011–2013 seismic series associated to the volcanic activity of El Hierro Island. J Geophys Res Solid Earth 120:7749–7770. https://doi.org/10.1002/2015JB012249

Domínguez Cerdeña I, del Fresno C, Moreno AG (2014) Seismicity patterns prior to the 2011 El Hierro eruption. Bull Seismol Soc Am 104(1):567–575

Driedger CL, Major JJ, Pallister JS, Clynne MA, Moran SC, Westby EG, Ewert JW (2020) Ten ways Mount St. Helens changed our world—the enduring legacy of the 1980 eruption. U.S. Geological Survey Fact Sheet 2020–3031, 6 p. http://doi.org/10.3133/fs20203031

Foshag WF, Gonzalez RJ (1956) Birth and development of Paricutin volcano. US Geol Surv Bull 965:355–489

García A, Fernández-Ros A, Berrocoso M, Marrero JM, Prates G, De la Cruz-Reyna S, Ortiz R (2014) Magma displacements under insular volcanic fields, applications to eruption forecasting: El Hierro, Canary Islands, 2011, 2013. Geophys J Int 179:322–334. https://doi.org/10.1093/gji/ggt505

González PJ, Samsonov SV, Pepe S, Tiampo KF, Tizzani P, Casu F et al (2013) Magma storage and migration associated with the 2011–2012 El Hierro eruption: implications for crustal magmatic systems at oceanic island volcanoes. J Geophys Res Solid Earth 118:4361–4377

Gorbatikov AV, Montesinos FG, Arnoso J, Stepanova MY, Benavent M, Tsukanov AA (2013) New features in the subsurface structure model of El Hierro Island (Canaries) from low-frequency microseismic sounding: an insight into the 2011 seismovolcanic crisis. Surv Geophys 34(4):463–489. http://doi.org/10.1007/s10712-013-9240-4

Gudmundsson M (2000) The eruption at Gjalp, Vatnajoekull in 1996—lessons on subglacial eruptions

Guilbaud M-N, Siebe C, Layer P, Salinas S, Castro-Govea R, Garduño Monroy V, Le Corvec N (2011) Geology, geochronology, and tectonic setting of the Jorullo Volcano region, Michoacán, México. J Volcanol Geoth Res 201:97–112 http://doi.org/10.1016/j.jvolgeores.2010.09.005

Hernández-Pacheco A (1982) Sobre una posible erupción en 1793 en la isla de El Hierro (Canarias). Estud Geol 38:15–25

Hincks TK, Komorowski JC, Sparks SR et al (2014) Retrospective analysis of uncertain eruption precursors at La Soufrière volcano, Guadeloupe, 1975–77: volcanic hazard assessment using a Bayesian belief network approach. J Appl Volcanol 3:3. https://doi.org/10.1186/2191-5040-3-3

Ibáñez JM, De Angelis S, Díaz-Moreno A, Hernández P, Alguacil G, Posadas A, Pérez N (2012) Insights into the 2011–2012 submarine eruption off the coast of El Hierro (Canary Islands, Spain) from statistical analyses of earthquake activity. Geophys J Int 191:659–670

Kereszturi G, Németh K (2012) Monogenetic basaltic volcanoes: genetic classification, growth, geomorphology and degradation. In: Updates in volcanology—new advances in understanding volcanic systems. IntechOpen. http://doi.org/10.5772/51387

Koulakov I, Gordeev E, Dobretsov N, Vernikovsky V, Senyukov S, Jakovlev AV, Kairly J (2013) Rapid changes in magma storage beneath the Klyuchevskoy group of volcanoes inferred from time-dependent seismic tomography. J Volcanol Geoth Res 263:75–91. https://doi.org/10.1016/j.jvolgeores.2012.10.014

Longpré MA (2021) Reactivation of Cumbre Vieja volcano. Science 374(6572):1197–1198. http://doi.org/10.1126/science.abm9423

Longpré MA, Felpeto A (2021) Historical volcanism in the Canary Islands; part 1: a review of precursory and eruptive activity, eruption parameter estimates, and implications for hazard assessment. J Volcanol Geotherm Res 419(art no 107363). http://doi.org/10.1016/j.jvolgeores.2021.107363

Longpré MA, Klügel A, Diehl A, Stix J (2014) Mixing in mantle magma reservoirs prior to and during the 2011–2012 eruption at El Hierro, Canary Islands. Geology 42(4):315–318

Longuet-Higgins MS (1950) A theory of the origin of microseisms. Philos Trans Roy Soc Lond Ser A Math Phys Sci 243:1–35

López C, Blanco MJ, Abella R, Brenes B, Cabrera Rodríguez VM, Casas B, Domínguez-Cerdeña I, Felpeto A, de Villalta MF, del Fresno C, García O, García-Arias MJ, García-Cañada L, Gomis Moreno A, González-Alonso E, Guzmán Pérez J, Iribarren I, López-Díaz R, Luengo-Oroz N, Meletlidis S, Moreno M, Moure D, de Pablo JP, Rodero C, Romero E, Sainz-Maza S, Sentre Domingo MA, Torres PA, Trigo P, Villasante-Marcos V (2012) Monitoring the volcanic unrest of El Hierro (Canary Islands) before the onset of the 2011–2012 submarine eruption. Geophys Res Lett 39. https://doi.org/10.1029/2012GL051846

López C, Martí J, Abella R, Tárraga M (2014) Applying fractal dimensions and energy-budget analysis to characterize fracturing processes during magma migration and eruption: 2011–2012 El Hierro (Canary Islands) submarine eruption. Surv Geophys. https://doi.org/10.1007/s10712-014-9290-2

López C, García-Cañada L, Martí J, Domínguez Cerdeña I (2017a) Early signs of geodynamic activity before the 2011–2012 El Hierro eruption. J Geodyn 104:1–14. https://doi.org/10.1016/j.jog.2016.12.005

López C, Benito MA, Martí J, del Fresno C, García-Cañada L, Albert H, Lamolda H (2017b) Driving magma to the surface: the 2011–2012 El Hierro eruption, G3

Martí J, Pinel V, López C, Geyer A, Abella R, Tárraga M, Blanco MJ, Castro A, Rodríguez C (2013) Causes and mechanisms of the 2011–2012 El Hierro (Canary Islands) submarine eruption. J Geophys Res Solid Earth 118:823–839. https://doi.org/10.1002/jgrb.50087

Martí J, Villaseñor A, Geyer A, López C, Tryggvason A (2017) Stress barriers controlling lateral migration of magma revealed by seismic tomography. Sci Rep 7:40757. https://doi.org/10.1038/srep40757

Mattsson H, Hoskuldsson A (2003) Geology of the Heimaey Volcanic Centre, South Iceland: early evolution of a central volcano in a propagating rift? J Volcanol Geoth Res 127. http://doi.org/10.1016/S0377-0273(03)00178-1

Melián G, Hernández PA, Padrón E, Pérez NM, Barrancos J, Padilla G, Dionis S, Rodríguez F, Calvo D, Nolasco D (2014) Spatial and temporal variations of diffuse CO_2 degassing at El Hierro volcanic system: relation to the 2011–2012 submarine eruption. J Geophys Res Solid Earth 119:6976–6991. https://doi.org/10.1002/2014JB011013

National Academies of Sciences, Engineering, and Medicine (2017) Volcanic eruptions and their repose, unrest, precursors, and timing. The National Academies Press, DC. http://doi.org/10.17226/24650

Padilla GD, Hernández PA, Padrón E, Barrancos J, Pérez NM, Melián G, Nolasco D, Dionis S, Rodríguez F, Calvo D, Hernández I (2013) Soil gas radon emissions and volcanic activity at El Hierro (Canary Islands): the 2011–2012 submarine eruption. Geochem Geophys Geosyst 14:432–447. https://doi.org/10.1029/2012GC004375

Padrón E, Melián G, Marrero R, Nolasco D, Barrancos J, Padilla G, Hernández PA, Pérez NM (2008) Changes in the diffuse CO_2 emission and relation to seismic activity in and around El Hierro, Canary Islands. Pure Appl Geophys 165:95–114. https://doi.org/10.1007/s00024-007-0281-9

Padrón E, Pérez NM, Hernández PA, Sumino H, Melián GV, Barrancos J, Nolasco D, Padilla G, Dionis S, Rodríguez F, Hernández Í, Calvo D, Peraza MD, Nagao K (2013) Diffusive helium emissions as a precursory sign of volcanic unrest. Geology 41:539–542. http://doi.org/10.1130/G34027.1

Pérez NM, Padilla GD, Padrón E, Hernández PA, Melián GV, Barrancos J, Dionis S, Nolasco D, Rodríguez F, Calvo D, Hernández Í (2012) Precursory diffuse CO_2 and H_2S emission signatures of the 2011–2012 El Hierro submarine eruption, Canary Islands. Geophys Res Lett 39. https://doi.org/10.1029/2012GL052410

Ponte A (1911) Cologán: Volcán del Chinyero. Memoria histórico-descriptiva de esta erupción volcánica acaecida en 18 de noviembre de 1909. Tipolit, Tenerife

Prates G, García A, Fernández-Ros A, Marrero JM, Ortiz R, Berrocoso M (2013) Enhancement of sub-daily positioning solutions for surface deformation surveillance at El Hierro volcano (Canary Islands, Spain). Bull Volcanol 75(6):1–9

Romero C (2003) El relieve de Lanzarote. Servicio de Publicaciones, Cabildo de Lanzarote

Romero Ruiz C (1990) Las manifestaciones volcánicas históricas del Archipiélago Canario. Universidad de La Laguna

Rueda Núñez JJ, Abella Meléndez R, Blanco Sánchez MJ, Díaz Suárez EA, Domínguez Cerdeña IF, Domínguez Valbuena J, Fernández de Villalta Compagni M, del Fresno Rodríguez-Portugal C, López Díaz R, López Moreno C, López Muga M, Muñoz Santamaría A, Sánchez Sanz C, Mezcua Rodríguez J (2020) Revisión del catálogo sísmico de las Islas Canarias (1341–2000), 1st edn. Administración General del Estado, Madrid. http://doi.org/10.7419/162.34.2020

Sainz-Maza S, Arnoso J, Montesinos FG, Martí J (2014) Volcanic signatures in time gravity variations during the volcanic unrest on El Hierro (Canary Islands). J Geophys Res Solid Earth 119. http://doi.org/10.1002/2013JB010795

Sánchez C (2014) Revisión del catálogo sísmico de las Islas Canarias. Universidad Politécnica de Madrid, Madrid

Sánchez-Pastor P, Obermann A, Schimmel M (2018) Detecting and locating precursory signals during the 2011 El Hierro, Canary Islands, submarine eruption. Geophys Res Lett 45:10288–10297. https://doi.org/10.1029/2018GL079550

Tárraga M, Martí J, Abella R, Carniel R, López C (2014) Volcanic tremors: good indicators of change in plumbing systems during volcanic eruptions. J Volcanol Geoth Res 273:33–40. https://doi.org/10.1016/j.jvolgeores.2014.01.003

Telesca L, Lovallo M, Martí J, López C, Abella R (2014) Using the Fisher–Shannon method to characterize continuous seismic signal during volcanic eruptions: application to 2011–2012 El Hierro (Canary Islands) eruption. Terra Nova 1–5. http://doi.org/10.1111/ter.12114

Torres-González P, Moure-García D, Luengo-Oroz N, Villasante-Marcos V, Iribarren I, Blanco MJ, Soler V, Jiménez-Abizanda A, García-Fraga J (2019) Geochemical signals related to the 2011–2012 El Hierro submarine eruption. J Volcanol Geoth Res 381:32–43. https://doi.org/10.1016/j.jvolgeores.2019.05.018

Torriani L (1592) Descripción e historia del Reino de las Islas Canarias, antes Afortunadas, con el parecer de sus fortificaciones (1592). Traducción de A. Cioranescu. Goya Ediciones. Santa Cruz de Tenerife, 1978

Villasante-Marcos V, Pavón-Carrasco J (2014) Paleomagnetic constraints on the age of Lomo Negro volcanic eruption (El Hierro, Canary Islands). Geophys J Int 199:1497–1514. https://doi.org/10.1093/gji/ggu346

Yamamoto T, Soya T, Suto S, Uto K, Takada A, Sakaguchi K, Ono K (1991) The 1989 submarine eruption off eastern Izu Peninsula, Japan: ejecta and eruption mechanisms. Bull Volcanol 53:301–308. https://doi.org/10.1007/BF00414526

Geomorphology of Tagoro Volcano Along Eruptive and Posteruptive Phases

7

Juan-Tomás Vázquez, Olga Sánchez Guillamón, Desirée Palomino, Luis Miguel Fernández Salas, Patricia Bárcenas, María Gómez-Ballesteros, María Olvido Tello, Nieves López-González, Carmen Presas-Navarro, and Eugenio Fraile-Nuez

Abstract

The Spanish Institute of Oceanography has realized nineteen oceanographic cruises in order to monitor the geomorphological changes during the submarine eruption of the Tagoro volcano and later evolution. The major geomorphological features were achieved fundamentally by the use of Multibeam EM710 echosounder data. Eruption was characterized by two main phases, the first one alternate stages of vertical growth and denudation by development of basal and southern flank collapses of the main edifice took place; the second phase was characterized by a fissure growth with a NNW-SSE trend. The eruption produced a main volcanic edifice rising from 400 to 88 m water depth. The edifice consists of four attached cones extended and at least fifteen emission vents. This edifice has a quasi-circular base and its final morphology was modulated by the activity of emission vents during the second phase which produced a NNW-SSE elongated summit line. Both vertical growth and instability phases were conditioned by preexistent southwestwards gradient of the seafloor slope and its initial

J.-T. Vázquez (✉) · O. Sánchez Guillamón · D. Palomino · P. Bárcenas · N. López-González
Centro Oceanográfico de Málaga, Instituto Español de Oceanografía, Puerto Pesquero s/n, 29640 Fuengirola, Spain
e-mail: juantomas.vazquez@ieo.csic.es

O. Sánchez Guillamón
e-mail: olga.sanchez@ieo.csic.es

D. Palomino
e-mail: desiree.palomino@ieo.csic.es

P. Bárcenas
e-mail: patricia.barcenas@ieo.csic.es

N. López-González
e-mail: nieves.lopez@ieo.csic.es

L. M. Fernández Salas
Centro Oceanográfico de Cádiz, Instituto Español de Oceanografía, Puerto Pesquero, Muelle de Levante, s/n, 11006 Cádiz, Spain
e-mail: luismi.fernandez@ieo.csic.es

M. Gómez-Ballesteros · M. O. Tello
Instituto Español de Oceanografía, Sede Central, C/ Corazón de María, nº 8, 28002 Madrid, Spain
e-mail: maria.gomez@ieo.csic.es

M. O. Tello
e-mail: olvido.tello@ieo.csic.es

C. Presas-Navarro · E. Fraile-Nuez
Centro Oceanográfico de Canarias, Instituto Español de Oceanografía, C/ Farola del Mar nº 22, Dársena Pesquera, 38180 Santa Cruz de Tenerife, Spain
e-mail: carmen.presas@ieo.csic.es

E. Fraile-Nuez
e-mail: eugenio.fraile@ieo.csic.es

P. J. González (ed.), *El Hierro Island*, Active Volcanoes of the World,
https://doi.org/10.1007/978-3-031-35135-8_7

location into a gully on the southern submarine island flank. In the proximal area, morphology is also characterized in by four ridges that correspond to semi-buried residual scars of different collapse phases. On the SW flank an apron of mixed lavas, pyroclastic and debris flows were deposited along more than 5 km length. These deposits were channeled throughout the previous gully and three parts are differentiated: proximal apron from the cone to an intermediate ravine located at 2.5 km away from the base of the main edifice where its maximum thickness occur in an accumulation front, the intermediated ravine and the distal apron fan deposits from the mouth of the ravine to 1800 m depth. The Tagoro volcano was built during a monogenetic eruption dominated by pyroclastic and lava balloon emissions, with lava emissions in the deepest vents. Its evolution alternating constructive and destructive stages and its morphology being similar to that of long-lived volcanoes located on a steep seafloor and dominated by pyroclastic emissions.

Keywords

Submarine geomorphology · Growth and destruction of submarine volcanic cone · Volcanic sediments · Seafloor mapping · Eruption · Posteruption dynamics

7.1 Introduction

Most of the volcanoes on the Earth are found on the seafloor. More than 5000 active underwater volcanoes have been identified, and their emitted products represent more than 75% of the total volcanic products that are ejected during a year (Huff and Owen 2013). Most of these active underwater volcanoes are located along mid-ocean ridges as the Axial volcano seamount (e.g. Clague et al. 2013; Chadwick et al. 2016; Le Saout et al. 2020), but they are also located in relation to subduction zones, especially in volcanic arc zones as the Kick-Em-Jenny, NW Rota and W Mata volcanoes (e.g. Schnur et al. 2017; Allen et al. 2018; Chadwick et al. 2019), back arc basins as the Kolumbo volcano (Rizzo et al. 2016), and on oceanic lithosphere domains in an intraplate context such the Loïhi seamount volcano (e.g. Schipper et al. 2011). The last type constitutes the hot spots systems which normally are related to the magmatic activity of upwelling mantle plumes (Siebert et al. 2015). Therefore, submarine volcanoes are probably one of the most important but often overlooked active geological processes, as is the case with much of the ocean floor. Different efforts are being made to improve this knowledge by monitoring their development (e.g. Chadwick et al. 2012, 2013; Fraile-Nuez et al. 2012, 2018; Bachèlery and Villeneuve 2013; Kelley et al. 2014; Clague et al. 2017); in this sense, the development of high-resolution bathymetric techniques and their analysis are essential for these studies.

Volcanoes are characterized by the emission of different types of products, lavas, hot rock fragments and gases that accumulate on the seafloor, although the mechanical characteristics of the accumulated materials, slope changes induced by ground dilations or the explosive nature of the eruption itself can also favor the formation of collapses of the volcanic edifice that is being generated, additionally landslides on its flanks that even can affect the adjacent areas to the volcano could also be favored along the eruption. Emissions from submarine volcanoes are conditioned by the presence of the mass of water above them, its cooling down effect, the geochemical interactions of water with volcanic products and the pressure that water column exerts on the surface of the seabed influence the character of the eruption (Bachèlery and Villeneuve 2013; Huff and Owen 2013). The most obvious changes that take place during an underwater eruption are those that occur on the surface of the seafloor that characterize the eruption style (Clague et al. 2011). Quantitative characterization of the size and shape of volcanic edifices is an essential step towards the understanding of factors controlling volcano growth and morphology (Grosse et al. 2012). The acquisition of high resolution bathymetric data repeatedly over time is a fundamental technique to control changes throughout the evolution of an active volcano. The development of multibeam bathymetry techniques greatly advanced the study of active underwater volcanoes (Caress et al. 2012; Allen et al. 2018; Chadwick et al. 2018; Tepp et al. 2019).

The El Hierro Island is the westernmost island of the Canary archipelago which extends along 500 km from east to west (Carracedo 1999) and forms part of the Canary Islands volcanic province constituted by volcanic islands and seamounts along more than one thousand kilometres from the NE to the SW of the archipelago (van den Bogaard 2013). The age of the beginning of sub-aerial volcanic activity decrease from 21 to 20 Ma at the eastern islands to less than 2 Ma at El Hierro and La Palma (Carracedo 1999). Several geodynamic models have been proposed to explain its origin starting from a context of anomalous hotspot systems, including anomalous mantle plumes and controls to the regional tectonic regime (Schmincke 1982; Holik et al. 1991; Hoernle and Schmincke 1993; Carracedo et al. 1998; Anguita and Hernán 2000; Acosta et al. 2003; Acosta Yepes et al. 2003). The islands are emplaced on a Jurassic oceanic lithosphere ($\sim$ 150–180 Ma; Hoernle 1998) which thickness thins westwards, from 35 km at Lanzarote (Martinez-Arevalo et al. 2013) to 12–15 km at El Hierro. The beginning of sub-aerial volcanic activity at El Hierro has been dated at 1.12 Ma (Guillou et al. 1996) and three main volcanic cycles have been defined (Carracedo et al. 2001; Becerril et al. 2015), that alternated with quiescence, structural deformation, and sector collapses (Masson 1996; Gee et al. 2001a, b; Masson et al. 2002; Longpré et al. 2011). The last cycle corresponds to the Rift Volcanism (0.158 Ma–present) characterized by cinder cones and thin lava flows (Carracedo et al. 2001).

Between 10 October 2011, and 5 March 2012, an underwater volcanic eruption took place approximately 1.8 km south of La Restinga port at the submarine southern flank of El Hierro Island (Fig. 7.1). It was emplaced in the western flank of the Southern Rift of the island (Carracedo et al. 2012; Rivera et al. 2013). As result of this eruption a new submarine volcano was formed, the Tagoro volcano, whose summit reached 88 mbsf. This eruption corresponds to a single submarine monogenetic eruption (e.g., López et al. 2012; Martí et al. 2013a; Longpré et al. 2014). It was characterized by the emission into the water column of gases and varied pyroclastic products, from ash to bombs. The most remarkable products were float xenopumice and basaltic lava balloons (Meletlidis et al. 2012; Carracedo et al. 2015). The Tagoro volcano corresponds to an eruption in a geodynamic intraplate context over an oceanic lithosphere and classified between intermediate to shallow eruption during their evolution (Ercilla et al. 2021), shallow water eruptions (< 200 m below sea level—mbsl) are normally characterized by higher explosiveness, especially if they rise above 50 m depth (Kokelaar 1986), meanwhile intermediate water eruptions (200–600 mbsl) have both lava and pyroclastic emissions and are

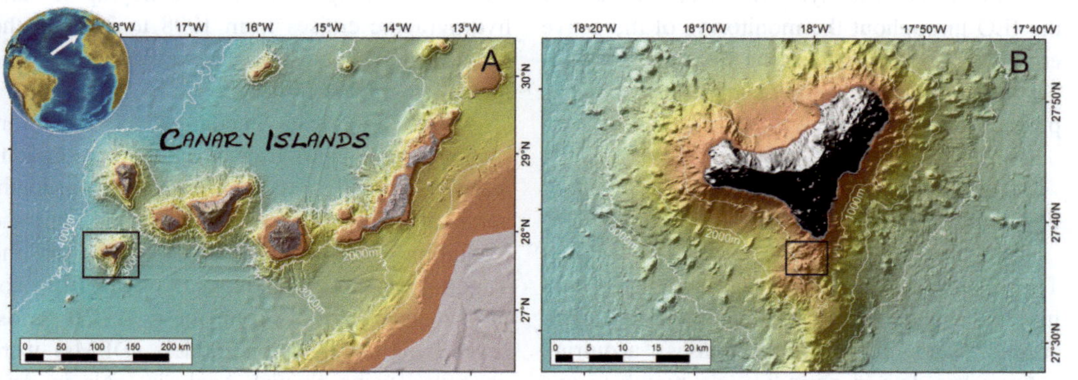

Fig. 7.1 Location of the study area. **a** Location of the Canary Islands in the West African continental margin. **b** Location of the study area in El Hierro Island continental slope

often characterized by lava balloon or pumice lava emissions (Gaspar et al. 2003; Kueppers et al. 2012; Conte et al. 2014; Kelly et al. 2014; Madureira et al. 2017).

Given the alert situation created by this eruptive process and its underwater characteristics, the Spanish government entrusted the Spanish Institute of Oceanography (IEO) to advise the PEVOLCA (Special Plan of Civil Protection and Emergency Attention for volcanic risk in the Autonomous Community of the Canary Islands) for the monitoring of this submarine eruption, investigating their geological characteristics, seafloor modifications, physical and chemical changes taking place in the water column and its impact on the marine ecosystem, to help establish the influence of the eruption on environment and its possible interaction with human populations and socioeconomic activities. The monitoring was led by the IEO projects and cruises: BIMBACHE (Fraile-Nuez et al. 2012; Rivera et al. 2013) and when the eruptive activity ceased, the IEO maintained the monitoring of the Tagoro volcano through two projects of the R + D + I Program of Spain (VULCANO-I and VULCANO-II projects) as well as through an IEO monitoring program (VULCANA-I, VULCANA-II and VULCANA-III) and additional cruises realized with different purposes (Somoza et al. 2017; Fraile-Nuez et al. 2018; González-Vega et al. 2020).

In this chapter a review of the morphological evolution of the Tagoro volcano is carried out from the different bathymetric surveys leaded by the IEO throughout the monitoring of the active eruptive episode. Finally, the final geomorphology of the volcanic building resulting from this process is analyzed.

7.2 Data and Methods

The Spanish Institute of Oceanography has invested an extensive effort in monitoring Tagoro eruption and post-eruptive phases, acquiring bathymetric data in periodic—six month to one year—consecutive oceanographic surveys over the last ten-year period (Table 7.1). During the eruption phase, detailed multibeam bathymetries of the Tagoro volcano and adjacent areas were obtained during eight BIMBACHE oceanographic surveys. These surveys were carried out along October 2011 to February of 2012, in order to study the rapid evolution of the submarine eruption and monitor the main morphological changes. On the post-eruptive and hydrothermal phase, twelve oceanographic surveys have been developed in the framework of different research projects (Raprocan, Vulcano-I, Vulcano-II and Vulcana-I and II) where detailed multibeam bathymetries of the Tagoro volcano and adjacent areas have continued being performed.

The bathymetry datasets were acquired using multibeam echosounder systems Kongsberg EM710 on board R/V Ramón Margalef and R/V Ángeles Alvariño, which have covered diverse study areas, mainly focusing on the site where Tagoro eruption took place, the adjacent western and eastern slopes and the whole Southern Rift of El Hierro island (Table 7.1). Datasets were processed with Caris HIPS and SIPS software, using CUBE algorithm and yielding the generation of bathymetric models of diverse 1–10 m resolution grid exported in different output formats (ASCII grid, geotiff, BAG, etc.) (Fig. 7.2). Since eruption, repeat bathymetric models have been collected during the 19 oceanographic surveys from 2011 to 2019. Before the eruption in 2011, a previous 30 m resolution bathymetry was compiled by the Instituto Hidrográfico de la Marina (Ministerio de Defensa, Spain), These data were collected during several oceanographic and hydrographic cruises from 1998 to 2008 in the framework of Spanish Exclusive Economic Zone and hydrographic projects.

ASCII (American Standard Code for Information Interchange) grids at different resolutions were individually exported depending on depth, study area and oceanographic survey, but the resolution used was mainly 1–2 m for the Tagoro volcano complex, and up to 10 m for the adjacent areas. From these ASCII grids, several digital elevation models (DEMs) were generated using Spatial Analysis tools in ArcGIS Desktop® software and the IDW (Inverse Distance Weighting) interpolation between

Table 7.1 Schematic table of the main datasets used for the eruption monitoring

Research project (year)	Oceanographic surveys	Research vessels (R/V)	Study areas	Depth (m)	Grid resolution (m)
ZEEE and Hidrography (1998–2008)	ZEEE_IHM-IEO and IHM	BIO Hespérides BH Malaespina	El Hierro island (previous data)	77–3975	30
BIMBACHE (2011–2012)	Bimbache_1011-1	Ramón Margalef	Western flank and Southern Rift	12–1612	10
	Bimbache_1011-2	Ramón Margalef		63–1756	10
	Bimbache_1011-3	Ramón Margalef	Tagoro volcano	105–1225	10
	Bimbache_1011-4	Ramón Margalef	Western flank and Southern Rift	62–2028	10
	Bimbache_1011-6	Ramón Margalef	Southern flank	7–1993	10
	Bimbache_1011-7	Ramón Margalef		46–1933	10
	Bimbache_1011-9	Ramón Margalef		59–1930	10
	Bimbache_1011-11	Ramón Margalef		8–1965	10
RAPROCAN (2012)	Raprocan_1212	Ramón Margalef		23–1918	5
VULCANO (2013–2015)	Vulcano_0313	Ramón Margalef	Western and eastern flanks and Southern Rift	12–2009	5
	Vulcano_1013	Ramón Margalef	Southern Rift	15–2164	5
	Vulcano_0314	Ramón Margalef	Southern Rift	68–1655	5
VULCANA (2015–2019)	Vulcana_0515	Ángeles Alvariño		89–1064	2
	Vulcana_1015	Ángeles Alvariño		89–1078	5
	Vulcana_0316	Ángeles Alvariño		89–1053	2
	Vulcana_0417	Ramón Margalef		89–2051	5
VULCANO II (2016–2018)	Vulcano_1016	Ángeles Alvariño		69–954	1
VULCANA II (2017–2019)	Vulcana_0318	Ángeles Alvariño		90–2475	1
	Vulcana_0319	Ángeles Alvariño	Tagoro volcano and Southern Rift	90–2724	1

Research project, oceanographic surveys, research vessels, study areas involved and limits of depths, grid resolution for each DEM (digital elevation models) and corresponding evolutive phase of Tagoro volcano

pixels (Fig. 7.2). Overall, the geomorphological analyses were based on the interpretation of these bathymetric models using ArcGIS extensions.

The analyses and further geomorphological interpretations have been carried applying a variety of techniques that are summarized in Fig. 7.2. As a starting point, shaded reliefs maps were created from the DEMs, that allow conduct a first subjective interpretation of geomorphologic features. From these DEMs, the first and second derivatives of depth, slope and curvature

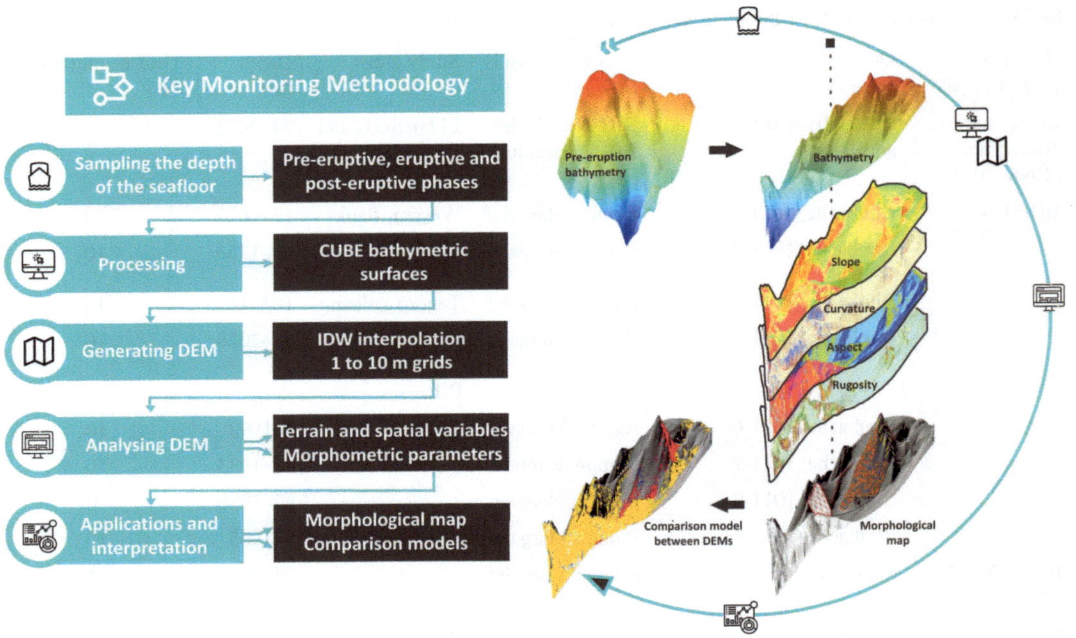

Fig. 7.2 Key monitoring methodological workflow for the acquisition, processing, analyses and applications of the digital elevations models (DEM) in this chapter for geomorphological analyses of Tagoro volcano

were created, showing relevant morphological changes, besides other indirect terrain variables such as aspect (to identify the orientation of pixels) and roughness of the seafloor (as the variation in three dimensional orientations of pixels within a neighborhood). The work realized, integrating all these variables contributed to carry out the geomorphological and comparative analyses in the study area during the study period and phases of the volcano. More specifically, to identify morphological changes the slope gradient maps and aspect maps in degrees, profile curvature map and plan curvature map were used in order to discern minor variations through time. The construction of a lot of bathymetric cross-sections with the use of profile graphs by sorting slope gradients within each flank for prefixed depth intervals have allowed to cartography morphological features.

For the characterization of main morphological types, several morphometric parameters have been used such as height, average basal diameter, slope gradient and main orientation for the main edifice whereas for the linear and minor morphologies its width, length, maximum slope, etc., have been calculated for understanding the formation and morphological evolution of Tagoro volcano from repeated bathymetric models. In order to complement the geomorphological analyses in this study, other specific tools have been used to compare Tagoro volcano with the surroundings areas. In this sense, the Benthic Terrain Modeler (BMT) toolbox for ArcGIS software has allowed us to create raster of Bathymetric Position Index (BPI) and define the location of specific features within the same bathymetric model such as other volcanic edifices situated in the southern ridge of El Hierro island, in order to characterize its similarity in terms of depth, roughness or slope gradients, making comparison among them and analyzing the morphological evolution of the whole area of study. Moreover, ArcGIS tools for volume calculation have been used in order to determine the difference of volume of the study areas through time using consecutive DEMs at the same resolution (Fig. 7.2).

Finally, the Kernel Density algorithm has been used to calculate the number of the volcanic edifices-per-unit area from centroids of the polygons of the volcanic edifices using a kernel function to fit a smoothly tapered surface to each point. To carry out morphometric computations, the spatial extent of each volcanic edifice was defined combining the bathymetric position index and slope obtaining a polygonal layer. A set of morphometric parameters was calculated as width, height, perimeter and mean slope.

7.3 Volcanism in the El Hierro Insular Slope

7.3.1 Physiographic Characteristics

El Hierro island has an estimated total volcanic edifice volume (emerged and submarine) of 5500 km^3 (Schmincke 1990). The highest point is located in the centre of the island (Pico de Malpaso at 1501 m altitude) rises about 5500 m from its base at 4000 mbsl. The island morphology is characterized by three large embayments interpreted as sector collapse scars with corresponding off-shore debris avalanche deposits (El Golfo, Las Playas y El Julán) separated by three ridges oriented at 120° to each other (NW ridge, NE ridge and S ridge) (Carracedo et al. 1999).

The Southern ridge, so called Southern Rift, has a different morphology from the other flanks (NW and NE) showing a slope excavated by deep gullies turning to a smoothly sedimentary seabed. This morphology is interpreted as the eroded remnant of a volcanic edifice (Gee et al. 2001b). The Southern Rift divides Las Playas (to the SE) and El Julán (to the SW) embayments that were formed as the result of huge lateral collapses of the flanks during the evolution of the island and subsequent submarine mass wasting to the SE and SW of the island respectively between 500 and 150 ky (Masson 1996; Carracedo et al. 2001). The NW and NE flanks show steep areas on the slopes, with excavated ridges and gullies. With the exception of the El Golfo valley, there are a large number of volcanic edifices distributed along the flanks. This exception is also observed in the El Julán valley, with both landslides showing a convex to concave profile in the sediment accumulation zone. The SE flank (Las Playas) is the widest of the three valleys. It shows a complex head wall region, which includes a narrow, steep-sided embayment associated whit a series of eroded strike-slip faults oriented perpendicular to the shoreline (Gee et al. 2001b). The SW flank (El Julan) has a smooth morphology, similar to that of the El Golfo valley, although with less well-defined lateral escarpments. It is also characterized by the presence of a few volcanic edifices.

The submarine area of the island (Fig. 7.3) is divided into a narrow insular shelf whose width ranges between 100 and 1200 m. This zone has a low slope not exceeding 2° and the shelf break is located between 120 and 150 mbsl. The insular slope extends from the edge of the shelf to approximately 4000 mbsl showing an eminently concave profile in the three slided zones and a profile more linear on the ridges. In the areas of concave slope profile, the gradient is steeper up to 500 mbsl, reaching values greater than 20° and then decreasing with depth until stabilizing between 5 and 10° on the lower slope above 2000 mbsl. In the other areas of the slope with a more linear profile, the slope oscillates between 40° and 15°, stabilizing at 5° at 3500 mbsl.

7.3.2 Spatial Distribution of the Volcanic Edifices

A set of morphometric parameters that characterize the size and shape of the volcanic buildings has been defined to interpret bathymetry in terms of volcanological processes. A total of two hundred eighty-seven volcanic edifices have been defined around of El Hierro island slope using the high-resolution bathymetric DEMs. The edifices are in a water depth ranging between 94 and 2000 mbsl. The distinguished volcanic edifices have variable shapes such as conical, subconical or composite and complex edifices or massifs with irregular shapes. The spatial density map of the volcanic edifices shows that there are two main areas with the highest concentration of volcanic edifices: The Southern Rift; and the western rift (Fig. 7.4). The areas with the lowest density of volcanic edifices logically coincide with the areas of previous landslides, such as the

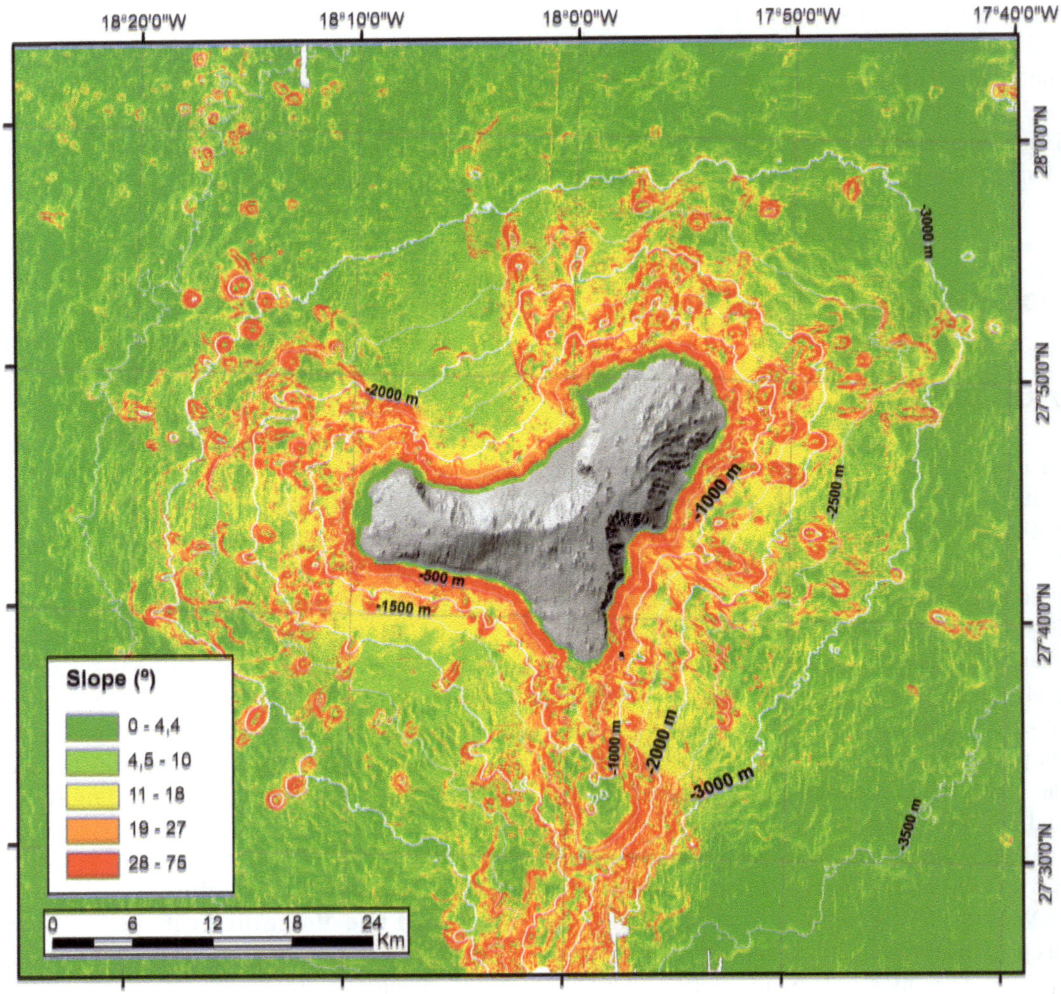

Fig. 7.3 Slope gradient map of El Hierro island shelf and flanks. The map is based on 30 m-resolution multibeam compilation bathymetric data of IHM previous data and IEO post-eruption data (VULCANO and VULCANA projects)

Julan, El Golfo and Las Playas areas where the morphological features prior to the landslide must have been eroded.

7.4 Tagoro Evolutive Geomorphology

7.4.1 Geomorphology Before Eruption

The Tagoro eruption took place at the submerged part of the Southern Rift that has a maximum length of 40 km and is extended to the south, where its distal segment curves slightly southwestward (Acosta et al. 2003). The minimum depth is located south of La Restinga town, and their flanks go deeper to the south reaching the maximum depth at 3700 mbsl. It has 20 km maximum wide in the proximal segment and is 8 km wide in the distal segment, being the flank gradient considerably more abrupt on the eastern side. The eruption began in its western flank from a fracture opened in the seafloor into a pre-existing submarine gully with a general ENE-WSW direction that extended in front of the La Restinga village, approximately at 250–350 mbsl (Fig. 7.5).

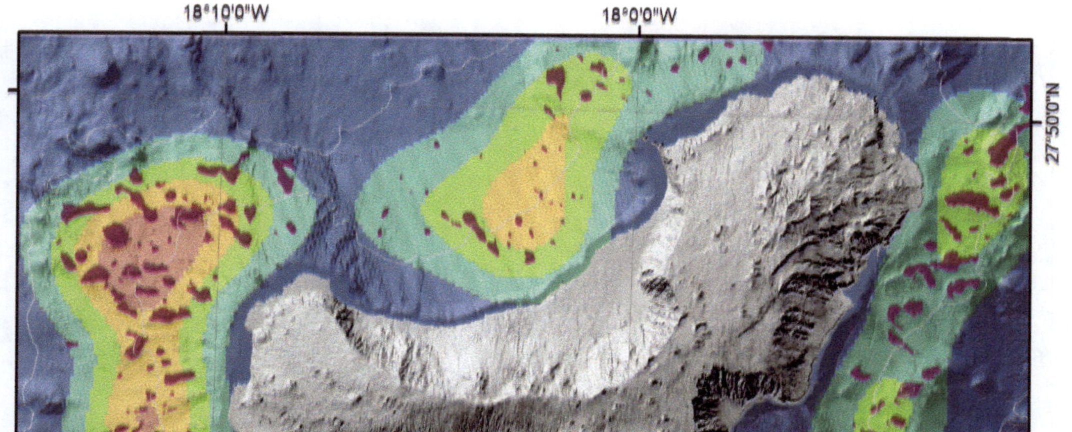

Fig. 7.4 Spatial density map of volcanic edifices around the island of El Hierro. In this study we have used the volcanic edifices that are located approximately at less than 2000 m depth

In the bathymetry preceding the eruption, in addition to gullies and valleys, the main geomorphological features were volcanic defined as cones, ridges including de Southern Rift axis, no conic positive reliefs (so called blocks), and bedrock outcrops. The Southern Rift consisted on multiple non-eroded volcanic cones located from the top of the rift to more than 2000 mbsl (Fig. 7.5). Up to 20 volcanic cones and blocks have been identified in the study zone; most of them were conical in shape although some present several different geometries as elongated, rectangular or rhombic shape. General distribution of the volcanic morphological features has a tendency to a NE-SW alignment. They presented a mean wide of 1 km and they rose 50 m above the surrounding seafloor although their bathymetric profile was very asymmetric as they were located on the steep flanks of the rift. They presented the lowest slope values in the summit (1–2°) and reached 30° along their flanks. Ridges have a NNE-SSW and NE-SW direction and they are 1–2 km length. They could be observed as a linear crest or with an irregular shape as result of the coalescence of small volcanic cones. They were 40 m high

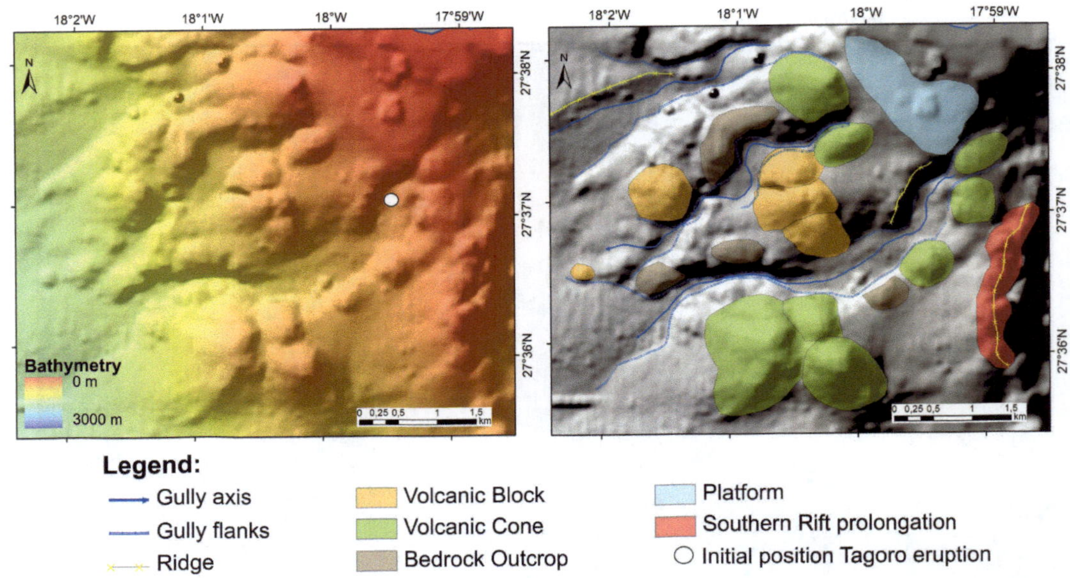

Fig. 7.5 Bathymetry previous to eruption (**a**) and **b** geomorphologic interpretation of the study zone before the eruption where different volcanic edifices, a platform, submarine gullies and ridges are shown

although the ridge located at the crest of the rift was higher ($\sim$ 100 m).

Submarine gullies were identified from the upper part of the rift to 4.5–7 km to the west and southwest, sinuously. They were flat-floored (< 5° slope values), and they were fitted between very steep lateral walls, up to 150 m high, and surrounded by the volcanic cones and blocks. Steep gullies predominate in the upper part of the submarine flank of the island, while broad valleys with smooth flanks predominate in the lower part of the flank. In the case of the gully where the growth of the Tagoro volcano would take place, the upper gully and lower valley sections are also connected by a marked ravine. The southwestern section is characterized by a fan-lobed 1.5 km wide that have been interpreted as recent volcaniclastic flows.

7.4.2 Morphological Evolution During Eruptive Phase

The bathymetric data obtained mainly during the BIMBACHE cruises carried out by the IEO (Rivera et al. 2013, 2014) has allowed us to monitor the bathymetry in the area (Fig. 7.6). The difficulties of conducting a bathymetric monitoring on a relatively explosive erupting underwater volcano did not always allow the exact location of the top of the edifice with the bathymetric acquisition techniques. This was a consequence of safety problems in ships navigation over the area subject to the emission of volcanic products. Therefore 100% coverage of the area was not always feasible. From the morphological changes observed in the bathymetry, the main stages of the Tagoro eruption interpreted are described below:

(1) The first surveys carried out by the IEO on October 25, 2011, allowed to locate a volcanic cone of 0.5 km of diameter whose top reached approximately up to 205 mbsl. The volcanic cone grew up constrained within the head of the preexistent submarine gully. In addition, an apron began to develop in favor of the gully slope towards the southwest (Figs. 7.6a and 7.7a).

(2) On October 29 a general collapse of the cone had taken place (Fig. 7.6b) that produced a subsidence of at least 30 m of the outer edge

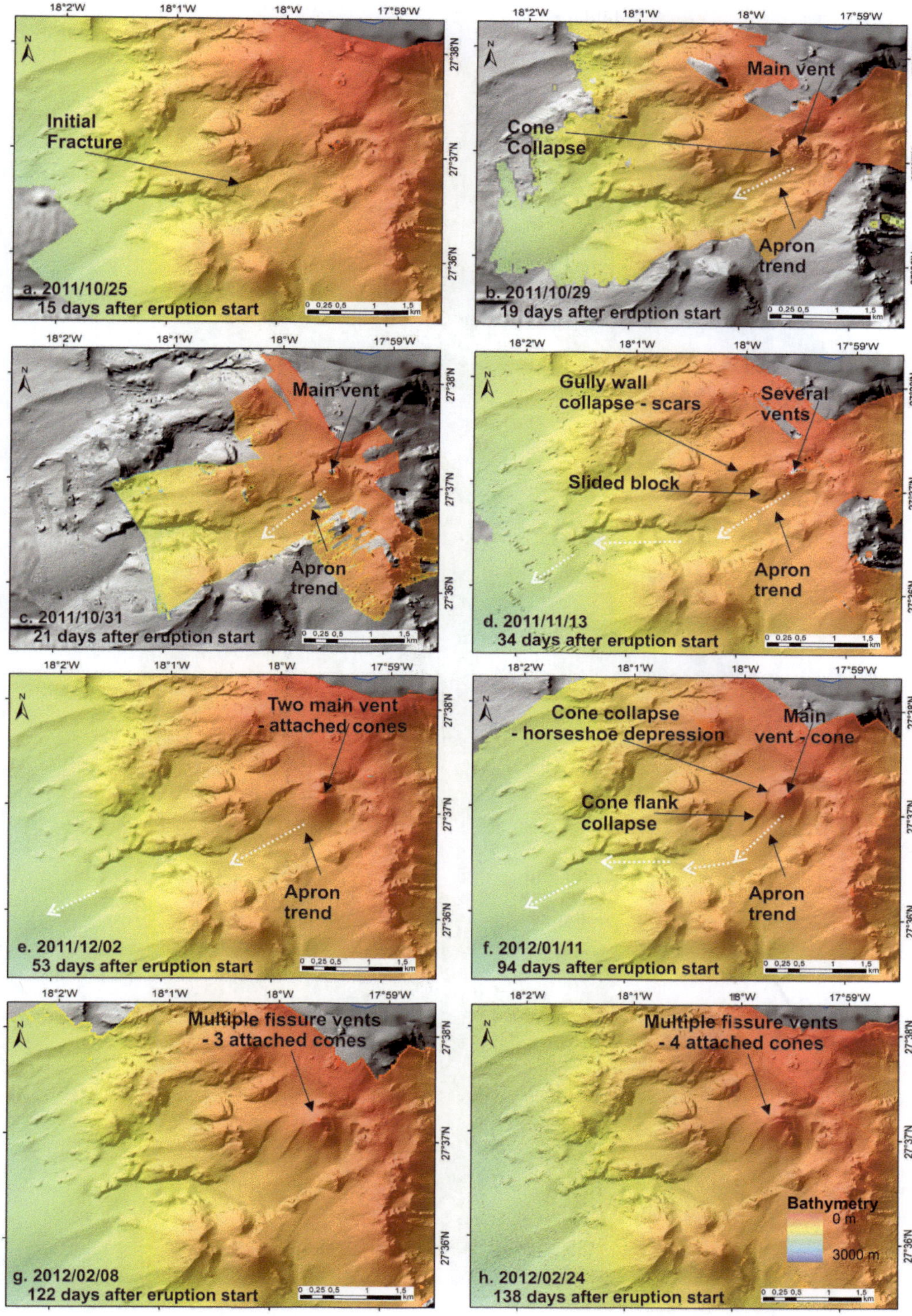

Fig. 7.6 Bathymetric DTM models made during the eight cruises dedicated to bathymetric monitoring during the eruptive phase of the Tagoro volcano. See Table 7.1. All bathymetric scales as in h. White arrows correspond to the apron trend evolution

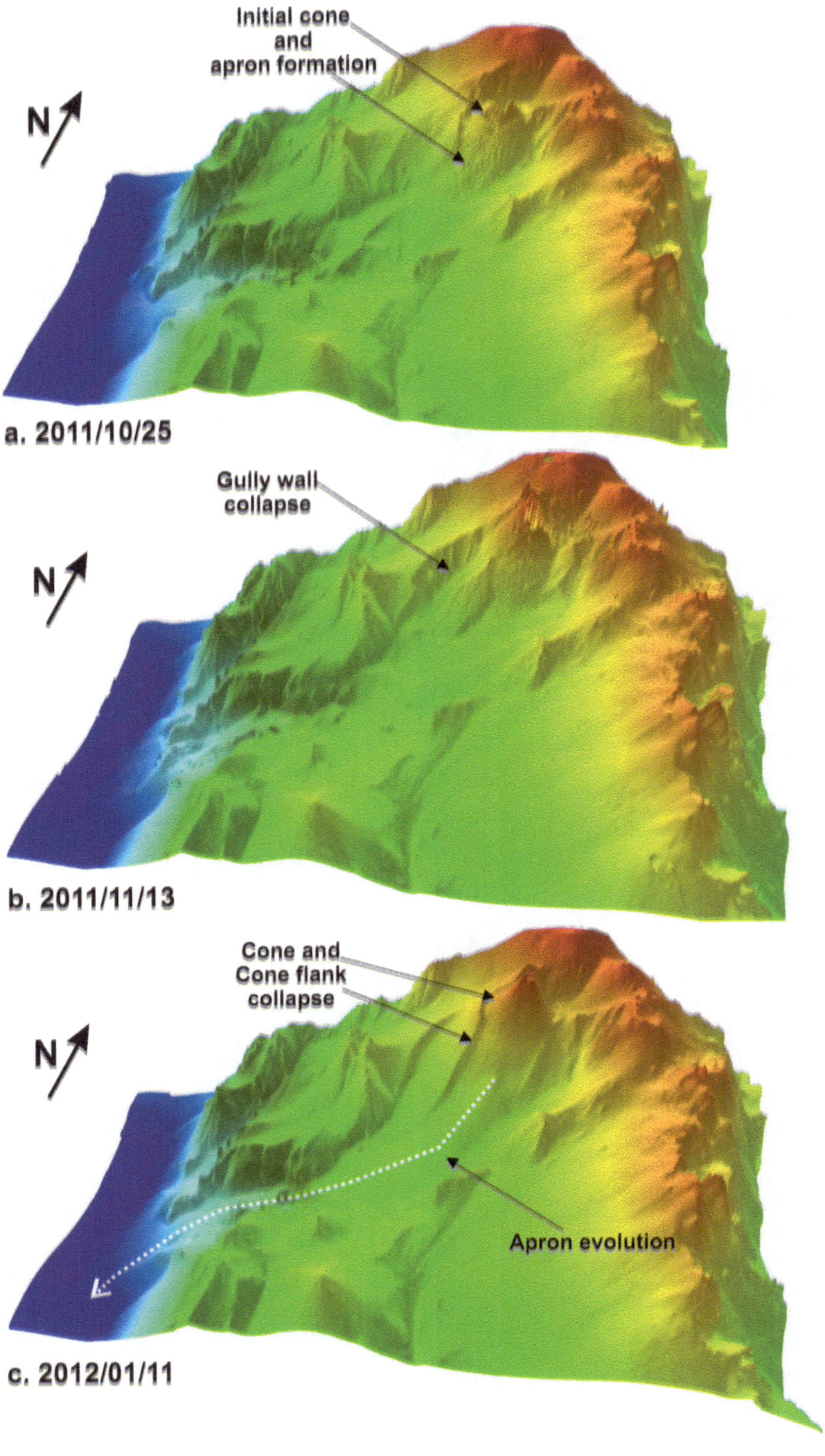

Fig. 7.7 3D bathymetric model of the most representative eruption stages monitored along its evolution during the main first phase. All the model views from the S

of the edifice whose flanks acquired concave profiles, although the width of the base remained relatively stable constrained by the gully and the summit could be kept at a similar height due to the continuous outflow of volcanic material. The southwestern apron continued to form in favor of the gully gradient. In the following days its rapid growth took place, recovering its regular conical geometry in the bathymetric survey on October 31 (Fig. 7.6c). At this time basal dimensions remains similar, the height of the summit reached approximately 170 mbsl, and the apron deposits filled the gully thalweg.

(3) According to the bathymetric mapping of November 13 (Figs. 7.6d and 7.7b), it was observed the formation of a second collapse event that significantly affected the cone, although its summit was conditioned by a high emission rate of volcanic materials in these stages (Rivera et al. 2013), and continued to grow until reaching around 150 mbsl. This episode also affected the western flank of the pre-existing gully; it suffered strong episodes of instability that generated the sliding of large blocks transported downslope towards the axis of the gully. This process allowed the increase in basal space for the growth of the volcanic cone. The base width was extended up to 0.65 km and the gully width was also increase in more than 0.5 km. In this context, the emission center shifted to the north and multiple vents developed that were activated on the southern flank of the volcano.

(4) Subsequently, the volcano again had a stage of vertical growth of the conical building that produced a strong expansion of the basal width up to 0.9 km. The level of the summit did not increase significantly, reaching around 135 mbsl the 2nd of December. Although there was a significant pile-up of material, generally on the flanks that even buried the slided blocks on the western flank of the gully (Fig. 7.6e). During this phase, the emission was produced by means of two main sources that were creating two secondary cones and progressively coalesced to form a main one.

(5) In December, the growing cone again suffered a basal collapse, leaving the level of the summit associated with the emitting center around 175 mbsl. Basal width remains with similar dimensions and around the cone a rim horseshoe depression was formed for depths between 200 and 280 mbsf around the northern flank. Simultaneously, the southern flank of the volcano suffers a slide in favor of the gully slope (Figs. 7.6f and 7.7c) and a ridges complex was formed in the outer boundaries of the slide, this ridges have initially a NE-SW direction. This landslide favored the flow and transport of flank materials downslope to the southwest.

(6) After the third collapse, there was a last progressive phase of growth that would last until the final of the eruption. This last phase was characterized by the development of several attached cones with multiple emission vents forming the main volcanic edifice. Both attached cones and multiple vents are aligned with a NNW-SSE trend related with a volcanic fissure with the same trend. In the bathymetric mapping of February 8, 2012 (Figs. 7.6g and 7.8a) there were three attached cones that reached from NNW to SSE 103, 102 and 206 mbsl respectively and distances between them of 79 and 221 m respectively, and at least nine emission vents. The ENE-WSW ridge between the main peaks and the NNW-SSE ridge were also formed. The northern cone is characterized at least by three active vents with peaks at 103 and 109 mbsl to the east and 122 mbsl to the west. While in the February 24, 2012 bathymetric mapping (Figs. 7.6h and 7.8b), four attached cones had developed, located from north to south successively at 88, 115, 204 and 246 mbsl, with distances between them of 82, 252 and 124 m, and there were at least 12 emission vents. This stage was characterized by the growth both of the highest Tagoro summit from the vent located to the west of the northern cone, and the deepest attached cone located to the SSW of

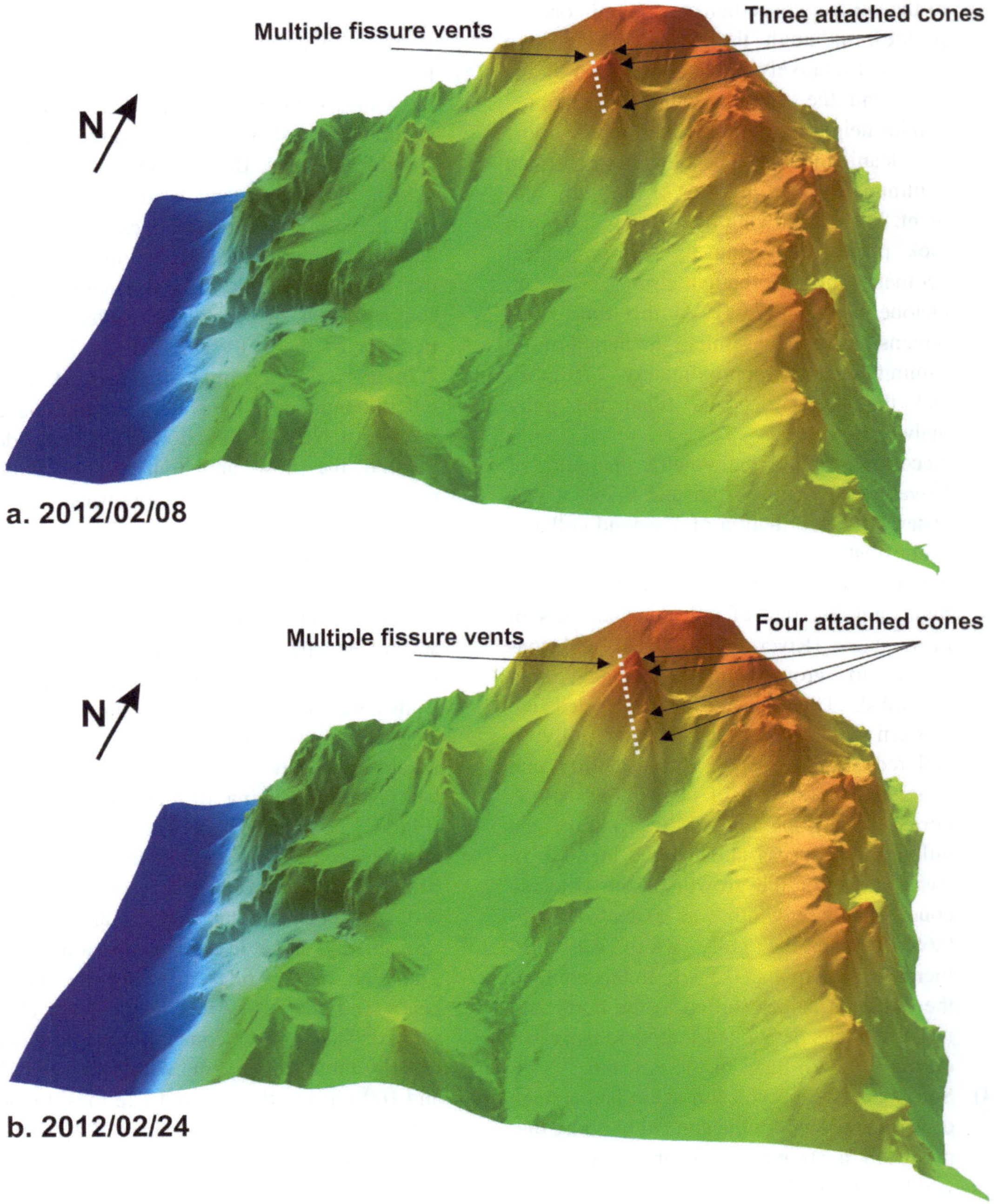

Fig. 7.8 3D bathymetric model of the most representative eruption stages monitored along its evolution during the main second phase. Both model views from the S

the main edifice, in addition the ENE-WSW ridge between the two main cones finished its formation in relation with the sinking of the second cone.

7.4.3 Morphological Evolution During Post-eruptive Phase

The improvement of DEM after the eruption has allowed us to discern minor morphological changes in the main features of the volcano that have took place during the hydrothermal phase.

The main edifice of Tagoro volcano has reduced its height at 89 mbsl at the 2019 bathymetric data, one meter below the highest point (88 mbsl) measured during the eruptive and initial post-eruptive phases (2012 to date). The resolution of the bathymetric data was increased from the Vulcana_0515 cruise (Table 7.1) up to 2 or 1 m allowing the detailed difference analysis between successive DTMs. In this sense, the main summit (northern cone) does not show significant changes since 2015 in depth. Meanwhile the heights of the other three attached cones peaks have deepening. Their peaks reached respectively, 116, 204 and 247 mbsl during 2013, reducing their heights at 118, 205 and 248 mbsl, between 1 and 2 m below the initially registered. The high resolution monitoring since 2015 allow to identified a decay of these peaks approximately from about 30 cm/yr covering the period of the last 5 years. In addition, the ENE-WSW ridge located at the main edifice extends at least 20 m from their northeast end, and has deepening between 0.2 and 1.2 m. The cones flanks gradients have gradually been higher and reaching up to 39° in the proximal areas of the main and attached cones. These cones are at present characterized by at least 15 conical vents presented as small mounded features with heights from 1 to 20 m and slopes up to 35°.

Finally, the most remarkable feature generated within the main edifice along this phase corresponds to a depression-like crater and other minor depressions associated to these secondary cones. The main circular depression was first mapped during Vulcano 0313 cruise on the top of an arc-like shape ridge of NE-SW orientation, 80 m length and 127.5 mbsl in depth, and originally had 15 × 6 m of dimension and 0.3 m depth. The DEMs obtained in 2014 and 2015 reveal the depression has started enlarged its width up to 8 m and its depth in 0.7 m reaching a deepest point of 128.2 mbsl. The latest bathymetric DEM also reveals its deepening and extending to 16 × 22 m in diameter, reaching a deepest point of 129.7 mbsl, with a current depth of 2 m. The subsidence of this crater-like depression is therefore approximately 0.36 cm/yr similar to that observed at the cone peaks.

7.4.4 Current Geomorphology of the Tagoro Volcano Complex

The current geomorphological characterization of the Tagoro volcano is based on the 2019 bathymetric model (Fig. 7.9a) and has allowed differentiating several morphological units (Fig. 7.9b):

(1) Old volcanic cones. In the vicinity of the Tagoro volcano there are several volcanic edifices previously described (Fig. 7.5). These are conical structures with a circular or oval base that are in different degrees of degradation, mainly due to the development of erosive and gravitational processes that produce changes in the initial morphology. Its diameters oscillate between 0.7 and 0.9 km and their summits are located between 125 and 520 mbsl. The relief between the base and the top varies between 95 and 395 m and the slopes of its flanks between 20 and 50°. They usually have a single top, but are occasionally characterized by several minor cones or small ridges on top. The increase of erosion is characterized by development of scarps that affect the flanks of these features.

(2) Volcanic Blocks. This morphologic type corresponds to individual outcrops of

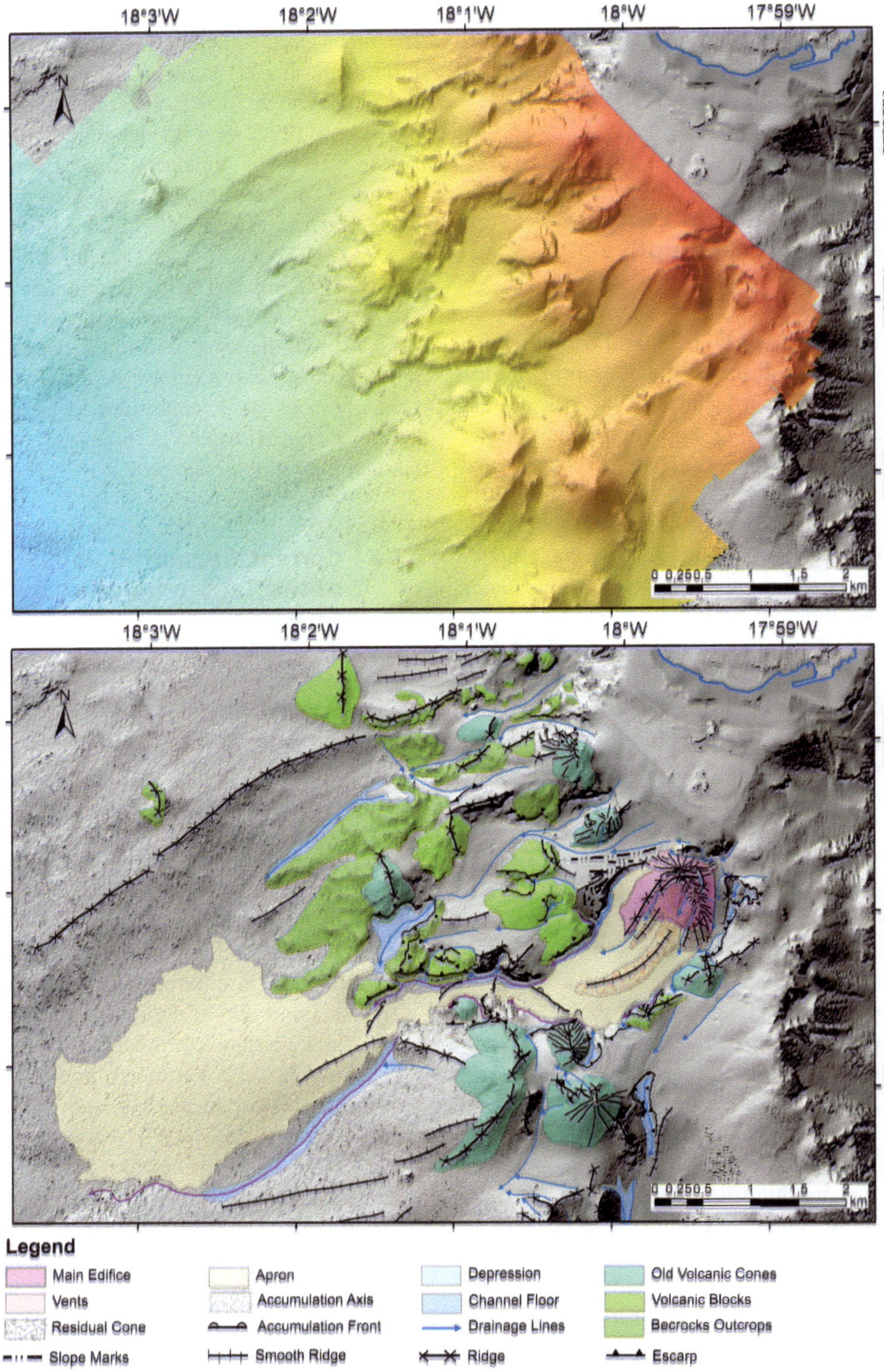

Fig. 7.9 **a** Bathymetric data used to map de current geomorphology of Tagoro volcanic complex acquired by the IEO in 2019. **b** Geomorphological map. Maps are represented in 50% transparency on the shade relief map

volcanic materials with several geometries, from longitudinal ridges to rectangular or rhombic in basal shape internally marked by ridges (Fig. 7.9). Outer borders are relatively irregular, with boundaries characterized both by irregular or longitudinal escarps. In the study area they are more frequent in the upper flank.

(3) Bedrock outcrops. It corresponds to areas where irregular outcrops of volcanic rocks occur. These outcrops occur in areas where there are no large volcanic cones or dominant blocks in the seafloor, although the roughness of the bathymetric response allows us to interpret these landforms as wide fields of rocky outcrops regardless of their origin. In the area near Tagoro (Fig. 7.9), these fields are found in the upper part of the island's flank and are related to the identified blocks.

(4) The main edifice had its top at 88 mbsl after the eruption, with slopes between 20 and 34°, extending approximately to a depth of 400 mbsl where the slope decreases below 20°. The base of this edifice is irregular to quasi-circular with diameters that vary between 0.9 and 1.2 km in length (Figs. 7.9 and 7.10). The irregularities are cause by two sets of morphologic elements:

(a) Four ridges, two of them with a NE-SW direction located on the west flank of the main building, a third with an ENE-WSW trend that crosses the second peak of the main edifice, and the fourth with a NNW-SSE orientation located on the eastern flank of the main edifice.

(b) Four attached cones that compartmentalize the main edifice and decrease in height and size from the NNW to the SSE, the

Fig. 7.10 Detail of the geomorphological map of proximal Tagoro area. Legend as in Fig. 7.9. Geomorphology represented in 50% transparency on the shade relief map made from the 2019 bathymetric data

height of their current peaks reached respectively 89, 118, 205 and 248 mbsl. On these attached cones, the presence of at least 15 vents centers has also been recognized. Usually, these vents have heights between 1 and 5 m on the seafloor, but they reach up to 10 m height in the case of vent related to the main cones.

(5) Volcaniclastic Apron. The apron extends from the SW flank of the main edifice (Fig. 7.9) and it channels the flow of materials from all the flanks of the volcano except the northern one. It has been divided into three parts that have been defined as distal and proximal pyroclastic apron by Somoza et al. (2017), connected by a marked ravine sector of the preexistent gully.

   (a) The proximal apron extends for 2.5 km southwestwards reaching 850 mbsl of depth. It is characterized by slopes gradients that decrease from 20° to 8° to the southwest and by the development of an accumulation front in its final zone of WNW-ESE trend. This sector of the apron is constrained to the interior of the preexistent gully and has a major accumulation body in its central part which produces a smooth longitudinal ridge.
   (b) A central steep ravine in the seafloor connects the two main parts of the Tagoro Apron. It is located between 850 and 1050 mbsl. It has been partially filled and is characterized by the development of two accumulation fronts located successively at 950 and 1030 mbsl, the first of N-S direction and the second of NE-SW trend (Fig. 7.10).
   (c) The distal apron extends in a general SW direction at the mouth of the ravine, along the slope of the island for approximately 3.5 km, forming a fan-like lobe geometry for depths between 1150 and 1800 mbsl, and slopes between 5° and 15°.

   Finally, at the northern area of the Tagoro volcano a minor lobate deposit produced a secondary volcaniclastic apron north-westwards into another second previous gully developed between volcanic reliefs.

(6) Depressions are low areas in the seafloor ground surrounded by higher ground in all directions, as a hole in the seafloor. It has been defined at least two depressions (Fig. 7.10). A NNW-SSE longitudinal depression is located in the eastern flank of the main edifice in relation to the same trend ridge. It extended along 0.26 km; it has a 0.04 km of width and a maximum of 2 m of negative relief, its course is controlled by a ridge in this flank. The second one is a minor circular depression, located in the summit of the second cone in relief. It has a maximum diameter around 0.02 km and a maximum negative relief of 2 m.
(7) Insular flank Drainage System. The area has a southwestwards slope gradient that favors the development of a series of gullies and valleys of different size with a dominant NE-SW trend (Fig. 7.9). Gullies characterize the upper section of the island flank. They have higher slopes (10–20°); their course could be irregular by the interference with volcanic edifices and varies between 2.5 and 4 km. Along the eruption the NW and SE walls of the initial gully were Tagoro took place have increased in width as much as 110 m and increased in depth as much as 0.8 m. Valleys are dominant in the lower section of the flank. They have lowers slopes (5–10°), more linear courses with 5.5 up to 8 km in length along the study area. They are related to fan lobe development as part of the general apron of the island.

7.5 Discussion

7.5.1 Tagoro Eruptive Evolution Geomorphology

The bathymetric monitoring carried out on the Tagoro volcano has made it possible to establish the different phases of its evolution which correspond to two different phases during the eruption and a third phase after it. Along the two active eruption phases the Tagoro volcano has formed throughout four growth stages which have alternated with three destructive stages controlled by the development of collapse episodes (Fig. 7.6). Subsequently, the third phase corresponds to evolution once the eruption ceases and consists of only one stage in which the maintenance of the monitoring has also made it possible to control the minor-scale changes in the morphology of the volcano, generated by processes of lower speed that have affected this volcano. Therefore, the evolution of the Tagoro volcano çan be summarized as follows

Phase 1 from October 10th to December 2011

In this phase the changes occurred rapidly corresponding to the highest rates of volcanic emissions (Rivera et al. 2013), the bathymetric monitoring has allowed us to identify this great variability of different stages.

Stage 1. On October 10th, 2011, the eruption began with the growth of a volcanic cone favored by a ENE-WSW fault into a submarine gully and the development of an initial apron down-gully.
Stage 2. At the end of October there was a basal collapse of the volcanic edifice increased apron materials down-gully.
Stage 3. Vertical growth predominance of the volcanic edifice.
Stage 4. Second collapse before November 13th that affected both the main edifice and the northwestern wall of the gully which generate significant slided blocks to the gully thalweg.
Stage 5. Vertical growth with development at least of two simultaneous emission vents that evolved to a main cone as well as the continuous growth of the apron, reaching to bury the blocks slided in the previous stage.
Stage 6. Third basal collapse and development of a major slide affecting the southern flank of the cone.

Phase 2, from December 2011 to March 2012

In this phase the growth of the volcano was more homogeneous, characterized only by a stage with the development of a progressive fissure edifice (Vázquez et al. 2016).

Stage 7. Vertical growth with several active vents that progressively generate up to four attached cones. Both vents and attached cones aligned in a NNW-SSE trend to responding to a fissure evolution. The last attached cone growth after February 8th 2012 in the last episodes of the eruption.

Phase 3, from March 2012 to Present

This phase corresponds to the Tagoro evolution after the magmatic eruption, it constituted by the eighth and longest stage that corresponds to the post-eruptive phase of degassing and hydrothermalism.

Stage 8. The end of the eruption took place in early March 2012, after that the degassing and hydrothermalism processes dominated the building which has been characterized by development of minor vents and the progressive crater-like depression formation. Minor subsidence effect has also been detected.

This evolution of the volcanic eruption coincides with the proposal of Martí et al. (2013a, b) who, based on geophysical and petrological data, have divided the eruption into two main phases. The first phase took place from October 10th to mid December 2011 and the second phase from this date to March 2012. This evolution was confirmed by the interpretation of Somoza et al. (2017) who observed a change in the morphological style of the volcanic construction from the cone collapse event that took place in mid December 2011.

The style of the eruption was predominantly explosive, with numerous emissions of pyroclasts ranging in sizes from ash to bombs (Somoza et al. 2017), although it maintained some mixed characteristics between shallow and intermediate marine eruptions such as correspond to the extensive emissions of lava balloons, that reached the surface, or pillow-lavas and lava ponds (González et al. 2020) observed in relation to the latest and deepest attached cone built up in the second phase. In any case, during the eruption the simultaneous activity of several emission vents located at different depths was frequent and they could have emitted different types of volcanic products, mainly pyroclastic but also lava emissions.

7.5.2 Tagoro Geomorphology and Processes

The previous morphological characteristics and physiography existing in the eruption zone was a strong constraint along the evolution of the volcano growth. These characteristics are marked by several volcanic features NE-SW alienated that have been related to igneous dykes radiating from the island (Gee et al. 2001b) and they had different stages of conservation. Two physiographic features controlled the initial morphology of the volcano. First, the emission center was initially installed near the head of a submarine gully in relation to a ENE-WSW fault which location would indicate that the fracture was close to the boundary between the axis of the valley and its northern wall. The diameter of the volcano base was strongly constrained by the characteristics of this gully. The first stage of basal collapse of the cone, at the end of October 2011, did not produce the widening of its base; instead it forced the development of a flow of materials towards the southwest along the thalweg of the gully. In November 2011, the evolution of the eruption induced the instability of the gully which head and northwestern wall were eroded and enlarged; these processes mainly produced a strong setback of its NW wall and allowing from this moment the increase of the cone base and its vertical growth.

Second, the volcano from a physiographic point of view was emplaced on the island's submarine flank that is characterized by a steep morphological context (ranging between 10 and 40°), it can be described as a hillside volcano. This context conditioned the asymmetric geometry of the volcanic edifice as well as its building evolution in favor of the pre-existing depositional gradients in the area. This aspect introduces an instability factor and favored the development of collapse events, affecting both the main building base and flanks that facilitated the development of apron deposits.

Two new large morphological units have been generated by this eruption. They correspond basically to the main edifice and to the debris-volcaniclastic apron that extends to the southwest of the volcano. Tagoro volcano began to grow approximately at 350 mbsl, although it quickly grew and the top remained between 200 and 88 mbsl. The location of emission centers varieds during eruption generating different volcanic cones along time. This produced that the summit of the edifice had different locations throughout the volcano evolution. The Tagoro volcano morphology and nearby areas influenced by the eruption were changing rapidly throughout its development as a function of the rates of materials emitted at each moment (Rivera et al. 2013), but also of the instability processes that affected both the growing volcano itself and the surrounding areas. These changes occurred more quickly during the first phase of the eruption, where the eruption had a high intensity and the growing volcanic edifice had in consequence a greater instability.

A flow of volcanic material was generated from the initial stages to constitute the volcaniclastic apron. This flow was generated both by effusive emissions, volcaniclastic deposits and mass movement of volcano and gully wall collapses in favor of the gully gradient and drove along the gully thalweg. The flow of materials extended from the SW flank of the main edifice and it was mainly channeled down-gully from all volcano flanks except the northern one. This flow was increased during the collapse and landslide events. The apron is divided in three sectors, the proximal sector is characterized by a smooth longitudinal ridge that could be directly related to the collapses of the southwest volcano flank and transversal accumulation fronts, the central sector functioned as a passage zone of the deposits proceed from the eruption, but has two transversal accumulation fronts, the distal sector has a fan lobe shape that is frequent in adjacent sectors of the island flank, especially to the west. This morphologic type corresponds to previous volcaniclastic accumulations related with volcanic eruptions, erosion and mass movement processes proceed from upper flank sectors, including the emerged island.

Current geomorphology is controlled by the evolution of the eruption itself, although most of the morphological features observed today were mainly formed during the second phase of the eruption that largely obliterated the previous features, such as the rim/horseshoe depression and ridge system related to the collapse and slide events. Some of the morphological features of the first phase which current remain, partially buried, are:

- The scar that affected the NW gully wall during the gully wall collapse of November 2011. Now it is represented by the most western ridge with a NE-SW direction subparallel to the previous gully walls direction;
- The second NE-SW ridge is likewise the residue of the collapse structure occurred in December 2011 that affected both the main volcanic edifice and its southwestern flank.

The direction of these ridges is similar both to the pre-existing gully axis and its western wall trends, therefore the development of these slides could be controlled by the previous morphostructure.

The current attached cones correspond to the progressive growth of the main edifice generated after the December collapse during the second phase of the eruption characterized by a fissure evolution. In this context the ENE-WSW ridge corresponds to the limit between the two main attached cones which constituted the main edifice, probably also related to a minor collapse of the second attached cone southeastwards whose peak went from 102 to 115 mbsl throughout February 2012. In addition, the NNW-SSE ridge in the eastern volcano flank could also formed in the second phase in relation to the instability of the edifice to the SSE. Finally, the rain of small sized pyroclastic materials occurred in a generalized way in the area, they were deposited on most of the morphological features described producing their partial buried and smoothing their characteristics (Vázquez et al. 2016).

A subsidence process has been observed in the Tagoro volcano along the post-eruptive phase that affect the general edifice and minor secondary morphological features. This subsidence could be caused by several processes, highlighting the consolidation of the fine-grained (pyroclastic material) accumulations and deposits into more coherent hard bedrocks, as evidenced in other zones of the Canary Islands (Hunt et al. 2011) and related to the post-emplacement thermal cooling of the surface in contact with the seawater, generating the contraction of the lava flows. These phenomena are observed at many active volcanoes, e.g., at Etna volcano in Italy (Stevens et al. 2001), Kilauea volcano in Hawaii (Dietterich et al. 2012) or those of the Iceland lava fields (Wittmann et al. 2017). Furthermore, the diameter and deep increase of the hydrothermal crater-like depression could be directly related to degasification processes of Tagoro started when it ceased to be active. This is supported by the water samples obtained into the depression that show the maximum physic-chemical anomalies in the area, comprising a release of heat and gases (mainly CO_2), by this depression and the minor secondary craters (Santana-Casiano et al. 2016), and the presence of hydrothermal chimneys for expelling fluids observed in situ since the first oceanographic cruises in 2017 (Sotomayor-García et al., this volume).

In the analysis of morphometric parameters, such as slope and height/width ratio, no significant differences were observed between the Tagoro volcano and the rest of the edifices identified at the El Hierro island submarine flanks (Fernández-Salas et al. 2014). Tagoro main edifice shows intermediate values of all parameters, indicating that this new cone does not have a special constructional morphology to other edifices generated at previous eruptive events. Although Tagoro reaches the highest values of slope, which may indicate the already mentioned physiographic control of it base that have not allowed to reach higher values of the height-width ratio, however it has achieved a steep slope that will decrease as the erosive processes act on the new volcanic edifice in the next years.

7.5.3 Bathymetric Monitoring

Bathymetric monitoring is essential to know the character of the changes in the relief of the seabed related to submarine eruptions. These changes can be positive and negative in height and, can be generated both by erosion and/or collapse processes, as well as by accumulation processes and/or inflation of the seafloor. The first systematic geological studies on active submarine volcanoes acquired bathymetric data (Moore and Peck 1965). The development of multibeam bathymetric echosounders from the 90s of the twentieth century has allowed obtaining detailed bathymetric maps with 100% coverage of submarine active volcanoes (e.g. Chadwick et al. 2001; Baker et al. 2002). The strategy of bathymetric monitoring is to acquire repeated seafloor bathymetric DEMs so that the changes in morphology can be verified. The morphological analysis is realized by means of GIS tools (e.g. Clague et al. 2011; Wormald et al. 2012), especially difference maps are used to define the morphological changes between two bathymetric DEMs acquired at different times and calculate affected areas and volumes involved as well as the processes dominant on the seafloor (e.g. Walker et al. 2008; Wright et al. 2008; Caress et al. 2012; Embley et al. 2014). Finally, the analysis of different morphological parameters, or rates, allows a better characterization of the volcanic forms and the eruptive processes (e.g. Chadwick et al. 2001; Grosse et al. 2012; Sánchez-Guillamón et al. 2018).

The Tagoro eruption was characterized first by different vents that were located in intermediate to shallow depths, and second by their proximity close to the coastline. These two peculiarities allowed the IEO to acquire a series of bathymetric DEMs with a high temporal frequency to monitoring the eruption, especially in its first phase. Therefore, the changes of the Tagoro morphology throughout its growth could be well defined in these bathymetric DEMs, with episodes of vertical growth, fissure growth alternating with episodes of denudation with basal or flanks collapses, or even collapses of the gully walls adjacent to the eruption area. This bathymetric monitoring has allowed a very well defined record of the processes throughout an individual eruptive episode. Unlike what normally happens in other regions of the world where it is usual to perform the bathymetric acquisition every some time periods or after a specific eruption ends to determine the changes associated with each specific eruption, that frequently also alternated episodes of growth and denudation, providing a good record of the changes generated by the different eruptions throughout the life of the active volcano (Chadwick et al. 2008, 2013, 2019; Wright et al 2008; Caress et al. 2012; Embley et al. 2014; Schnur et al. 2017; Allen et al. 2018).

The Tagoro volcano corresponds to a monogenetic eruption that has built during a singled volcanic episode, as usually happens in different eruptions that have taken place recently in the Canary Islands, such as the multiples volcanic edifices identified at El Hierro (Becerril et al. 2015), Tenerife (Dóniz et al. 2008) or Gran Canaria islands (Rodriguez-Gonzalez et al. 2012) and therefore a new eruption of this singled volcano is not expected. However, as can be seen in Figs. 7.3 and 7.4, there are two areas on the submarine flanks of the El Hierro island, the western promontory and the Southern Rift where numerous volcanic cones with similar morphological characteristics to Tagoro volcano can be located. Therefore, the possibility of a new eruption with similar characteristics, either on land or offshore in these areas, would not be ruled out.

Unlike the Tagoro volcano, the submarine volcanic areas usually studied correspond to long-lived active volcanoes that have a long-term activity alternating active episodes with quiescent periods. In addition, most of them correspond to deep volcanoes dominated by lava flows emissions and seafloor inflation/deflation deformation (e.g. Chadwick et al. 2013; Nooner and Chadwick 2016; Clague et al. 2019; Le Saout et al. 2020) or conversely coastal eruptions of emerged volcanoes dominated by flank slope processes and lava delta deposits (Casalbore et al. 2010; Di Traglia et al. 2018). Morphological features observed in the Tagoro volcano, mainly asymmetrical cone growth, cone collapses and flank slides, are comparable to those identified on volcanoes with summits located at intermediate to shallow depths such as the W Rota (Schnur et al. 2017) or Kick-'Em-Jenny (Allen et al. 2018) volcanoes, they are also characterized by higher volumes of pyroclastics materials. A close similarity exists with the Kick-'Em-Jenny volcano located on a steep physiographic domain and it is morphologically characterized by a cone collapse that have resulted in the formation of a horseshoe-shaped rhim depression and a flank collapse bounded by lateral ridges (Allen et al. 2018), so similar as the features described in the Tagoro volcano that were developed in December 2011 and observed in the bathymetric DEM obtained on January 11, 2012. In any case, the development of growth phases alternating with denudation phases, with collapses or flank landslides, is a fairly generalized process characteristic of active volcanoes, although gravitational landslides could also occur in the quiescent phases.

7.6 Recommendations for Further Geological Underwater Monitoring

Given the monogenetic nature of the last volcanic eruptions in the Canary Islands, it is not easy to carry out a geological monitoring strategy for submarine volcanism in these regions. A first conclusion is evident, it is necessary to have a good control of the bathymetric and morphological characteristics of the seabed of the Canary Islands environment. The areas that would have the highest priority are those in which, given their volcanic history, they may be more susceptible to a new eruption, especially around the islands of La Palma and El Hierro. Likewise, perhaps to a lesser extent but not less interesting, the surroundings of the islands of Tenerife and Lanzarote would also have to be studied. A good bottom bathymetry, with the best possible resolution, will make it possible to establish a geomorphological catalog of reliefs of volcanic origin and other geological characteristics of the seabed. This catalog of resulting morphological types can be analyzed and compared with volcanic reliefs from different areas, including those of emerged areas. In addition, this bathymetry will also allow extending the use of the analysis of morphometric parameters and their application as potential indicators of the action of constructive or destructive processes on submarine volcanoes, it will also be fundamental for the analysis of differences that will be carried out with subsequent bathymetric models when it will be necessary. This catalog can be complemented with studies of the alternating tephra layers in marine sediments, so that obtaining sediment cores would allow establishing the recent history of eruptions in a region (e.g. Portner et al. 2015).

The sequential acquisition of seafloor bathymetric data over time would also allow the location of potential volcanic areas to be identified from possible seafloor inflations that would be obtained from the analysis of the differences of different bathymetric models. In the same way, systematic sweeps with the water column module of multibeam bathymetric echosounders could allow the definition of possible active sources, at least hydrothermal vents that could also be considered as potential precursors. A new quantum leap in the acquisition of multibeam bathymetric DEMs consists in the possibility of bringing echosounders closer to the seabed using autonomous underwater vehicles (AUVs) and thus increasing resolution, especially in the case of deep-sea volcanoes. AUVs can be programmed to follow a predetermined itinerary independently of the mother ship. This technique has been applied for example in the Axial volcano where bathymetric DEMs of 1 m resolution were obtained at more than 1400 m depth (Caress et al. 2012). The use of these systems would be applied mainly to areas already classified as potentially active, which implies extensive previous bathymetric and geomorphological analysis work.

When eruptions do not affect the sea surface, it is also interesting to use the water column module of multibeam echosounders, it can allow to locate with greater accuracy the vents that are actives in a submarine volcano during the eruption (e.g. Baker et al. 2008; Chadwick et al. 2014), although its use is more frequent for studies of hydrothermal emissions that imply a lower risk to navigation and a lower load of emitted volcanic particles into the water masses.

Finally, the geological monitoring of submarine volcanoes should not be limited exclusively to the acquisition of bathymetric data, it would be important to go a step further with the installation of underwater monitoring stations in areas defined as potentially volcanic, connected to base stations onland by coaxial cables (Kelley et al. 2014), so that an online monitoring of parameters such as microseismicity, variation of the slope of the seafloor (tilting) or even the variation of the magnetic field related to changes in temperature in the subsoil, as well as the content could be carried out from water bodies into gases such as methane, carbon dioxide or hydrogen sulfide.

Acknowledgements Bathymetric monitoring of the Tagoro volcano eruption and subsequent evolution has been possible thanks to funding from the Spanish Institute of Oceanography (IEO) through the BIMBACHE (IEO-2011-2012), RAPROCAN-III (IEO-2010-2012), VULCANA-I (IEO-2015-2017) and VULCANA-II (IEO-2018-2020) projects, as well as through the Ministerio de Economía y Competitividad del Gobierno de España (MINECO) and FEDER through VULCANO-I (CTM2012-36317) and VULCANO-II (CTM2014-51837-R) projects. We would like to especially thank the crews of the Ramón Margalef and Ángeles Alvariño Research Vessels, as well as the participants in the different oceanographic cruises that have been carried out on this volcano.

References

Acosta J, Uchupi E, Smith D, Muñoz A, Herranz P, Palomo C, Llanes P, Ballesteros M, ZEE Working Group (2003a) Comparison of volcanic rifts on La Palma and El Hierro, Canary Islands and the Island of Hawaii. Mar Geophys Res 24:59–90. http://doi.org/10.1007/s11001-004-1162-6

Acosta Yepes J, Uchupi E, Muñoz A, Herranz P, Palomo C, Ballesteros M (2003) Geologic evolution of the Canarian Islands of Lanzarote, Fuerteventura, Gran Canaria and La Gomera and comparison of landslides at these islands with those at Tenerife, La Palma and El Hierro. Mar Geophys Res 24(1–2):1–40. https://doi.org/10.1007/s11001-004-1513-3

Allen RW, Berry C, Henstock TJ, Collier JS, Dondin FJY, Rietbrock A et al (2018) 30 years in the life of an active submarine volcano: a time-lapse bathymetry study of the Kick-'em-Jenny volcano, Lesser Antilles. Geochem Geophys Geosyst 19:715–731. https://doi.org/10.1002/2017GC007270

Anguita F, Hernán F (2000) The Canary Islands origin: a unifying model. J Volcanol Geotherm Res 103(1–4):1–26. https://doi.org/10.1016/S0377-0273(00)00195-5

Bachèlery P, Villeneuve N (2013) Hot spots and large igneous provinces. In: Shroder J (Editor in Chief), Owen LA (ed) Treatise on geomorphology, vol 5. Tectonic geomorphology. Academic Press, San Diego, CA, pp 193–233

Baker ET, Massoth GJ, de Ronde CEJ, Lupton JE, McInnes BIA (2002) Observations and sampling of an ongoing subsurface eruption of Kavachi volcano, Solomon Islands, May 2000. Geology 30(11):975–978

Baker ET, Embley RW, Walker SL, Resing JA, Lupton JE, Nakamura KI, de Ronde CEJ, Massoth GJ (2008) Hydrothermal activity and volcano distribution along the Mariana arc. J Geophys Res 113:B08S09. http://doi.org/10.1029/2007JB005423

Becerril L, Galindo I, Martí J, Gudmundsson A (2015) Three-armed rifts or masked radial pattern of eruptive fissures? The intriguing case of El Hierro volcano (Canary Islands). Tectonophysics 647–648:33–47. https://doi.org/10.1016/j.tecto.2015.02.006

Caress DW, Clague DA, Paduan JB, Martin JF, Dreyer BM, Chadwick WW Jr, Denny A, Kelley DS (2012) Repeat bathymetric surveys at 1-metre resolution of lava flows erupted at Axial Seamount in April 2011. Nat Geosci 5(7):483–488. https://doi.org/10.1038/ngeo1496

Carracedo JC (1999) Growth, structure, instability and collapse of Canarian volcanoes and comparisons with Hawaiian volcanoes. J Volcanol Geotherm Res 94:1–19

Carracedo JC, Day S, Guillou H, Rodríguez Badiola E, Canas JA, Pérez Torrado FJ (1998) Hotspot volcanism close to a passive continental margin: the Canary Islands. Geol Mag 135(5):591–604. https://doi.org/10.1017/S0016756898001447

Carracedo JC, Day SJ, Guillou H (1999) Quaternary collapse structures and the evolution of the western Canaries (Las Palmas and Hierro). J Volcanol Geotherm Res 94:169–190

Carracedo JC, Badiola ER, Guillou H, de la Nuez J, Perez Torrado FJ (2001) Geology and volcanology of La Palma and El Hierro, Western Canaries. Estud Geol 57:175–273

Carracedo JC, Torrado FP, González AR, Soler V, Turiel JLF, Troll VR, Wiesmaier S (2012) The 2011 submarine volcanic eruption in El Hierro (Canary Islands). Geol Today 28:53–58. http://doi.org/10.1111/j.1365-2451.2012.00827.x

Carracedo JC, Troll VR, Zaczek K, Rodrıguez-Gonzalez A, Soler V, Deegan FM (2015) The 2011–2012 submarine eruption off El Hierro, Canary Islands: new lessons in oceanic island growth and volcanic crisis management. Earth Sci Rev 150:168–200

Casalbore D, Romagnoli C, Chiocci F, Frezza V (2010) Morpho-sedimentary characteristics of the volcaniclastic apron around Stromboli volcano (Italy). Mar Geol 269(3–4):132–148. http://doi.org/10.1016/j.margeo.2010.01.004

Chadwick WW Jr, Scheirer DS, Embley RW, Johnson HP (2001) High-resolution bathymetric surveys using scanning sonars: lava flow morphology, hydrothermal vents, and geologic structure at recent eruption sites on the Juan de Fuca Ridge. J Geophys Res 106 (B8):16075–16099. https://doi.org/10.1029/2001jb000297

Chadwick WW Jr, Wright IC, Schwarz-Schampera U, Hyvernaud O, Reymond D, de Ronde CEJ (2008) Cyclic eruptions and sector collapses at Monowai submarine volcano, Kermadec arc: 1998–2007. Geochem Geophys Geosyst 9:Q10014. https://doi.org/10.1029/2008GC002113

Chadwick WW Jr, Dziak RP, Haxel JH, Embley RW, Matsumoto H (2012) Submarine landslide triggered by volcanic eruption recorded by in-situ hydrophone. Geology 40:51–54. https://doi.org/10.1130/G32495.1

Chadwick WW, Clague DA, Embley RW, Perfit MR, Butterfield DA, Caress DW et al (2013) The 1998 eruption of Axial Seamount: new insights on submarine lava flow emplacement from high-resolution mapping. Geochem Geophys Geosyst 14:3939–3968. https://doi.org/10.1002/ggge.20202

Chadwick WW Jr, Merle SG, Buck NJ, Lavelle JW, Resing JA, Ferrini V (2014) Imaging of CO_2 bubble plumes above an erupting submarine volcano, NW Rota-1, Mariana Arc. Geochem Geophys Geosyst 15:4325–4342. https://doi.org/10.1002/2014GC005543

Chadwick WW Jr, Paduan JB, Clague DA, Dreyer BM, Merle SG, Bobbitt AM et al (2016) Voluminous eruption from a zoned magma body after an increase in supply rate at Axial Seamount. Geophys Res Lett 43:12063–12070. https://doi.org/10.1002/2016GL071327

Chadwick WW Jr, Merle SG, Baker ET, Walker SL, Resing JA, Butterfield DA, Anderson MO, Baumberger T, Bobbitt AM (2018) A recent volcanic eruption discovered on the central Mariana back-arc spreading center. Front Earth Sci 6(art no 172). http://doi.org/10.3389/feart.2018.00172

Chadwick WW Jr, Rubin KH, Merle SG, Bobbitt AM, Kwasnitschka T, Embley RW (2019) Recent eruptions between 2012 and 2018 discovered at West Mata Submarine Volcano (NE Lau Basin, SW Pacific) and characterized by new ship, AUV, and ROV data. Front Mar Sci 6:495. https://doi.org/10.3389/fmars.2019.00495

Clague DA, Paduan JB, Caress DW, Thomas H, Chadwick WW Jr, Merle SG (2011) Volcanic morphology of West Mata Volcano, NE Lau Basin, based on high-resolution bathymetry and depth changes. Geochem Geophys Geosyst 12:QOAF03. http://doi.org/10.1029/2011GC003791

Clague DA, Dreyer BM, Paduan JB, Martin JF, Chadwick WW, Caress DW et al (2013) Geologic history of the summit of Axial Seamount, Juan de Fuca Ridge. Geochem Geophys Geosyst 14:4403–4443. https://doi.org/10.1002/ggge.20240

Clague DA, Paduan JB, Caress DW, Chadwick WW Jr, Le Saout M, Dreyer BM, Portner RA (2017) High-resolution AUV mapping and targeted ROV observations of three historical lava flows at Axial Seamount. Oceanography 30(4). http://doi.org/10.5670/oceano.2017.426

Clague DA, Paduan JB, Caress DW, Moyer CL, Glazer BT, Yoerger DR (2019) Structure of Lō'ihi seamount, Hawai'i and lava flow morphology from high-resolution mapping. Front Earth Sci 7(art no 58):1–17. http://doi.org/10.3389/feart.2019.00058

Conte AM, Martorelli E, Calarco M, Sposato A, Perinelli C, Coltelli M, Chiocci FL (2014) The 1891 submarine eruption offshore Pantelleria Island (Sicily Channel, Italy): identification of the vent and characterization of products and eruptive style. Geochem Geophys Geosyst 15:2555–2574. http://doi.org/10.1002/2014GC005238

Di Traglia F, Nolesini T, Solari L, Ciampalini A, Frodella W, Steri D et al (2018) Lava delta deformation as a proxy for submarine slope instability. Earth Plan Sci Lett 488:46–58. http://doi.org/10.1016/j.epsl.2018.01.038

Dietterich HR, Poland MP, Schmidt DA, Cashman KV, Sherrod DR, Espinosa AT (2012) Tracking lava flow emplacement on the east rift zone of Kilauea, Hawai'i, with synthetic aperture radar coherence. Geochem Geophys Geosyst 13:Q05001. https://doi.org/10.1029/2011GC004016

Dóniz J, Romero C, Coello E, Guillén C, Sánchez N, García-Cacho L, García A (2008) Morphological and statistical characterisation of recent mafic volcanism on Tenerife (Canary Islands, Spain). J Volcanol Geotherm Res 173(3–4):185–195. http://doi.org/10.1016/j.jvolgeores.2007.12.046

Embley RW, Merle SG, Baker ET, Rubin KH, Lupton JE, Resing JA, Dziak RP, Lilley MD, Chadwick WW Jr, Shank T, Greene R, Walker SL, Haxel J, Olson E, Baumberger T (2014) Eruptive modes and hiatus of volcanism at West Mata seamount, NE Lau basin: 1996–2012. Geochem Geophys Geosyst 15:4093–4115. https://doi.org/10.1002/2014GC0

Ercilla G, Casas D, Alonso B et al (2021) Marine geological hazards: charting the course of progress and future directions. Oceans (In press)

Fernández-Salas LM, Vázquez JT, Díaz del Río V, López-González N, Rueda JL, Fraile-Nuez E (2014)

Preliminar morphometric characterization of the submarine volcanic cones in the southern promontory of El Hierro Island. In: IV international symposium of marine sciences. Book of abstracts, pp 217–218

Fraile-Nuez E, González-Dávila M, Santana-Casiano JM, Arístegui J, Alonso-González IJ, Hernández-León S, Blanco MJ, Rodríguez-Santana A, Hernández-Guerra A, Gelado-Caballero MD, Eugenio F, Marcello J, De Armas D, Domínguez-Yanes JF, Montero MF, Laetsch DR, Vélez-Belchí P, Ramos A, Ariza AV, Comas-Rodríguez I, Benítez-Barrios VM (2012) The submarine volcano eruption at the island of El Hierro: physical-chemical perturbation and biological response. Sci Rep 2(art no 486). http://doi.org/10.1038/srep00486

Fraile-Nuez E, Santana-Casiano JM, González-Dávila M, Vázquez JT, Fernández-Salas LM, Sánchez-Guillamón O, Palomino D, Presas-Navarro C (2018) Cyclic behavior associated with the degassing process at the shallow submarine Volcano Tagoro, Canary Islands, Spain. Geosciences (Switzerland) 8(12)(art no 457)

Gaspar JL, Queiroz G, Pacheco JM, Ferreira T, Wallenstein N, Almeida MH, Coutinho R (2003) Basaltic lava balloons produced during the 1998–2001 Serreta submarine ridge eruption (Azores). Geophys Monogr Ser 140:205–212. https://doi.org/10.1029/140GM13

Gee MJR, Watts AB, Masson DG, Mitchell NC (2001a) Landslides and the evolution of El Hierro in the Canary Islands. Mar Geol 177:271–293

Gee MJR, Masson DG, Watts AB, Mitchell NC (2001b) Offshore continuation of volcanic rift zones, El Hierro, Canary Islands. J Volcanol Geotherm Res 105:107–119

González FJ, Rincón-Tomás B, Somoza L, Santofimia E, Medialdea T, Madureira P, López-Pamo E, Hein JR, Marino E, de Ignacio C, Reyes J, Hoppert M, Reitner J (2020) Low-temperature, shallow-water hydrothermal vent mineralization following the recent submarine eruption of Tagoro volcano (El Hierro, Canary Islands). Mar Geol 430(art no 106333). http://doi.org/10.1016/j.margeo.2020.106333

González-Vega A, Fraile-Nuez E, Santana-Casiano JM, González-Dávila M, Escánez-Pérez J, Gómez-Ballesteros M, Tello O, Arrieta JM (2020) Significant release of dissolved inorganic nutrients from the Shallow Submarine Volcano Tagoro (Canary Islands) based on seven-year monitoring. Front Mar Sci 6(art no 829)

Grosse P, van Wyk de Vries B, Euillades PA, Kervyn M, Petrinovic IA (2012) Systematic morphometric characterization of volcanic edifices using digital elevation models. Geomorphology 136:114–131

Guillou H, Carracedo JC, Torrado FP, Badiola ER (1996) K-Ar ages and magnetic stratigraphy of a hotspot-induced, fast grown oceanic island: El Hierro, Canary Islands. J Volcanol Geotherm Res 73:141–155. https://doi.org/10.1016/0377-0273(96)00021-2

Hoernle KAJ (1998) Geochemistry of Jurassic oceanic crust beneath Gran Canaria (Canary Islands): implications for crustal recycling and assimilation. J Petrol 39:859–880

Hoernle K, Schmincke HU (1993) The role of partial melting in the 15-MA geochemical evolution of Gran Canaria: a blob model for the Canary hotspot. J Petrol 34(3):599–626. https://doi.org/10.1093/petrology/34.3.599

Holik JS, Rabinowitz PD, Austin JA Jr (1991) Effects of Canary hotspot volcanism on structure of oceanic crust off Morocco. J Geophys Res 96(B7):12039–12067. https://doi.org/10.1029/91jb00709

Huff WD, Owen LA (2013) Volcanic landforms and hazards. In: Shroder J (Editor in Chief), Owen LA (ed) Treatise on geomorphology, vol 5. Tectonic geomorphology. Academic Press, San Diego, CA, pp 148–192

Hunt JE et al (2011) Sedimentological and geochemical evidence for multistage failure of volcanic island landslides: a case study from Icod landslide on north Tenerife, Canary Islands. Geochem Geophys Geosyst 12:12

Kelley DS, Delaney JR, Juniper SK (2014) Establishing a new era of submarine volcanic observatories: cabling Axial Seamount and the endeavour segment of the Juan de Fuca Ridge. Mar Geol 352:426–450. https://doi.org/10.1016/j.margeo.2014.03.010

Kelly JT, Carey S, Pistolesi M, Rosi M, Croff-Bell KL, Roman C, Marani M (2014) Exploration of the 1891 Foerstner submarine vent site (Pantelleria, Italy): insights into the formation of basaltic balloons. Bull Volcanol 76(7)(art no 844):1–18. http://doi.org/10.1007/s00445-014-0844-4

Kokelaar P (1986) Magma-water interactions in subaqueous and emergent basaltic. Bull Volcanol 48:275–289. https://doi.org/10.1007/bf01081756

Kueppers U, Nichols ARL, Zanon V, Potuzak M, Pacheco JMR (2012) Lava balloons—peculiar products of basaltic submarine eruptions. Bull Volcanol 74:1379–1393. https://doi.org/10.1007/s00445-012-0597-x

Le Saout M, Bohnenstiehl DR, Paduan JB, Clague DA (2020) Quantification of eruption dynamics on the north rift at Axial Seamount, Juan de Fuca Ridge. Geochem Geophys Geosyst 21:e2020GC009136. https://doi.org/10.1029/2020GC009136

Longpré MA, Chadwick JP, Wijbrans J, Iping R (2011) Age of the El Golfo debris avalanche, El Hierro (Canary Islands): new constraints from laser and furnace 40Ar/39Ar dating. J Volcanol Geotherm Res 203:76–80

Longpré MA, Klügel A, Diehl A, Stix J (2014) Mixing in mantle magma reservoirs prior to and during the 2011–2012 eruption at El Hierro, Canary Islands. Geology 42:315–318

López C, Blanco MJ, Abella R, Brenes B, Cabrera Rodríguez VM, Casas B, Domínguez Cerdeña I, Felpeto A, de Villalta MF, del Fresno C, García O, García-Arias MJ, García-Cañada L, Gomis Moreno A, González-Alonso E, Guzmán Pérez J, Iribarren I, López-Díaz R, Luengo-Oroz N, Meletlidis S,

Moreno M, Moure D, de Pablo JP, Rodero C, Romero E, Sainz-Maza S, Sentre Domingo MA, Torres PA, Trigo P, Villasante-Marcos V (2012) Monitoring the volcanic unrest of El Hierro (Canary Islands) before the onset of the 2011–2012 submarine eruption. Geophys Res Lett 39. http://doi.org/10.1029/2012GL051846

Madureira P, Conceição P, Silva P, Relvas J, Barriga FJAS (2017) The 1998–2001 submarine lava balloon eruption at the Serreta ridge (Azores archipelago): constraints from volcanic facies architecture, isotope geochemistry and magnetic data. J Volcanol Geoth Res 329:13–29. https://doi.org/10.1016/j.jvolgeores.2016.11.006

Martí J, Castro A, Rodríguez C, Costa F, Carrasquilla S, Pedreira R, Bolos X (2013a) Correlation of magma evolution and geophysical monitoring during the 2011–2012 El Hierro (Canary Islands) submarine eruption. J Petrol 54:1349–1373

Martí J, Pinel V, López C, Geyer A, Abella R, Tárraga M, Blanco MJ, Castro A, Rodríguez C (2013b) Causes and mechanisms of the 2011–2012 El Hierro (Canary Islands) submarine eruption. J Geophys Res Solid Earth 118:823–839. https://doi.org/10.1002/jgrb.50087

Martinez-Arevalo C, Mancilla FD, Helffrich G, Garcia A (2013) Seismic evidence of a regional sublithospheric low velocity layer beneath the Canary Islands. Tectonophysics 608:586–599

Masson DG (1996) Catastrophic collapse of the volcanic island of Hierro 15 Ka ago and the history of landslides in the Canary Islands. Geology 24:231–234

Masson DG, Watts B, Gee MJR, Urgeles R, Mitchell NC, Le Bas TP, Canals M (2002) Slope failures on the flanks of the western Canary Islands. Earth Sci Rev 57:1–35

Meletlidis S, Di Roberto A, Pompilio M, Bertagnini A, Iribarren I, Felpeto A, Torres PA, D'Oriano C (2012) Xenopumices from the 2011–2012 submarine eruption of El Hierro (Canary Islands, Spain): constraints on the plumbing system and magma ascent. Geophys Res Lett 39(17)(art no L17302)

Moore JG, Peck DL (1965) Bathymetric, topographic, and structural map of the south-central flank of Kilauea Volcano, Hawaii. Miscellaneous geologic investigations map I-456. Department of the Interior United States Geological Survey

Nooner SL, Chadwick WW (2016) Inflation-predictable behavior and co-eruption deformation at Axial Seamount. Science 354(6318):1399–1403. https://doi.org/10.1126/science.aah4666

Portner RA, Clague DA, Helo C, Dreyer BM, Paduan JB (2015) Contrasting styles of deep-marine pyroclastic eruptions revealed from Axial Seamount push core records. Earth Planet Sci Lett 423:219–231. https://doi.org/10.1016/j.epsl.2015.03.043

Rivera J, Lastras G, Canals M, Acosta J, Arrese B, Hermida N, Micallef A, Tello O, Amblas D (2013) Construction of an oceanic island: Insights from the El Hierro (Canary Islands) 2011–2012 submarine volcanic eruption. Geology 41(3):355–358

Rivera J, Hermida N, Arrese B, González-Aller D, Sánchez de Lamadrid JL, Gutiérrez de la Flor D, Acosta J (2014) Bathymetry of a new-born submarine volcano: El Hierro, Canary Islands. J Maps 10(1):82–89

Rizzo A, Caracausi A, Chavagnac V et al (2016) Kolumbo submarine volcano (Greece): an active window into the Aegean subduction system. Sci Rep 6:28013. https://doi.org/10.1038/srep28013

Rodriguez-Gonzalez A, Fernandez-Turiel JL, Perez-Torrado FJ, Paris R, Gimeno D, Carracedo JC, Aulinas M (2012) Factors controlling the morphology of monogenetic basaltic volcanoes: the Holocene volcanism of Gran Canaria (Canary Islands, Spain). Geomorphology 136(1):31–44. http://doi.org/10.1016/j.geomorph.2011.08.023

Sánchez-Guillamón O, Fernández-Salas LM, Vázquez JT, Palomino D, Medialdea T, López-González N, Somoza L, León R (2018) Shape and size complexity of deep seafloor mounds on the Canary Basin (West to Canary Islands, Eastern Atlantic): a DEM-based geomorphometric analysis of domes and volcanoes. Geosciences 8(2):37. https://doi.org/10.3390/geosciences8020037

Santana-Casiano JM, Fraile-Nuez E, González-Dávila M, Baker ET, Resing JA, Walker SL (2016) Significant discharge of CO_2 from hydrothermalism associated with the submarine volcano of El Hierro Island. Sci Rep 6

Schipper CI, White JDL, Houghton BF (2011) Textural, geochemical, and volatile evidence for a Strombolian-like eruption sequence at Lō'ihi Seamount, Hawai'i. J Volcanol Geotherm Res 207:16–32

Schmincke HU (1982) Volcanic and chemical evolution of the Canary Islands. In: VonRad U, Hinz K, Sarnthein M, Seibol E (eds) Geology of the northwest African continental margin. Springer, New York, pp 273–306

Schmincke HU (1990) Geological field guide Gran Canaria. With contributions from Freundt A, Ferriz H, Kobberger G, Leat P. Excursion 7BI International Volcanological Congress, Mainz, FRG, Pluto Press, Witten, 210 p

Schnur SR, Chadwick WW Jr, Embley RW et al (2017) A decade of volcanic construction and destruction at the summit of NW Rota-1 seamount: 2004–2014. J Geophys Res Solid Earth 122:1558–1584. https://doi.org/10.1002/2016JB013742

Siebert L, Cottrell E, Venzke E, Andrews B (2015) Earth's volcanoes and their eruptions: an overview. In: The encyclopedia of volcanoes, Ch 12. Elsevier, pp 239–253. http://doi.org/10.1016/B978-0-12-385938-9.00012-2

Somoza L, González FJ, Barker SJ, Madureira P, Medialdea T, de Ignacio C, Lourenço N, Leon R, Vázquez JT, Palomino D (2017) Evolution of submarine eruptive activity during the 2011–2012 El Hierro event as documented by hydroacoustic images and remotely operated vehicle observations. Geochem Geophys Geosyst 18:3109–3137. https://doi.org/10.1002/2016GC006733

Sotomayor-García A, Rueda JJ, Sánchez-Guillamón O, Urra J, Martín-Arjona A, González-Porto M, Vazquez JT, Palomino D, López-González N, Fernández-Salas LM, Santana-Casiano JM, González-Dávila M, Fraile-Nuez E (this volume) Impact of Tagoro Volcano formation on benthic and associated biota: a review

Stevens NF, Wadge G, Williams CA (2001) Post-emplacement lava subsidence and the accuracy of ERS InSAR digital elevation models of volcanoes. Int J Remote Sens 22(5):819–828. https://doi.org/10.1080/01431160051060246

Tepp G, Chadwick WW Jr, Haney MM, Lyons JJ, Dziak RP, Merle SG, Butterfield DA, Young CW III (2019) Hydroacoustic, seismic, and bathymetric observations of the 2014 submarine eruption at Ahyi Seamount, Mariana Arc. Geochem Geophys Geosyst 20(7):3608–3627. https://doi.org/10.1029/2019GC008311

van den Bogaard P (2013) The origin of the Canary Island Seamount Province—new ages of old seamounts. Sci Rep 3:1–7. https://doi.org/10.1038/srep02107

Vázquez JT, Sánchez-Guillamón O et al (2016) Preliminary geomorphological analysis of the Tagoro Volcano underwater eruption (submarine slope of El Hierro Island). In: Abstract book V Simposium Internacional de Ciencias del Mar, Universidad de Alicante, Alicante, Spain

Walker SL, Baker ET, Resing JA, Chadwick WW Jr, Lebon GT, Lupton JE, Merle SG (2008) Eruption-fed particle plumes and volcaniclastic deposits at a submarine volcano: NW Rota-1, Mariana Arc. J Geophys Res 113:B08S11. http://doi.org/10.1029/2007JB005441

Wittmann W, Sigmundsson F, Dumont S, Lavallée Y (2017) Post-emplacement cooling and contraction of lava flows: InSAR observations and a thermal model for lava fields at Hekla volcano, Iceland. J Geophys Res 122(2):946–965. https://doi.org/10.1002/2016JB013444

Wormald SC, Wright IC, Bull JM, Lamarche G, Sanderson DJ (2012) Morphometric analysis of the submarine arc volcano Monowai (Tofua–Kermadec Arc) to decipher tectono-magmatic interactions. J Volcanol Geotherm Res 239–240:69–82

Wright IC, Chadwick WW Jr, de Ronde CEJ, Reymond D, Hyvernaud O, Gennerich HH, Stoffers P, Mackay K, Dunkin MA, Bannister SC (2008) Collapse and reconstruction of Monowai submarine volcano, Kermadec arc, 1998–2004. J Geophys Res 113:B08S03. http://doi.org/10.1029/2007JB005138

Part III

The 2011–2014 Volcanic Unrest and Submarine Eruption: Marine Environment

Ten Years of Intense Physical–Chemical, Geological and Biological Monitoring Over the Tagoro Submarine Volcano Marine Ecosystem (Eruptive and Degassing Stages)

8

Eugenio Fraile-Nuez, J. Magdalena Santana-Casiano, Melchor González-Dávila, Alba González-Vega, Juan Tomás Vázquez, Ana Sotomayor-García, Isabel Ferrera, Carolina Santana-González, Francisco Eugenio, Javier Marcello, Santiago Hernández-León, Evangelos Bakalis, José L. Rueda, María Gómez-Ballesteros, Antonio M. Álvarez-Valero, Olga Sánchez-Guillamón, Desirée Palomino, Olvido Tello, Carmen Presas-Navarro, José Escánez-Pérez, Marcos González-Porto, María Luz Fernández de Puelles, Anna Olivé-Abelló, Beatriz Vinha, Francisco Machín, Juan Pablo Martín-Díaz, and Jesús M. Arrieta

E. Fraile-Nuez (✉) · A. González-Vega · C. Presas-Navarro · J. Escánez-Pérez · M. González-Porto · A. Olivé-Abelló · B. Vinha · J. M. Arrieta
Centro Oceanográfico de Canarias, Instituto Español de Oceanografía (IEO), Consejo Superior de Investigaciones Científicas (CSIC), 38180 Santa Cruz de Tenerife, Spain
e-mail: eugenio.fraile@ieo.csic.es

J. M. Santana-Casiano · M. González-Dávila · C. Santana-González · F. Eugenio · J. Marcello · S. Hernández-León
Instituto de Oceanografía y Cambio Global (IOCAG), Universidad de Las Palmas de Gran Canaria (ULPGC), 35250 Las Palmas de Gran Canaria, Spain

J. T. Vázquez · I. Ferrera · J. L. Rueda · O. Sánchez-Guillamón · D. Palomino
Centro Oceanográfico de Málaga, Instituto Español de Oceanografía (IEO), Consejo Superior de Investigaciones Científicas (CSIC), 29640 Fuengirola, Spain

A. Sotomayor-García · A. Olivé-Abelló
Instituto de Ciencias del Mar, ICM-CSIC, 08003 Barcelona, Spain

E. Bakalis
Dipartimento di Chimica "G. Ciamician", Università di Bologna, Bologna BO, Italy

M. Gómez-Ballesteros · O. Tello
Centro Oceanográfico de Madrid, Instituto Español de Oceanografía (IEO), Consejo Superior de Investigaciones Científicas (CSIC), 28080 Madrid, Spain

P. J. González (ed.), *El Hierro Island*, Active Volcanoes of the World,
https://doi.org/10.1007/978-3-031-35135-8_8

Abstract

The shallow Tagoro submarine volcano monitoring represents a unique opportunity not only for improving our sparse understanding of submarine volcanic processes in specific scientific fields as physical and chemical oceanography or marine geology but also its interactions over the marine biology in one of the richest marine ecosystems in Europe. This chapter aims to summarize the most relevant physical–chemical, geological and biological changes that occurred in the marine ecosystem of El Hierro island, at the Marine Reserve Punta de La Restinga—El Mar de Las Calmas, due to the genesis of the new underwater volcano Tagoro (27°37′07″N–017°59′28″W) in October 2011. During the first six months of the eruption, extreme physical–chemical perturbations caused by this event, comprising thermal increase from up to + 18.8 °C, water acidification with a pH decrease of 2.9 units, deoxygenation to anoxic levels and extremely high metal enrichment among others, resulted in significant and dramatic alterations of the marine ecosystem. After March 2012, once the eruptive phase was finished, the new submarine volcano entered an active hydrothermal phase involving the release of heat with smaller but still significant and important thermal anomalies of up to + 2.55 °C around the craters, density decrease of − 1.43 kg m^{-3}, pH decrease of − 1.25 units, and high concentrations of metals and inorganic nutrients similar to upwelling zones. These enrichments are still active up to date, producing clear signs of marine recovery not only in the benthonic strata but also in the whole water column compared with pre-eruptive data. Since its eruption ten years ago, an unprecedented monitoring effort has turned into the longest and most complete multidisciplinary time-series for the study of a shallow submarine volcano, with the realization of 31 oceanographic multidisciplinary expeditions that systematically measure more than 40 different physical–chemical and biological variables. All this information and the results obtained during the evolution of the process could serve as a baseline for better understanding future or similar submarine eruptions worldwide.

8.1 Introduction

About 80% of Earth volcanism occurs on the ocean floor which has greatly hindered our understanding of the natural outgassing of volatiles from the Earth's interior and its impact on the environment (Crisp 1984). A significant percentage of new submarine volcanoes have been discovered in the last decades together with a high number of geochemical studies to investigate the current status of submarine hydrothermal systems (Rizzo et al. 2019). Although it is well-known that submarine hydrothermal systems are instrumental in transferring heat, gases, acidic and reduced metal-rich fluids relative to ambient seawater in the global ocean (Butterfield et al. 1990; Fraile-Nuez et al. 2012; Tassi et al. 2009), the chemical composition of hydrothermal fluids change as deep and shallow systems are affected by variations in permeability and fluid source processes. Shallow hydrothermal vents are located in different tectonic settings, in particular related to recent subaerial and submarine

A. M. Álvarez-Valero
Departamento de Geología, Universidad de Salamanca, 37008 Salamanca, Spain

M. L. Fernández de Puelles
Centro Oceanográfico de Baleares, Instituto Español de Oceanografía (IEO), Consejo Superior de Investigaciones Científicas (CSIC), 07015 Palma de Mallorca, Spain

B. Vinha
Dipartimento di Scienze E Tecnologie Biologiche Ed Ambientali (DiSTeBA), Università del Salento, Lecce, Italy

A. González-Vega · F. Machín
Universidad de Las Palmas de Gran Canaria (ULPGC), 35250 Las Palmas de Gran Canaria, Spain

J. P. Martín-Díaz
Universidad de La Laguna (ULL), Tenerife, Santa Cruz de Tenerife, Spain

volcanic activity, along arcs, mid ocean ridges and in areas presenting intraplate oceanic volcanism (Botz et al. 1999; Cardigos et al. 2005; Dando et al. 1999; Fraile-Nuez et al. 2018; Fricke et al. 1989; Furushima et al. 2009).

Although hydrothermal emissions significantly affect the properties of seawater and marine sediments (Damm 1990; Thompson 1983), changing the geochemical processes and mineralization patterns (Rona et al. 2006) that dramatically affect the surrounding marine environments, it is unclear whether or not the emissions produced by these systems could have a clear effect on the primary producers inhabiting nearby photic waters. Nevertheless, several authors have shown that shallow inputs of hydrothermal fluids lead to an increase in the availability of nutrients and bioavailable iron which could lead to an increase in oceanic productivity in photic waters (Buck et al. 2018; Guieu et al. 2018; Santana-Casiano et al. 2013; Wilson et al. 2019). Yet, there are also other studies finding no significant evidences on that sense (Gómez-Letona et al. 2018). Moreover, recent studies over Tagoro submarine volcano have revealed an increase in zooplankton abundance in waters affected by the volcano, indicating that volcanic inputs have a significant effect on these organisms (Fernández de Puelles et al. 2021). On the other hand, submarine organisms such as black corals have also been revealed as a sink of volcanic elements (magmatic isotopes that are far to be nutrients), thus being excellent fingerprints of eruptive events (Álvarez-Valero et al. 2018).

This chapter aims to summarize the most relevant physical–chemical, geological and biological changes that occurred in the marine ecosystem of El Hierro island, at the Marine Reserve Punta de La Restinga—El Mar de Las Calmas, due to the genesis of the new underwater volcano "Tagoro" (27º37′07″N–017º59′28″W) in October 2011. Since its eruption ten years ago, a continuous and intense multidisciplinary monitoring has been carried out in the area with the realization of 31 oceanographic expeditions that could serve as a baseline for future or similar submarine eruptions.

8.2 Ten Years of Intense Physical–Chemical, Geological and Biological Monitoring

In July 2011, the youngest island of the Canary Archipelago, El Hierro, began to register an unusual seismic activity (see Chap. 6). In less than three months, the National Geographic Institute (IGN), the national institution in charge of monitoring the seismic activity in Spain, registered through its seismic network more than 12,000 low-magnitude earthquakes located at first in the northern part of the island and then progressively more towards the south, crossing the island practically by its half. These earthquakes had magnitudes that varied between 2 and 4.3 on the Richter scale and a ground deformation that exceeded 5 cm in its vertical component (Domínguez-Cerdeña et al. 2014). On October 10th, 2011, a substantial decrease of seismicity together with continuous volcanic tremor, indicated the beginning of an eruptive phase. The magma finally reached the Earth's surface underwater, at 1.8 km south of the fishing village of La Restinga, at a depth of 355 m below the ocean surface (Fraile-Nuez et al. 2018).

The Tagoro submarine volcano was entirely generated during the eruptive underwater phase that took place between October 10th, 2011 and March 5th, 2012 (Fraile-Nuez et al. 2018). During these six months, extreme physical–chemical perturbations, comprising thermal changes, water acidification, deoxygenation and metal-enrichment, resulted in significant and dramatical alterations of the marine ecosystem (Fig. 8.1) (Fraile-Nuez et al. 2018, 2012; Santana-Casiano et al. 2013). After March 2012, once the eruptive phase was over, the Tagoro underwater volcano entered an active hydrothermal phase which remains active, with a diffuse release of heat, gases, inorganic nutrients and metals from the main and secondary craters (Fraile-Nuez et al. 2018; Santana-Casiano et al. 2013).

Since the beginning of the eruption and to this day, the Spanish Institute of Oceanography (IEO) together with both Canary Islands'

Fig. 8.1 **a** A true color, high-resolution RapidEye satellite image featuring a gigantic stain visible on the surface of "El Mar de Las Calmas" marine reserve, El Hierro island, Spain (11–26–2011). **b** R/V Ramón Margalef, from the Spanish Institute of Oceanography, involved in the monitoring of the volcanic event. **c, d** "Guardia Civil" helicopter flight images over the Tagoro submarine volcano. **e–g** Pyroclastic rocks floating with gas release images (Sasemar, Nov.2011, photo: Jesús Pérez). **h** Bubbling over the main edifice (RTVE, 11–08–2011)

Universities (University of Las Palmas de Gran Canaria and University of La Laguna) and other national and international marine research centers and private foundations have dedicated funding, resources and effort for developing multidisciplinary monitoring and studies of the physical–chemical, geological and biological processes that took place over the Tagoro submarine volcano. To date, more than 31 different multidisciplinary oceanographic expeditions have been carried out in the area in the context of nine different scientific projects: Bimbache (IEO-2011–2012, Leg1-12), Raprocan-III (IEO-2010–2012), Cethobaph (CGL2009-1,311,218), Vulcano-I (CTM2012-36,317), Vulcano-II (CTM2014-51,837-R), Vulcana-I (IEO-2015–2017), Vulcana-II (IEO-2018–2020), Cyclovent (2019SP18), Vulcana-III (IEO-2021–2023) (Table 8.1; Fig. 8.2). These institutions have been able to collect quite an enormous amount of multidisciplinary data and water samples, including CTD stations (~1300), dissolved oxygen (~4500), inorganic nutrients (~5500), turbidity (~1400), salinity (~200), pH (~4000), inorganic total carbon (~4000), alkalinity (~4000), potential redox (~400), Fe (II) (~2200), metals (~700), phytoplankton abundance (~800), bacterial abundance (~1600), picoplankton diversity (~250), particulate organic carbon (~800), dissolved organic carbon (~1500), dissolved inorganic carbon (~700), photosynthetic pigments (~700), electron transport system (~400), plankton WP2-3 (~500), rocks and microbial mats (~300), ROV images (~90 h recordings), physical–chemical mooring (2), surface multiparametric buoy (1), among others.

This unprecedent monitoring effort has resulted in the longest and most complete multidisciplinary time-series for the study of the Tagoro submarine volcano, one of the most recent shallow monogenetic submarine volcanoes in Europe. This invaluable dataset has been the base for a large number of scientific publications that evidence the enormous potential for the new discoveries over the Tagoro submarine volcano

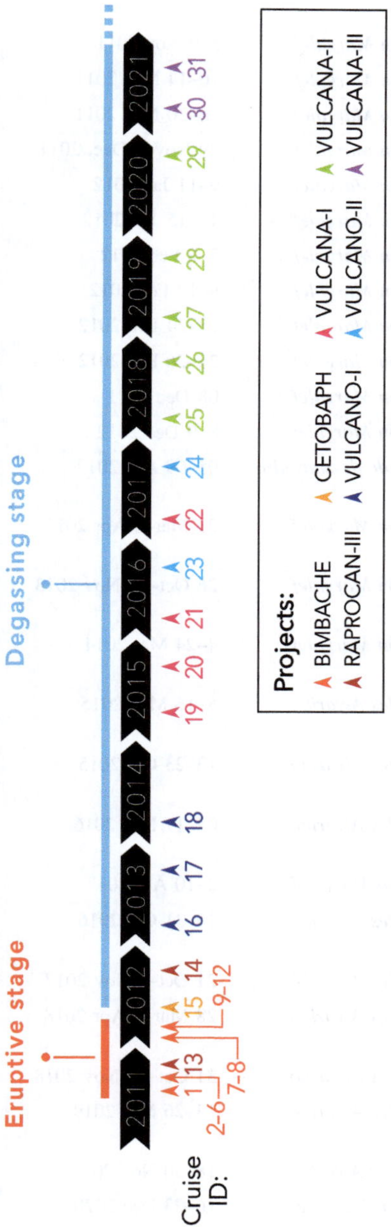

Fig. 8.2 Timeline of the multidisciplinary oceanographic cruises carried out for the monitoring of the submarine volcano Tagoro during both the eruptive and degassing stages in the context of different scientific projects. Each cruise has been assigned an ID number corresponding to the order presented in Table 8.1

(Ariza et al. 2014; Álvarez-Valero et al. 2018; Blanco et al. 2015; Coca et al. 2014; Danovaro et al. 2017; Eugenio et al. 2014; Fernández de Puelles et al. 2021; Ferrera et al. 2015; Fraile-Nuez et al. 2012, 2018; García-Davis et al. 2021; González-Vega et al. 2020; Olivé Abelló et al. 2021; Rivera et al. 2013; Santana-Casiano et al. 2013, 2016, 2018; Santana-González et al. 2017; Sotomayor-García et al. 2019, 2020).

8.3 Dramatic Effects Over the Marine Environment

8.3.1 Ocean Warming: A Continuous Flux of Energy into the Environment

The eruptive phase of the Tagoro submarine volcano involved the release of a huge amount of heat into the ocean. Physical observations in the affected area of the volcano revealed temperature and salinity anomalies of up to + 3 °C and − 0.3, respectively, at 80 m depth in the surrounding waters of the Tagoro edifice. Moreover, expendable bathythermograph (XBTs) probes were launched during November 2011 and February 2012 in the proximity of the main crater in order to quickly measure the vertical temperature structure (Fig. 8.3). Temperature values of 22.76, 28.80 °C were registered in the proximity of the main crater with maximum of 36.10 °C in the center of a crater at a depth of ~ 120–130 m. With respect to the reference temperature value, a maximum temperature anomaly of + 18.8 °C was observed (Fraile-Nuez et al. 2012).

During the degasification phase, the temperature anomalies at Tagoro volcano were only observed close to the seabed in the surrounding of the main and secondary craters. These 10 years of temperature anomalies could be registered thanks to the *Tow-yo* technique, which involves continuously lowering and raising the rosette between 1 and 40 m above the seabed and with the ship moving at 0.2–0.4 kn in dynamic positioning, obtaining a sawtooth-shaped dataset with high spatial resolution and closer to the source of the emissions (González-Vega et al.

Table 8.1 Detailed information about all the oceanographic cruises carried out in the last ten years for the multidisciplinary monitoring of Tagoro submarine volcano

ID	Project	Cruise	Type	Research vessel	Date
1	BIMBACHE (2011–2012)	Bimbache leg 1	Bathymetry	*Ramón Margalef*	22–26 Oct 2011
2		Bimbache leg 2	Oceanography—ROV	*Ramón Margalef*	27 Oct–3 Nov 2011
3		Bimbache leg 3	Oceanography	*Ramón Margalef*	5–9 Nov 2011
4		Bimbache leg 4	Bathymetry	*Ramón Margalef*	10–14 Nov 2011
5		Bimbache leg 5	Oceanography	*Ramón Margalef*	16–20 Nov 2011
6		Bimbache leg 6	Bathymetry	*Ramón Margalef*	29 Nov–3 Dec. 2011
7		Bimbache leg 7	Bathymetry	*Ramón Margalef*	9–11 Jan 2012
8		Bimbache leg 8	Oceanography	*Ramón Margalef*	13–15 Jan 2012
9		Bimbache leg 9	Bathymetry	*Ramón Margalef*	7–8 Feb 2012
10		Bimbache leg 10	Oceanography	*Ramón Margalef*	9–12 Feb 2012
11		Bimbache leg 11	Bathymetry	*Ramón Margalef*	23–24 Feb 2012
12		Bimbache leg 12	Oceanography	*Ramón Margalef*	24–26 Feb 2012
13	RAPROCAN-III (2010–2012)	Raprocan-1211	Oceanography	*Ramón Margalef*	08 Dec 2011
14		Raprocan-1212	Oceanography	*Ramón Margalef*	8–9 Dec 2012
15	CETOBAPH (2010–2012)	Cetobaph-0412	Oceanography	*Cornide de Saavedra*	05–09 Apr 2012
16	VULCANO-I (2013–2014)	Vulcano-0313	Oceanography-Bathymetry	*Ramón Margalef*	23 Mar–6 Apr 2013
17		Vulcano-1013	Oceanography-Bathymetry	*Ramón Margalef*	26 Oct–11 Nov 2013
18		Vulcano-0314	Oceanography-Bathymetry	*Ramón Margalef*	4–24 Mar 2014
19	VULCANA-I (2015–2016)	Vulcana-0515	Oceanography - Bathymetry	*Ángeles Alvariño*	5–16 May 2015
20		Vulcana-1015	Oceanography-Bathymetry	*Ángeles Alvariño*	13–23 Oct 2015
21		Vulcana-0316	Oceanography-Bathymetry	*Ángeles Alvariño*	07–17 Mar 2016
22		Vulcana-0417	Oceanography-ROV	*Ramón Margalef*	2–10 Apr 2017
23	VULCANO-II (2016–2017)	Vulcano-1016	Oceanography-Bathymetry	*Ángeles Alvariño*	15–31 Oct 2016
24		Vulcano-1017	Oceanography	*Ángeles Alvariño*	21 Oct–6 Nov 2017
25	VULCANA-II (2018–2020)	Vulcana-0318	Oceanography-Bathymetry	*Ángeles Alvariño*	28 Mar–9 Apr 2018
26		Vulcana-1118	Oceanography-ROV	*Ángeles Alvariño*	31 Oct–14 Nov 2018
27		Vulcana-0319	Oceanography-Bathymetry	*Ángeles Alvariño*	13–26 Mar 2019
28		Vulcana-1119	Oceanography	*Ángeles Alvariño*	18–30 Nov 2019
29		Vulcana-1220	Oceanography	*Ramón Margalef*	16–23 Dec 2020
30	VULCANA-III (2021–2023)	Vulcana-0421	Oceanography	*Ángeles Alvariño*	4–14 Apr 2021
31		Vulcana-1021	Oceanography-Bathymetry	*Ángeles Alvariño*	13–31 Oct 2021

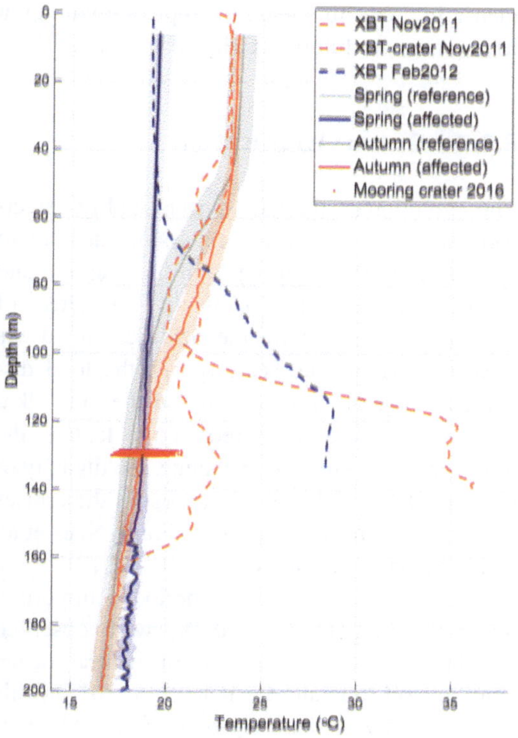

Fig. 8.3 Vertical temperature profiles from Expendable Bathythermograph (XBT) probes over the Tagoro submarine volcano in November 2011(dashed orange lines) and February 2012 (dashed blue line). Average vertical profiles of temperature in the unaffected waters as a reference (gray lines with shading) and in affected stations (solid lines with shading) around the volcano, in the upper 200 m for Spring (blue) and Autumn (orange). The shaded areas around each line represent the standard deviation. Temperature data extracted from a mooring installed in October 2016 in the interior of the main crater at 127 m depth (indicated in red)

2020). In that sense, in autumn, when the thermocline was located around 60 m depth of the water column, significant temperature anomalies of up to + 1.2 °C were found in a range of 60–120 m depth and only 1–30 m above the seabed. Conversely, during spring, the thermocline was found at a deeper depth of about 140 m, and a temperature anomaly of up to + 0.8 °C was registered in the range of 140–200 m depth (Fig. 8.3).

In October 2016, five years after the end of the eruptive phase, a physical mooring for thermohaline observations was installed in the interior of the main crater at a depth of 127 m. This instrument registered important physical anomalies over the seabed as compared to the background temperature values of the surrounding waters (Fig. 8.3). We have corroborated that the hydrothermal system has generated a physical environment that continuously supports maximum potential temperature anomalies of up to + 2.55 °C, a decrease in salinity of 1.02, and a decrease in density of 1.43 kg m^{-3} (Fraile-Nuez et al. 2018). The combined effect of these anomalies could generate a buoyant plume over the environment that returns downward waters closing the circulation of the advection cells, which generate horizontal flow due to the effect of rotation (Speer 1989). Theoretically, this vortex is of sufficient magnitude to trap water in a local recirculation, which could have a crucial effect on the expected dispersal of properties, including minerals, tracers, and organisms (Speer 1989).

From the previous temperature anomalies, we computed the amount of heat that emanates from the crater of the Tagoro volcano needed to produce the observed temperature anomaly. The procedure followed is consistent with previous publications (Baker and Massoth 1987; Lupton et al. 1985), using the following expression:

$$Q = \rho C_p A_r \Delta T v$$

where Q is the emitted power or heat flux, ρ is the density, C_p is the specific heat, A_r is the area affected by the emission, ΔT is the temperature anomaly, and v is the current speed. The Tagoro volcano environment is featured by the next values: $\rho = 1026$ kg m^{-3}, $Cp = 3.99 \times 10^3$ J kg^{-1} $°C^{-1}$, A_r comprises an area of about 10,000 m^2, with a reference anomaly for the temperature of 1 °C and a speed for the Canary Current of 0.1 m s^{-1}. In this way, the power or heat flux associated with the heat emission is in the order of 4000 MW. This value is the same order of magnitude as those reported in similar vents, where a heat flux of the order of 1000 MW is emitted (Baker and Massoth 1987). To put this value in context, the integral hydro-wind power plant that provides more than 90% of the electrical power demanded by the island of El Hierro has a peak production capacity of 11 MW.

The outgassing materials directly affect the conductivity and the temperature of the surrounding environment, which results in a high fluctuation of these parameters (Bakalis et al. 2017; 2018; Olivé-Abelló et al. 2021). In the water column above hydrothermal vents temperature and conductivity were recorded, among other properties, in the form of equidistant time series. Typically, these measurements can be made at single spots (using ROVs) (Bakalis et al. 2018, 2017) or can have a larger spatial coverage as was the case for the measurements done over Tagoro submarine volcano with CTD (Olivé-Abelló et al. 2021). The recorded time series were analyzed with the Generalized Moments Method (GMM), (see Bakalis et al. 2017, 2018; Olivé-Abelló et al. 2021 and references therein for details of the method), which reformulates the actual time of observation as time between consecutive measurements, in statistical sense, and returns the moments of the formed distributions at different times. The scaling of the moments provides the structure function whose form entails insights on the underlying stochastic processes.

The analysis of the time series of temperature and conductivity by GMM showed that Tagoro submarine volcano releases energy to its surroundings and slows down equilibration of temperature with the overlying aquatic environment. Temperature and conductivity are classified as multifractals whose structure functions present a crossover between two distinct regimes. Tagoro volcano can be characterized as an open system since temperature fluctuations do not correspond to stationary process. The behavior of these two properties can serve as an indicator of the state of activity (conductivity) and the physical setting, open or closed, of the submarine volcano (temperature). The present methodology can be useful especially when insights are needed for poorly-known deep-sea volcanoes, where in situ measurements by means of submarine robotic technologies are either too expensive or even technically impossible. The applicability of the method as universal geohazard warning predictor requires its application to other volcanic systems. The moments used by GMM may be used as a metric in a machine learning approach aiming at forecasting the future activity of volcanoes.

8.3.2 Ocean Acidification

The evolution of the CO_2 system and its effects were studied in the seawater around the Tagoro submarine volcano during both the eruptive and degasification stages. To quantify the effect of the gasses emission on the acidification of the area, the pH in total scale (pH_T), the total dissolved inorganic carbon (C_T) and the total alkalinity (A_T) were measured. To calculate the atmosphere–ocean CO_2 exchange, the fugacity of CO_2 (fCO_2) was measured in most of the cruises or calculated from C_T and pH_T (Fraile-Nuez et al. 2012; Santana-Casiano et al. 2018, 2013).

During the eruptive period, important anomalies in pH_T, C_T and A_T were observed which generated an episode of natural ocean acidification (Santana-Casiano et al. 2013). This acidification was a consequence of the CO_2, SO_2 and H_2S/HS^- volcanic emissions that decreased the pH from 8 to values lower than 6, with pH anomalies (ΔpH) larger than −2.9 in the first month. The C_T changed from the normal surface values for the area of 2100 µmol kg^{-1} to 7682 µmol kg^{-1} and the A_T decreased from 2430.1 µmol kg^{-1} to 1338 µmol kg^{-1} due to different chemical reactions. The $pCO2_{(pH,CT)}$ increased to values as high as 155,000 µatm. The surface seawater in the southwest of the island registered $pCO_{2(pH,CT)}$ values in the range of 11,000–19,000 µatm. pCO_2 in the areas unaffected by the volcanic emissions was of 414 ± 2 µatm. The air-sea CO_2 exchange calculated in one of the most active weeks of the volcano, between November 4th–9th, was 5×10^{10} g d^{-1} (Santana-Casiano et al. 2013). It was estimated that the volcano eruption released at least 1.3–2.1 Tg CO_2 in the first 147 days (Longpré et al. 2016).

The degassing period began in March 2012 and the area affected was concentrated in 0.5 km around the main crater of the volcano. The composition of the gases emitted was mainly CO_2 (Santana-Casiano et al. 2013). These

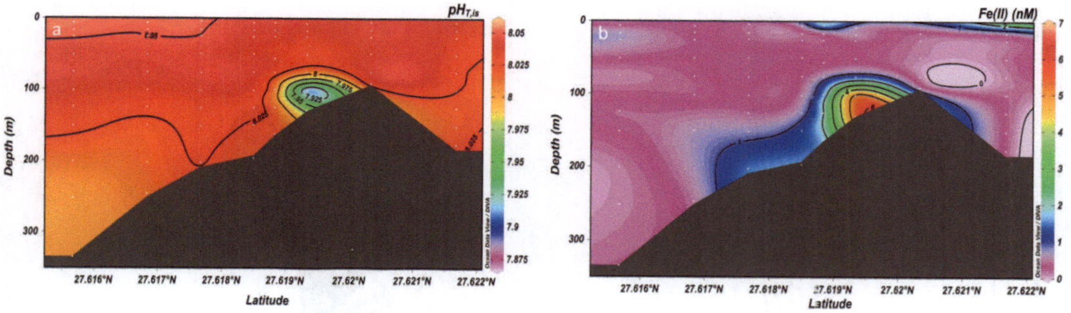

Fig. 8.4 Meridional transect with depth for the variation of **a** $pH_{T,is}$ and **b** Fe(II) variable over Tagoro submarine volcano (October 2013). Figure modified of Santana-González et al. (2017)

emissions generated a pH_T value of 6.1 (1.9 units below the normal values) and a C_T of 4.191 μmol kg^{-1} in the water column just over the crater.

During the post-eruptive stage, the study of the chemical properties continued. The $pH_{T,is}$ and Fe(II) variation in October 2013 is shown in Fig. 8.4 along the volcano. In 2014, a new methodology was introduced using *Tow-yo* studies (Santana-Casiano et al. 2016). The reduced and acidic characteristics of the Tagoro emissions allowed to detect anomalies related to changes in the chemical potential (pE) and proton concentration (pH, Fig. 8.3a) and to accurately define the affected area of emissions (13,000 ± 800 m^2) and the volume occupied (206,000 ± 50,000 m^3). The pE and pH changes were observed only around the main crater of the volcano. From the measurements of the CO_2 system variables (pH_T, A_T and C_T) in the water column and considering both the affected volume and the current speed, a CO_2 flux estimation of 6.0 10^5 ± 1.1 10^5 kg d^{-1}, principally transported to the south-west, was obtained for the Tagoro volcano. This emission of CO_2 reduced the pH increasing the seawater acidity of the affected area by ~ 20%. Maxima values of negative pH anomaly of ~ 0.30 units close to the crater duplicated the acidity of the waters (100% of increase).

In 2019, anomalies were also observed only above the crater. The anomaly in $pCO_{2(pH,CT)}$ was 80 μatm and − 0.05 units for the pH, which represented an increase of 12% in the acidity in the vicinity of the crater.

8.3.3 Deoxygenation

The eruption of the Tagoro submarine volcano released remarkable amounts of reduced chemical compounds into the surrounding waters, such as reduced sulfur species and Fe(II) (Santana-Casiano et al. 2013). The mixing of these reduced fluids with seawater caused a severe oxygen depletion and even anoxic episodes (González-Vega et al. 2022). The strongest episode occurred in late November 2011 (Fig. 8.5a). In a zone where dissolved oxygen concentrations under normal conditions are typically around 200 μmol kg^{-1}, concentrations as low as 7.71 μmol kg^{-1} were measured due to the volcanic eruption, which represents a 96% decrease in oxygen respect to normal conditions. The deoxygenation was observed in the first 250 m of the water column, with minimum values found around 100 m depth.

This patch of deoxygenated waters was transported to a wide area around the volcano, particularly south-west. Figure 8.5b shows an estimation of the area where oxygen concentrations were below normal levels at 100 m depth during November 2011. This area represents 378 km^2 and is in accordance with the observed transport of the volcanic decolorated plume from satellite images (Fig. 8.5c). This was the general distribution of the transported plume; however, sporadic processes also caused further transport outside of this area. For example, a low-oxygen plume was observed north-west of El Hierro island on late November 2011 due to changes in the local currents. Additionally, an anticyclonic

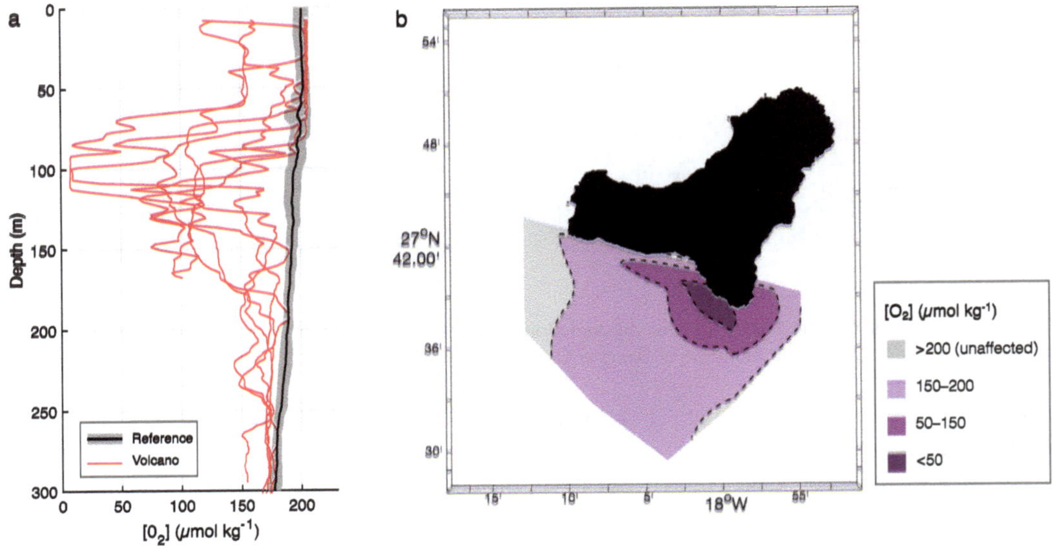

Fig. 8.5 **a** Vertical profiles of dissolved oxygen during the strongest episode of the eruptive stage (November 2011). Black line represents a reference profile averaged from stations outside of the affected area, with standard deviation (grey shade). **b** Horizontal section of dissolved oxygen at 100 m depth during November 2011

eddy on early November 2011 transported deoxygenated waters from the volcanic area as far as 80 km south of the island.

8.3.4 Reduction of the Epipelagic Stocks and A Diel Vertical Migration of Mesopelagic Organisms

Diel (diurnal) vertical migration is the synchronized movement of zooplankton and fish up and down in the water column over a daily cycle. In the eruption of Tagoro volcano, zooplankton and micronekton were heavily affected. During the days after the onset of the eruption, the use of zooplankton nets (WP-2) near the volcano was not possible due to the amount of ash clogging the net mesh. Micronekton organisms assessed using acoustics during those days displayed different densities and vertical distribution. Surveys performed in the area not affected by the volcano showed a deep scattering layer (DSL) between 400 and 600 m depth, which is the normal pattern is the waters surrounding the Canary Islands during the annual cycle (Uiblein and Bordes 1999). Near the volcano, however, organisms were affected by changes in the physical and chemical scenarios. Particulate material expelled by the volcano (measured as sea surface reflectance, SSR) decreased light penetration in the water column affecting the depth of the DSL (Ariza et al. 2014). While the normal position of the upper bound of the DSL is normally located at 400 m depth, the areas showing decreased light penetration displayed this layer between 200 and 300 m depth (Fig. 8.6).

Associated to the volcano plume, the oxygen, temperature, as well as chemicals and their effect (e.g., acidification) also disrupted the diel vertical migration of micronekton. These migrants around the islands are mainly decapods, cephalopods, and mesopelagic fishes (mainly myctophids; Ariza et al. 2016). These organisms diminished by 73% (± 18) in areas affected by the volcano plume during night in the epipelagic layer as observed from the 38 kHz frequency of the echosounder (Ariza et al. 2014). This decrease was consistent with anomalies related to rather high levels of turbidity and the decrease in dissolved oxygen.

During the eruptive phase (from October 2011 to February 2012), we recorded low chlorophyll

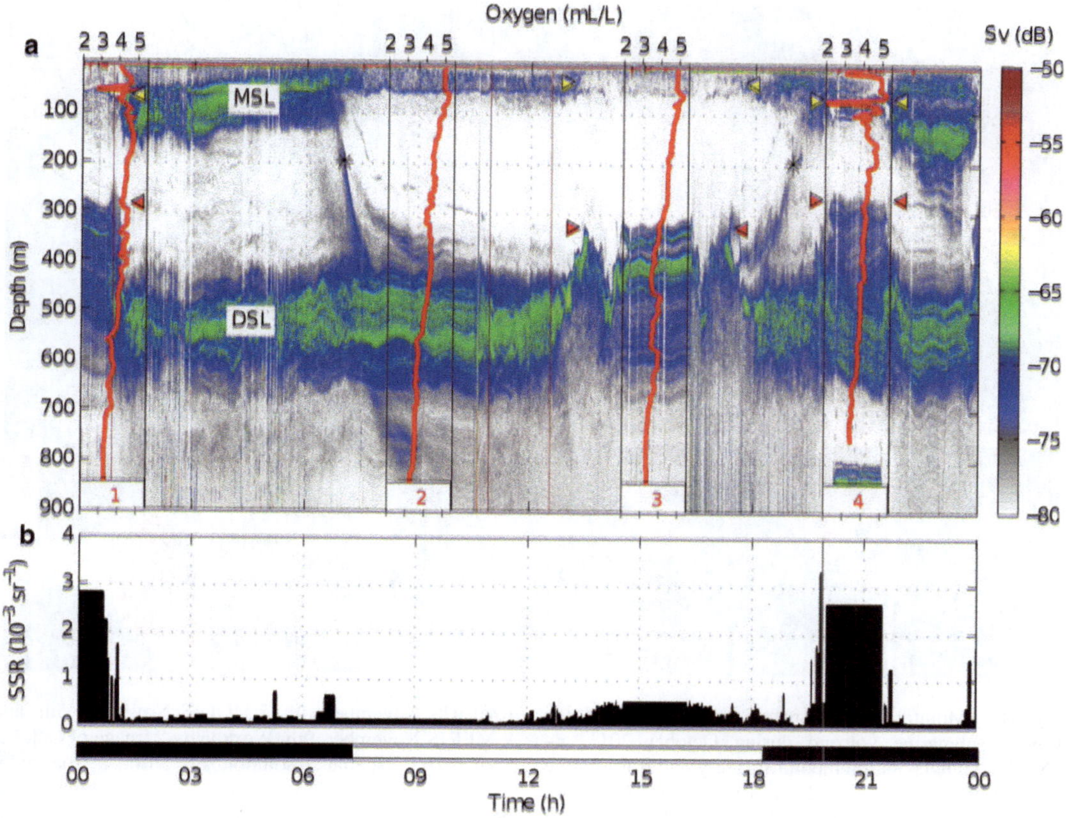

Fig. 8.6 38 kHz echogram **a** and Sea surface reflectance **b** on November 7th, 2011 along the acoustic transects indicated in Fig. 8.3b. The color scale refers to backscattering strength (Sv). Migrant and Deep Scattering Layers (MSL and DSL) as well as both diel migrations (*) are indicated. Acoustic anomalies are also indicated: the MSL weakening (yellow triangles) and the elevations of the DSL (red triangles). Red lines depict the dissolved oxygen profiles established during the acoustic track. The black color in the time scale refers to nighttime and white to daytime. *Source* https://doi.org/10.1371/journal.pone.0102354.g003

values (< 0.1 mg m^{-3}) near the plume and in the control station outside the waters affected by turbidity. These low values also coincided with low values of zooplankton (< 2 mg dry mass m^{-3}) and the proxy for this community (Mean Volume Backscattering Strength, MVBS, at 200 kHz, (Ariza et al. 2014)). Values of MVBS at 38 kHz used as a proxy for the migrant biomass was not so heavily affected during the eruptive and post-eruptive phases. Zooplankton biomass and proxies for zooplankton and micronekton biomass increased at the end of the eruption by February showing the highest values found during the study. These higher biomass records coincided with the so-called Late Winter Bloom in the Canary Current System (Hernández-León et al. 2007) as the effect of convective mixing and availability of nutrients in the euphotic zone, fostering higher biomass in upper trophic levels. This seasonal bloom hid a possible fertilization effect on chlorophyll (and therefore zooplankton) promoted by the important amount of iron expelled by the volcano.

8.3.5 Ocean Discoloration: Remote sensing of Volcanic Processes

Satellite remote sensing is a powerful tool to monitor and map the extent of natural hazards. Nowadays, there is wide availability and variety

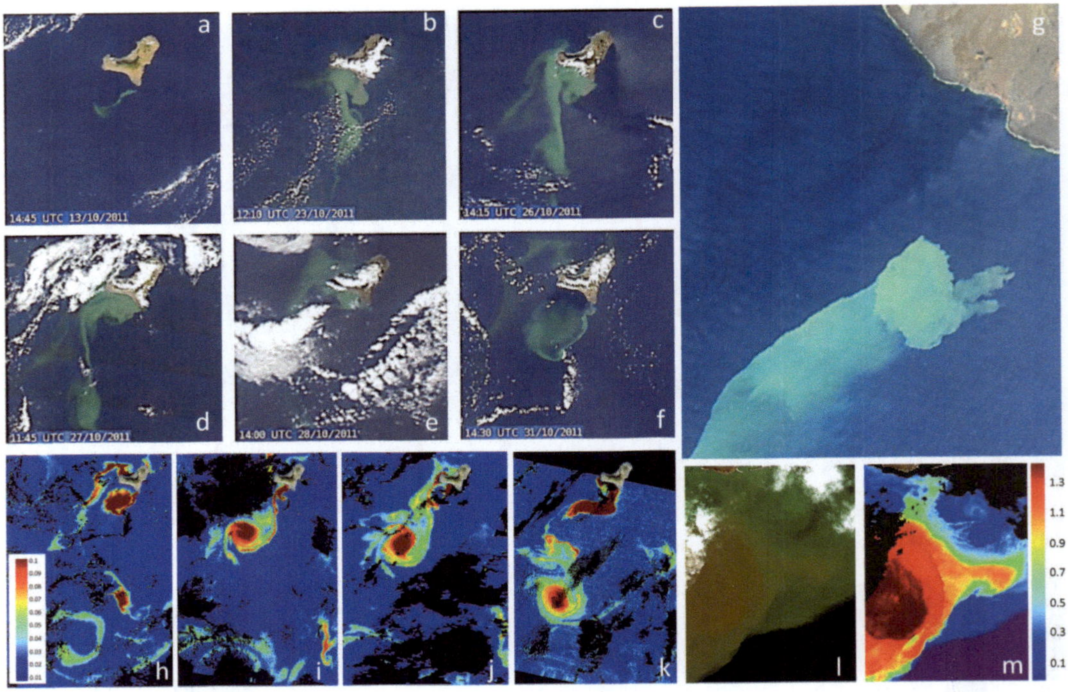

Fig. 8.7 Monitoring the surface perturbation of the Tagoro submarine volcano during October 2011, **a–f** NASA-MODIS multitemporal imagery and **g** RapidEye image of the October 13th, 2011. MODIS and MERIS K_d product (m^{-1}) for some days of 2011 (spatial resolution of 1 km and 300 m, respectively): **h** MODIS October, 31st, **i** MERIS November, 4th, **j** MODIS November 5th, and **k** MERIS November, 9th. Worldview-2 image of October 26th, 2011 (spatial resolution of 2 m): **l** true color composite (Bands 5–3–2), and **m** K_d product (m^{-1}) derived using the corrected bands

of sensors, with enhanced capabilities, providing data at higher spectral, spatial and temporal resolutions. To analyze the impacts of El Hierro submarine volcanic eruption, low and high spatial resolution imagery were used (Fraile-Nuez et al. 2012) (Fig. 8.7a–g). In particular, two fundamental water quality parameters were thoroughly studied: chlorophyll *a* concentration (Chl-*a*) and the diffuse attenuation coefficient, which allows the estimation of suspended and dissolved constituents in turbid waters, K_d.

In the last decades, different semi-analytical and empirical algorithms have been proposed to recover water quality indicators in coastal an open ocean waters. Unfortunately, these algorithms failed in El Hierro due to the substantial alterations in the water composition caused by the submarine eruption. Thus, accurate models were specifically developed for such singular scenario (Eugenio et al. 2014).

8.3.5.1 Low Resolution Data

MERIS and MODIS imagery were the main sources of remote sensing data used to assess the progress of the volcanic discharges, and to supply information of chemical, biological and physical parameters during the event. Such data played an essential role directing the oceanographic research vessels to the correct sampling sites during the field campaigns.

As mentioned, spacecraft estimations of Chl-*a* and K_d during these abnormal circumstances were evaluated. Comparing in situ measurements with derived MERIS and MODIS estimations, it was demonstrated that open ocean Chl-*a* algorithms completely fail, whereas the ALGAS-2 coastal model, developed for the MERIS instrument, achieves better accuracy but just in regions with lower-moderate turbidity. Likewise, the existing K_d algorithms for MERIS and MODIS sensors, in the surrounding waters around the volcanic eruption, were not acceptable, underestimating the real values. Accordingly, improved algorithms were developed for the retrieval of Chl-*a* and K_d in such turbid waters, demonstrating good correlation with the sampled measurements (Eugenio et al. 2014). Figure 8.7h–k display the volcanic plume evolution due to underwater volcanic activities using the diffuse attenuation coefficient parameter with low resolution MODIS and MERIS data.

Although the Canary Archipelago is a very active region in the generation of eddies, these structures are commonly studied and tracked using satellite sea level anomaly products that have a coarse spatial resolution of tens of kilometres and, thus, they are suitable for global studies but not very appropriate for coastal areas or detailed analysis at submesoscale levels. In that sense, the submarine eruption of Tagoro volcano provided an outstanding source of optical tracer that enabled the study of such oceanographic structures. Taking advantage of this tracer by the eruption and accomplishing ocean colour data, a comprehensive study at a much higher resolution allowed a better understanding of the physical process of axisymmetrization and filamentation predicted by theoretical studies and numerical modelling (Marcello et al. 2015).

8.3.5.2 Very High Resolution Data

Concerning very high resolution data, Worldview-2 imagery with 8 spectral bands at 2 m' resolution was used and the critical pre-processing steps were improved. Specifically, enhanced atmospheric correction and deglinting techniques were adapted to such unusual scenario. For WorldView-2, after applying the critical pre-processing steps, novel algorithms were developed with the aim to retrieve high resolution maps of the diffuse attenuation coefficient, total suspended matter and chlorophyll-*a* concentration. Algorithms were validated using in situ data, obtained during the oceanographic research cruises, and the results achieved confirmed the excellent performance under such challenging situation. Figure 8.7-l presents a true color composite image, using the WorldView-2 atmospheric corrected bands, and the diffuse attenuation coefficients at 490 nm. Green and brown colors in Fig. 8.7-m correspond to high volcanic emissions areas.

As a concluding remark, it is important to highlight that multitemporal and ultisensory satellite imagery, applying the appropriate processing techniques, becomes a powerful approach to monitor and understand submarine volcanic processes.

8.3.6 New Substrates: Annihilation of the Pre-eruptive Biological Communities and Establishment of Chemosynthesis-Based Communities

Baseline information on the marine ecosystem before the eruption and formation of Tagoro is essential to understand and evaluate the changes on the different components of the Tagoro ecosystem. Sotomayor-García et al. (see Chap. 4.3 of this book) provide an extensive review on baseline information on the marine habitats and associated biota of the area were Tagoro was formed as well as similar information on the current state of the marine habitats and associated benthic and demersal species. This baseline information is based on studies carried out on the same areas or nearby some years before the eruption.

Prior to reaching the present morphology, the Tagoro submarine volcano went through several eruptive events, overcoming two deflation main cone collapses (Fig. 8.8a; see Chap. 7). The eruption accumulated a total volume in the main edifice area of 102 × 10^6 m^3 of non-dense rock equivalent, and 227 × 10^6 m^3 in the pyroclastic apron and main edifice (Rivera et al. 2013). Such amount of new magmatic material covered the pre-existing substrate. This also caused the annihilation of the biota, mainly of benthic organisms and some demersal ones (e.g., fishes) (Fraile-Nuez et al. 2012), as well as the burial of complex habitats, including aggregations of gorgonians, antipatharians and sponges as the

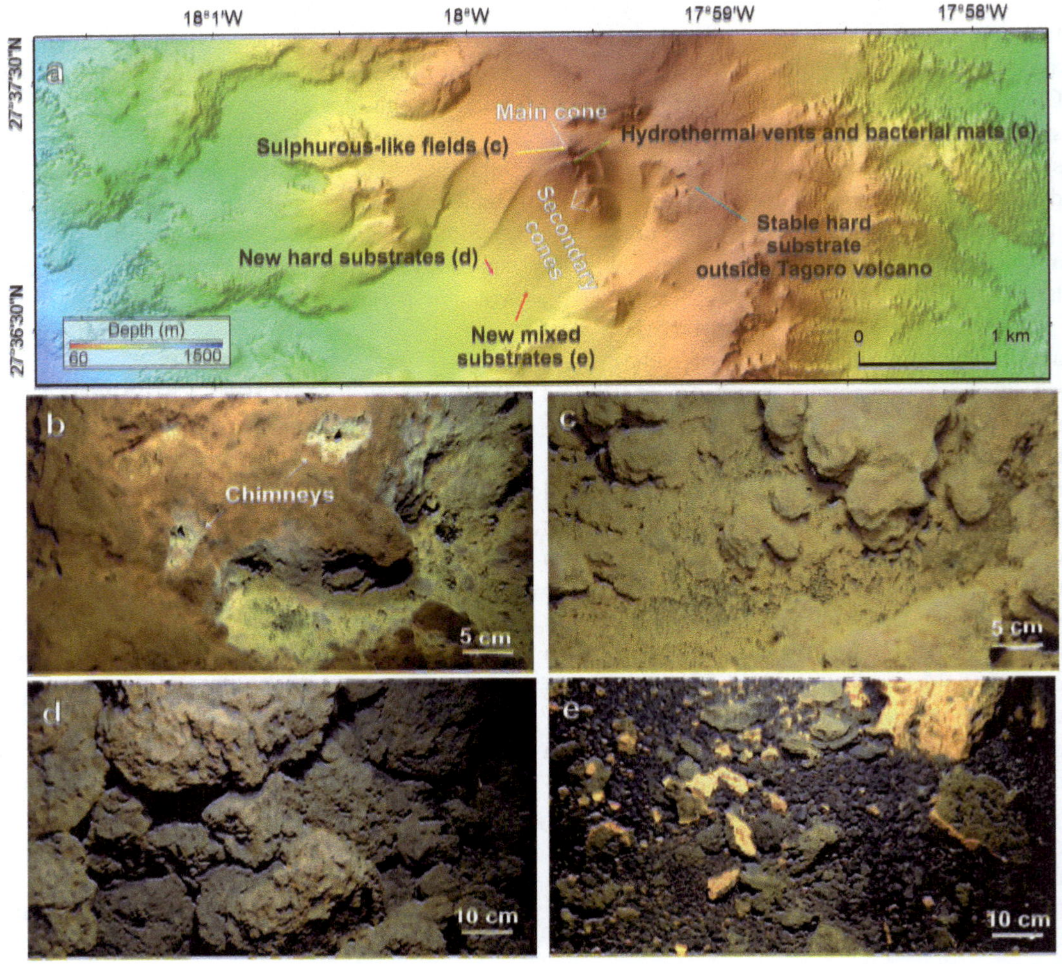

Fig. 8.8 **a** Bathymetric map off the southern insular slope of the El Hierro island with location of the Tagoro submarine volcano showing the main edifice and secondary volcanic cones. In this map the location of the main different types of new habitats are highlighted; **b–e** In situ photographs taken from VOR-Aphia 2012 remotely operated vehicle from the Instituto Español de Oceanografia showing the main habitats within the Tagoro volcano; **b** hydrothermal chimney-like structures, **c** the sulphurous–like fields, **d** the new hard substrates and **e** the new mixed substrates

ones detected in unburied areas located close by at similar depths (Sotomayor-García et al. 2019).

In specific areas of Tagoro, the new erupted materials as well as the fluid venting, facilitated the settlement and formation of new communities and habitats. In this context, basaltic lava blocks, pyroclastic deposits, and ash deposited in the new grounds, creating a new seabed (Danovaro et al. 2017; Somoza et al. 2017; Sotomayor-García et al. 2020). The new seabed was almost free of metazoans and some hydrothermal chimneys were formed in those areas where fluid venting was more acute. Sotomayor-García (2020) described different types of habitats according to their proximity to the main or secondary cones of the volcano. At the main edifice, two habitats with extremophile conditions were distinguished, including "hydrothermal vents with bacterial mats" (Fig. 8.8b), and "Sulphurous-like fields" (Fig. 8.8c). Both habitats were detected in the vicinities of the main and secondary cones, from 89 to 300 m depth, over hard substrates in areas with high slope (from 40° to 50°). The former habitat was characterized at the moment of the eruption by maximum extreme water temperature anomaly of + 18.8 °C and chemical anomalies around the hydrothermal vents (e.g., ΔpH up to − 2.9, oxidation reduction potential, ORP from 0.2 to − 0.03 V) (Fraile-Nuez et al. 2012; Santana-Casiano et al. 2013), which harbored some chimney-like structures (Fig. 8.8b). In areas with continuous fluid venting, extremophile bacterial communities have also been detected, including a recently described novel bacterium (Venus's Hair, Candidatus *Thiolava veneris*) (Danovaro et al. 2017). Then, surrounding the aforementioned habitat and with a lower fluid venting influence, vast fields of volcanic ash and fine yellowish sediment were detected and named as "Sulphurous-like fields" by Sotomayor-García et al. (2020). Oxidized deposits of S and Fe covered the area and characterized this habitat with lower fluid venting activity (Fig. 8.8c). These fields usually extend over large areas and sometimes cover basaltic blocks and pyroclasts. Those deposits are composed of fragile and easily resuspended material, and may create a barrier to a large number of hard substrate colonizers.

Two extra habitats were described further deep (between 300 and 900 m depth) in areas with a lower slope (10–20°): "Newly formed hard bottoms" (Fig. 8.8d) and "Newly formed mixed bottoms" (Fig. 8.8e). Both habitats were characterized by a seabed that was poorly colonized by metazoans, except for specific areas with few black corals (Stichopates cf. setacea) and hydrozans among other organisms, that were considered as survivors after the eruptions. These areas with survivors were detected at the distal part of the pyroclastic apron, where the arrival of erupted material would be less aggressive. The former habitat (Fig. 8.8d) was characterized by large pyroclastic rocks of tens of centimeters to meters in size, mostly volcanic bombs and lava-balloons, which are distributed chaotically creating a crumble of rocks with abundant caves and crevices, providing shelters and new substrates to colonizers. The latter habitat (Fig. 8.8e) was characterized by a combination of small to medium-sized rocks, gravel and fine sediment, and was mainly detected at the pyroclastic apron, but also around the former habitat.

8.3.7 Significant Changes in the Surrounding Bacterioplankton Communities

Bacterioplankton communities (i.e., bacteria and archaea) are major contributors to marine biodiversity and are a dominant component of the aquatic biota in terms of biomass and activity. These microbes, which represent the smallest biotic components of plankton, are capable of responding rapidly to alterations in the ecosystem′s conditions. After the eruption that gave rise to the Tagoro shallow submarine volcano, a series of oceanographic cruises allowed the study of the changes produced on the bacterioplankton communities (see Chap. 4.2 of this book). Initial surveys indicated, by means of flow cytometry, that the abundance of cyanobacteria, i.e., *Prochlorococcus* and *Synechococcus*, declined at depths near the main cone of the volcano as compared to unaffected stations (Fraile-Nuez et al. 2012). On the contrary, the abundance of heterotrophic prokaryotes increased with depth at stations affected by the volcanic emissions (Fraile-Nuez et al. 2012). Further exploration combining flow cytometry and high-throughput sequencing of 16S rRNA gene amplicons, a typical marker for prokaryotic diversity analysis, allowed to monitor the area around the volcano from the eruptive phase until it had ceased (November 2011 to April 2012). In particular, the effects of the eruption on the abundance, activity, diversity, and community structure of bacterioplankton communities were evaluated (Ferrera et al. 2015).

Bacterioplankon abundance data gathered from seven cruises showed that the number of prokaryotes in the surrounding waters of the volcano was generally higher during the eruption than in the post-eruption stages. Likewise, the percentage of cells with high nucleic acid content (HNA) was greater during the eruptive phase in the affected zone. Cells with HNA content tend to be more active (Gasol et al. 1999), thus, overall these results indicate that the eruption promoted an increase in the abundance and likely in the activity of heterotrophic prokaryotes. Nevertheless, the effect was observed during the eruptive phase and values returned to normal levels in the post-eruptive stages (Ferrera et al. 2015). In terms of diversity, as seen through the sequencing of the 16S rRNA gene in samples from four hydrographic cruises, the eruption promoted a decrease in bacterial richness, that is, the number of distinct bacterial taxa retrieved in samples collected during the eruption was lower than in samples collected in the following months (Fig. 8.9). This trend was however not observed for archaeal diversity. Similar observations were compiled in relation to the structure of bacterioplankton communities; the volcano induced changes in bacterial community structure while no major shifts were observed for Archaea. Yet, the limited amount of data generated for Archaea as compared to Bacteria could have resulted in an underestimation of the effect of the eruption on their diversity and community structure (Ferrera et al. 2015). Besides, changes in bacterioplankton composition were also observed; the percentage of major marine bacterial taxa such as Alphaproteobacteria and Bacteroidetes was lower in samples collected during the eruptive phase. Moreover, the presence of certain bacterial groups not commonly found in marine planktonic communities but that typically dominate marine hydrothermal environments, such as the Epsilonproteobacteria, was notorious at that stage. No archaeal groups associated with eruptive processes were retrieved from these samples. Nevertheless, most archaeal groups from hydrothermal systems are thermophilic or hyperthermophilic which limits the possibility that, even if they were released during the eruption, they could have developed in the water column.

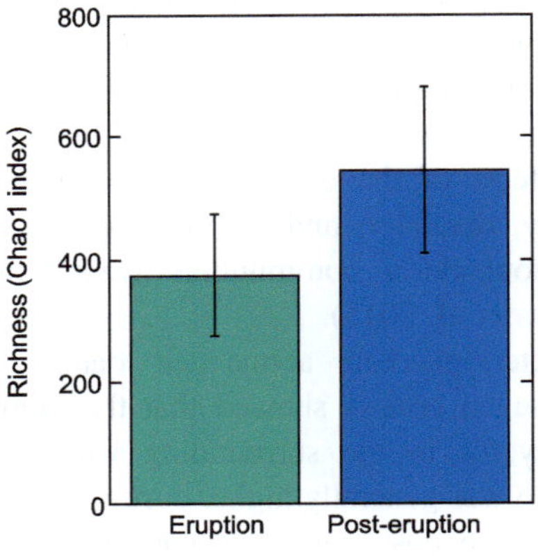

Fig. 8.9 Bacterial richness estimates (Chao1 index) in pelagic samples collected during the eruption and post-eruption stages (modified from Ferrera et al. 2015)

8.4 An Oasis of Life: Clear Signs of a Marine Ecosystem Recovery

8.4.1 Ocean Fertilization

The evolution of inorganic nutrients (Fe(II), $NO_3^- + NO_2^-$, PO_4, $Si(OH)_4$ and NH_4^+) were studied in the seawater around the Tagoro submarine volcano from the eruption to the degasification period (Fraile-Nuez et al. 2012; González-Vega et al. 2020; Santana-Casiano et al. 2018; 2013; Santana-González et al. 2017; see Chap. 9).

During the eruption period, high concentrations of Fe(II) and inorganic nutrients were measured in the area close to the volcano. The emissions of Fe(II) produced dissolved concentrations as high as 50 μmol kg^{-1} in the seawater compared to normal values in the area lower than 0.2 nmol l^{-1}. Maximum concentrations of macronutrients during the eruptive stage were also registered in the water column between 75–100 m depth with values up to 8.06 μmol kg^{-1} for $NO_3^- + NO_2^-$ (reference: 1.75 μmol kg^{-1}), 0.79 μmol kg^{-1} for PO_4 (reference: 0.11 μmol kg^{-1}), and 21.76 μmol kg^{-1} for $Si(OH)_4$ (reference: 1.23 μmol kg^{-1}); moreover, substantial concentration of these macronutrients often reached shallower layers of the water column between 0–50 m depth (González-Vega et al. 2020). This increase in Fe(II) concentrations together with the important input of macronutrients produced an enrichment in the water column. They provided the conditions for a fertilization event caused by the volcano emissions that have contributed to the regeneration of productivity in the area (González-Vega et al. 2020; Santana-Casiano et al. 2013).

Moreover, the diminution in pH decreased the Fe(II) oxidation rate and favoured the persistence of this bioactive metal in the medium. The oxidation of Fe(II) in seawater depends on the pH and the O_2 and H_2O_2 concentrations (González-Dávila et al. 2006; Santana-Casiano et al. 2005). Therefore, the decrease in pH, because of the volcano emissions, produced slower rates of Fe (II) oxidation in the area. However, the oxidation of Fe(II) for samples at the volcano affected area occurred more rapidly than for samples under normal conditions at the same pH, due to the presence of other components, in particular silicates, that accelerated the oxidation of Fe(II) in the area (Santana-González et al. 2017). The decrease in O_2 and H_2O_2 concentrations also favored the persistence of Fe(II) in the seawater (González-Dávila et al. 2005).

During the post-eruptive stage, the high resolution studies using *yo-yo* and *Tow-yo* techniques (Santana-Casiano et al. 2016; González-Vega et al. 2020) revealed positive anomalies in the vertical distribution of TdFe(II) concentration, which were also coincident with important negative anomalies in the pH signal (Fig. 8.4b), limited to an area and depth close to the main cone (Santana-González et al. 2017). The TdFe (II) concentrations measured in a close non-affected area and in surface waters were lower than 0.3 $nmolL^{-1}$, with some exceptions at the deep chlorophyll maxima (DCM) that presented an increase in TdFe(II) concentration in some of the stations, in the order of 1–2 $nmolL^{-1}$ (Santana-González et al. 2017). Using the *yo-yo* technique during October 2013 cruise, a TdFe(II) concentration as high as 48.92 $nmolL^{-1}$ and a pH of 7.91 were measured just above the main emitting vent. In March 2014, during the *yo-yo* study on the top of the secondary crater, a maximum value in TdFe(II) concentration was 1278.9 $nmolL^{-1}$ at around 100 m depth but decreased to 281 $nmolL^{-1}$ when were sampled only one and half hour later. The pH values changed from 7.76 to 7.84 at the same station. These variations could be due to small displacement in the sampling position between casts or due to changes in the amount of dissolved species emitted along the day by the hydrothermal vents, related to tidal effects as it was observed in the *yo-yo* studies (Santana-Casiano et al. 2016) or by changes in the periodicity of the emissions (Fraile-Nuez et al. 2018). In 2015, the maximum TdFe(II) measured was 44.6 $nmolL^{-1}$ with a pH of 8.0.

In the same way as Fe(II), remarkable differences in the macronutrients concentrations were observed with regard to the sampling methodology applied (González-Vega et al. 2020). Maximum phosphate and silicate concentrations were found in the samples collected directly from the vents with the use of the Liropus 2000 ROV from the Spanish Institute of Oceanography, with maxima of 1.26 and 189.40 $\mu mol\ kg^{-1}$, respectively; whereas for nitrate the maximum (6.21 $\mu mol\ kg^{-1}$) was found in the most mixed (water-column) samples and decreased when approaching the source (González-Vega et al. 2020). From these vent samples, ammonium concentrations were analyzed, finding maximums of 1.97 μM (reference: 0.02 μM), indicating that nitrogen is likely emitted in the form of ammonium rather than nitrate or nitrite. Average enrichments were obtained for all these nutrients, with water-column maxima of 8.8-fold (nitrate), 4.0-fold (phosphate) and 16.3-fold (silicate), and higher vent maxima of 10.5-fold (phosphate) and 325.4-fold (silicate) (González-Vega et al. 2020). These important emissions of macronutrients along with an enrichment and longer persistence of Fe(II) in the waters affected by the volcanic emissions contributed to the biological recovery of the system (Santana-Casiano et al. 2013; González-Vega et al. 2020).

8.4.2 The New Arrival: First Benthonic and Demersal Macro-colonizers

Despite the enormous amount of magmatic erupted material, and the unfavorable conditions in specific areas, some of the pre-eruptive habitats and associated biota managed to thrive in specific areas (Sotomayor-García et al. 2020; 2019). These habitats were locally impacted by the eruptive process through burial, mainly those conformed by antipatharians (*Stichopathes*, *Antipathes*, and *Tanacetipathes*) and hydrozoans. These were considered survivors of the eruption and also of the water column anomalies caused by the eruption (Sotomayor-García et al. 2020). Those organisms can be considered as relicts and may have acted as habitat-recovery promoters in those specific areas (Sotomayor-García et al. 2020). Nevertheless, pre-existing populations of most of the organisms diminished during and after the eruption and large areas were filled by erupted materials lacking metazoans. However, a large bulk of new colonizers settled and colonized the new substrates and habitats, once environmental conditions were similar to the pre-eruptive stage (Sotomayor-García et al. 2020). These new colonizers were mainly organisms with strong capacities, high mobility, fast growth rates and capability to adapt and benefit from scarce competition.

Around the main crater and the secondary cones, thermal and chemical anomalies were still detected a few years after the onset of the eruption (Fraile-Nuez et al. 2012; González-Vega et al. 2020), essentially due to fluid venting from specific hydrothermal vents. These vents provide unfavorable conditions to most benthic and demersal metazoans of the Macaronesian region, and no information on the presence of the typical vent invertebrates of the Atlantic Ocean is still available for Tagoro's extremophile habitats. Nevertheless, those anomalies generated the ideal conditions for extremophile bacteria to develop (Danovaro et al. 2017). Far away from those extremophile habitats, water masses without significant anomalies and big basaltic pyroclastic rocks, volcanic bombs, gravel and fine sediment, arranged in irregular fields. The ideal conditions were provided for the typical macaronesian fauna to settle and develop at this bottom type and depth range.

A study of the colonizing macro-fauna was published by Sotomayor-García (2019), based on video imagery and samples obtained with a benthic-dredge, which provided a list of 116 taxa from nine different phyla occurring in the area two years after the first eruption. From this bulk of species, it was possible to generate a picture of the organisms with the strongest capacities to adapt and develop in the erupted material, where shelter and symbiosis with other organisms was missing. A total of 93 different taxa were identified to be colonizers of these new habitats. Annelids, arthropods, cnidarians, and mollusks were the most diverse phyla, generally represented by small specimens (Fig. 8.10). Porifera was also present but in much lower representation than un-impacted areas located away from Tagoro. The most frequent taxa were similar among the different newly formed habitats, i.e., rocky and mixed bottoms, but the abundance was notably higher at the mixed bottoms due to the combination of typical organisms from hard and soft bottoms. Annelids were mainly represented by small sessile and mobile species (*Chloeia* cf. *venusta*—Fig. 8.10b, *Onuphis* sp.). *Monodaeus couchii* (Fig. 8.10a) and *Alpheus* sp. were the dominant arthropod species, and small hydrozoans (e.g., *Sterularella* cf. *tenella*, Fig. 8.10c) constituted one of the few cnidarians in these new substrates. On the other hand, mollusks were vastly represented by the ostreid bivalve *Neopycnodonte cochlear* (Fig. 8.10d), reaching densities of 100 individuals m^{-2} in some specific areas. Certain organisms, typical from undisturbed hard bottoms, represented the demersal community including pandalid shrimps (e.g., *Plesionika* spp.) and fishes that sometimes shelter in caves and crevices (e.g., *Anthias anthias*, *Conger conger*, *Gymnothorax* spp.,) (Sotomayor-García et al. 2019).

Colonization and initial recovery of the disturbed area by macro-fauna is viable thanks to the arrival of larvae or developed adult organisms from adjacent areas (Shanks 2009; Treml et al. 2012), and their settlement success in newly formed substrates (Gaines and Roughgarden 1985). Some of the identified species in Tagoro are mobile and, moreover, have planktotrophic development, which are important strategies for triggering a slow recovery process. Most of the observed individuals were juveniles (mainly

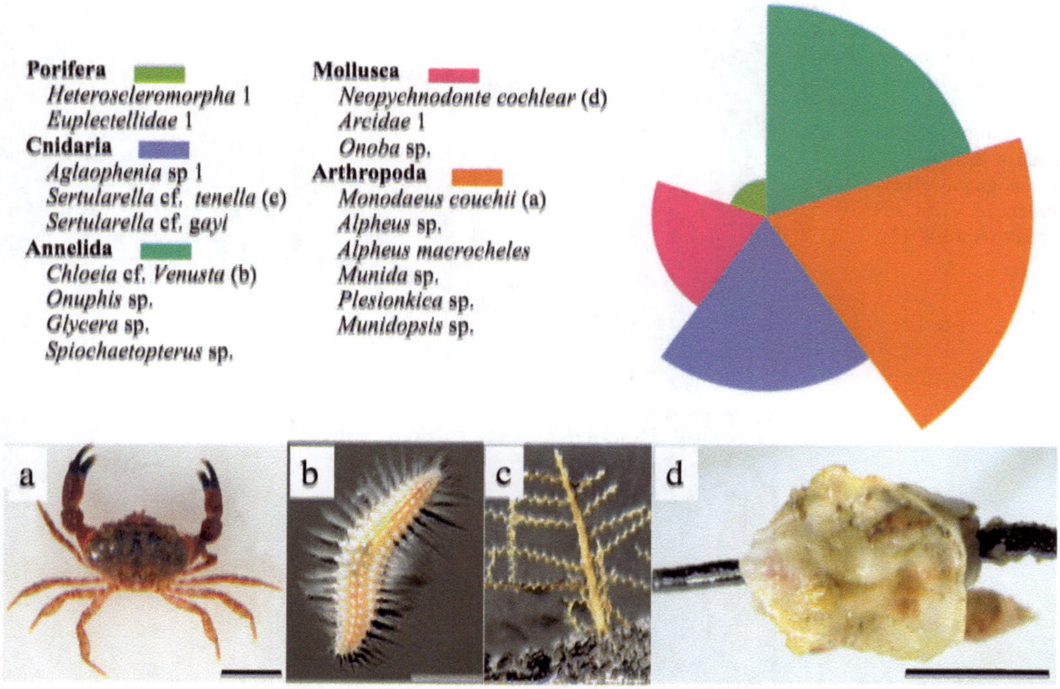

Fig. 8.10 At the top-right, the pie-chart visually shows the relative abundance of the phyla identified in the benthic dredge samples of Vulcano 0313. At the top-left, a list of the most abundant taxa of the identified phyla. At the bottom, the images illustrate the most representative species of the four most abundant phyla, arthopoda (**a**, Monodaeus couchii), annelida (**b**, Chloeia cf. venusta), cnidaria (**c**, Sertularella cf. tenella) and mollusc (**d**, Neopycnodonte cochlear)

from larvae drifting arrival), however, post-eruptive recruitment might have occurred for some species (e.g., *Glycera* sp. and *Chloeia* cf. *venusta*), as adult and juvenile individuals were observed in the newly formed bottoms.

8.4.3 Microbial Mats: The Discovery of a New Bacterium (*Thiolava Veneris*)

In November 2014, three years after the eruption, an expedition documented the presence of a whitish microbial mat massively covering the Tagoro submarine volcano. Microscopic analyses revealed that such colossal mat was produced by filaments made of bacterial trichomes enveloped within a sheath and colonized by epibiotic bacteria. Further investigations of these filaments resulted in the discovery of a new species of bacteria, commonly named Venus's Hair (Danovaro et al. 2017). Bacterial species are usually formally described after isolation in pure culture. Attempts to grow this filamentous species failed but nowadays powerful genetic tools allow to obtain the genomic information of uncultured microbes. Metagenomic analyses of the filaments allowed to identify not only a new species but also a new genus of the order Thiotrichales, within the Gammaprotobacteria class, named *Thiolava veneris*. This new species presents an unusual metabolic repertoire including the potential to utilize organic and inorganic carbon as well as the ability to uptake sulfur and nitrogen compounds. Perhaps this metabolic architecture explains why this organism was able to colonize the seafloor after the destruction that followed the submarine volcanic eruption, leading to a novel microbial habitat. What is unclear is the origin of these new colonizers; Venus's

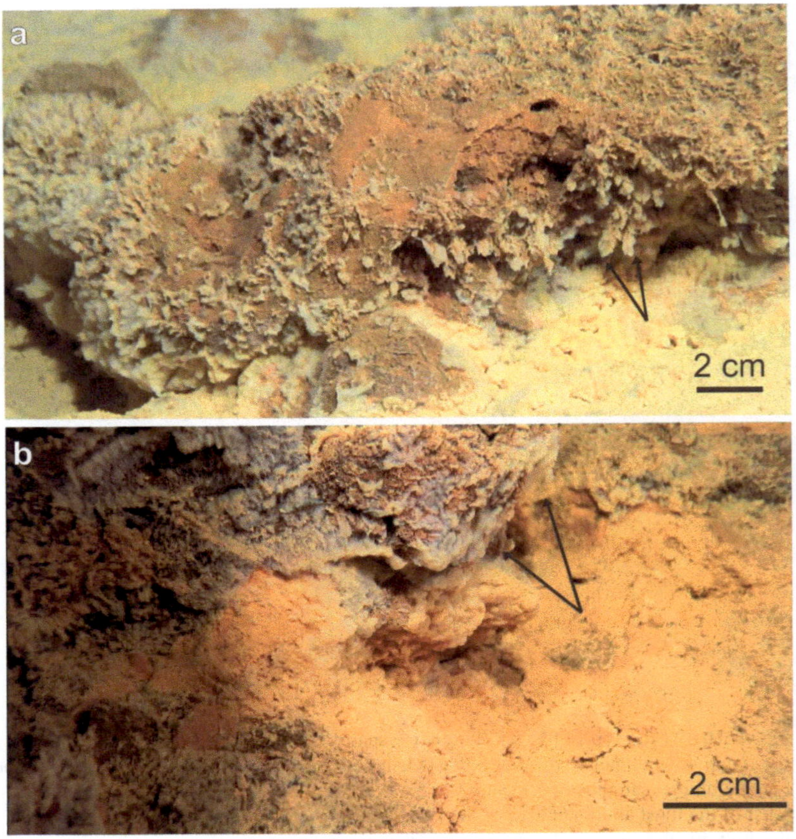

Fig. 8.11 a, b Bacterial mat likely formed by Thiolava veneris (white patches marked with black arrows) over hard substrates. Image captured by ROV Liropus 2000 from the Spanish Institute of Oceanography (IEO) in the context of VULCANA-II project (April 2017)

Hair was not detected in the surrounding seawater and, intriguingly, is macroscopically, phylogenetically and ecologically different from other known vent assemblages. How this organism was surviving before the eruption and how it spread remain a mystery. In any case, this discovery illustrates how despite the destruction that accompanies a volcanic eruption, a complex ecosystem eventually emerges (Witt et al. 2017). Nowadays the main crater area continues to be covered with the presence of long extension of mats likely formed by *Thiolava veneris* mats (Fig. 8.11).

8.5 Recommendations for Further Monitoring

The shallow Tagoro submarine volcano monitoring represents a unique opportunity not only for improving our sparse understanding of submarine volcanic processes in diverse fields such as physical and chemical oceanography or marine geology, but also the interactions among all of them and their synergic effects over the marine ecosystem.

Although our research over these ten years has addressed a broad number of studies mainly focused on the quantification, evolution and interactions of a relevant number of physical–chemical properties of the seawater in active submarine ambients (Ariza et al. 2014; Álvarez-Valero et al. 2018; Blanco et al. 2015; Coca et al. 2014; Danovaro et al. 2017; Eugenio et al. 2014; Fernández de Puelles et al. 2021; Ferrera et al. 2015; Fraile-Nuez et al. 2012, 2018; García-Davis et al. 2021; González-Vega et al. 2020, 2022; Olivé Abelló et al. 2021; Rivera et al. 2013; Santana-Casiano et al. 2013, 2016, 2018; Santana-González et al. 2017; Sotomayor-García et al. 2019, 2020), there are still many important questions that should be addressed in order to better understand this complex and interesting system. In a close future, all our physical–chemical monitoring effort should be focused on the long-term data acquisition with the use of multiparametric moorings, landers or even floats that could transmit data in real or quasi-real time. Nevertheless, the traditional data acquisition with the use of oceanographic vessels in the area are highly recommended and cannot be replaced. This includes the collection of large amounts of continuous CTD data and other physical–chemical sensors, in situ water samples collected with the oceanographic rosette for the analysis of numerous parameters, in situ experiments, tow-yo and yo-yo studies, etc.

From a microbiological perspective, the Tagoro volcano offers a unique opportunity to approach the long-standing question of microbial biogeography. The studies carried out up to date indicate that some of the taxa near the volcano were not present in the surrounding waters and it is unclear where Candidatus *Thiolava veneris* came from. A larger effort should be directed towards understanding whether the diversity of Tagoro's microorganisms are identical to those of other hydrothermal systems or the systems harbors its own local diversity. Further, investigating the presence of symbiotic bacteria in the biota that colonized the newly formed ecosystem would further contribute to this question since symbiotic bacteria can be selected by the environmental conditions but also by their hosts, which likewise can display strong biogeographical patterns.

Moreover, the Tagoro submarine volcano represents a unique opportunity for improving our knowledge on some key processes of marine benthic and demersal organisms of the Macaronesia, such as their colonization potential or resilience to ocean acidification. In order to improve the information on biological colonization, further monitoring should be done using non-invasive methods (ROV) in order to assess how the newly formed habitats are evolving towards complex habitats located nearby. This could be done in a 2–3 year periodicity in order to detect the arrival of those slow-growing organisms (e.g., gorgonians, antipatharians, scleractinians), because the information on the small opportunistic and fast-growing taxa is already available after the 3-years monitoring after the eruption. Moreover, our findings over Tagoro submarine volcano highlight the potential role of this eruptive process as a natural ecosystem-scale experiment for the study of extreme effects of global change stressors on marine environments.

Acknowledgements This ten years monitoring program over Tagoro submarine volcano has been possible thanks to the funding of several institutions and organizations: (i) Ministerio de Economía y Competitividad del Gobierno de España (MINECO) and FEDER through VULCANO-I (CTM2012-36317), VULCANO-II (CTM2014-51837-R), CETHOBATH (CGL2009-1311218), EXPLORA-CIENCIA (CGL2014- 61775-EXP), ECOFEMA (CTM2010-19517), EACFe (CTM2014-52342-P) projects. (ii) The Spanish Institute of Oceanography (IEO-CSIC) through BIMBACHE (IEO-2011-2012), RAPROCAN-III (IEO-2010-2012), VULCANA-I (IEO-2015- 2017), VULCANA-II (IEO-2018-2020) and VULCANA-III (IEO-CSIC-2021-2023) projects. (iii) Special funding from IEO-CSIC for the use of the ROV Liropus 2000 in several cruises. (iv) Fundación Caja-Canarias through CYCLOVENT (2019SP18) project. (v) The European project CARBOCHANGE-264879. We would also like to thank the extraordinary collaboration with the Secretaría General de Pesca (MAPA) from the Spanish Ministry, Cabildo de El Hierro, Ayuntamiento de El Pinar, Marine Reserve of La Restinga-Mar de las Calmas, SASEMAR, and every single person in El Hierro island without its help and support, this long monitoring would never have been carried out. Finally, the authors are grateful to Pablo González, editor of this book, for his support during the writing process.

References

Álvarez-Valero AM, Burgess R, Recio C, de Matos V, Sánchez-Guillamón O, Gómez-Ballesteros M, Recio G, Fraile-Nuez E, Sumino H, Flores JA, Ban M, Geyer A, Bárcena MA, Borrajo J, Compaña JM (2018) Noble gas signals in corals predict submarine volcanic eruptions. Chem Geol 480:28–34. https://doi.org/10.1016/j.chemgeo.2017.05.013

Ariza A, Kaartvedt S, Røstad A, Garijo JC, Arístegui J, Fraile-Nuez E, Hernández-León S (2014) The submarine volcano eruption off El Hierro Island: effects on the scattering migrant biota and the evolution of the pelagic communities. PLoS ONE 9:e102354–e102411. https://doi.org/10.1371/journal.pone.0102354

Ariza A, Landeira JM, Escánez A, Wienerroither R, Aguilar de Soto N, Røstad A, Kaartvedt S, Hernández-León S (2016) Vertical distribution, composition and migratory patterns of acoustic scattering layers in the Canary Islands. J Mar Syst 157:82–91. https://doi.org/10.1016/j.jmarsys.2016.01.004

Bakalis E, Mertzimekis TJ, Nomikou P, Zerbetto F (2017) Breathing modes of Kolumbo submarine volcano (Santorini, Greece). Sci Rep 7:46515. https://doi.org/10.1038/srep46515

Bakalis E, Mertzimekis TJ, Nomikou P, Zerbetto F (2018) Temperature and conductivity as indicators of the morphology and activity of a submarine volcano: Avyssos (nisyros) in the south aegean sea, Greece. Geosciences (Switzerland) 8. https://doi.org/10.3390/geosciences8060193

Baker ET, Massoth GJ (1987) Characteristics of hydrothermal plumes from two vent fields on the Juan de Fuca Ridge, northeast Pacific Ocean. Earth Planet Sci Lett 85:59–73

Blanco MJ, Fraile-Nuez E, Felpeto A, Santana-Casiano JM, Abella R, Fernández-Salas LM, Almendros J, Díaz-del-Río V, Domínguez Cerdeña I, García-Cañada L, González-Dávila M, López C, López-González N, Meletlidis S, Vázquez JT (2015) Comment on "Evidence from acoustic imaging for submarine volcanic activity in 2012 off the west coast of El Hierro (Canary Islands, Spain)" by Pérez NM, Somoza L, Hernández PA, González de Vallejo L, León R, Sagiya T, Biain A, González FJ, Medialdea T, Barrancos J, Ibáñez J, Sumino H, Nogami K and Romero C [Bull Volcanol (2014) 76:882–896]. Bull Volcanol 77:335. https://doi.org/10.1007/s00445-015-0947-6

Botz R, Winckler G, Bayer R, Schmitt M, Schmidt M, Garbe-Schönberg D, Stoffers P, Kristjansson JK (1999) Origin of trace gases in submarine hydrothermal vents of the Kolbeinsey Ridge, north Iceland. Earth Planet Sci Lett 171:83–93. https://doi.org/10.1016/S0012-821X(99)00128-4

Buck NJ, Resing JA, Baker ET, Lupton JE (2018) Chemical fluxes from a recently erupted shallow submarine volcano on the Mariana Arc. Geochem Geophys Geosyst 19:1660–1673. https://doi.org/10.1029/2018GC007470

Butterfield DA, Massoth GJ, McDuff RE, Lupton JE, Lilley MD (1990) Geochemistry of hydrothermal fluids from Axial Seamount hydrothermal emissions study vent field, Juan de Fuca Ridge: subseafloor boiling and subsequent fluid-rock interaction. J Geophys Res 95:12895–12921. https://doi.org/10.1029/JB095iB08p12895

Cardigos F, Colaço A, Dando PR, Ávila SP, Sarradin PM, Tempera F, Conceição P, Pascoal A, Serrão Santos R (2005) Shallow water hydrothermal vent field fluids and communities of the D. João de Castro Seamount (Azores). Chem Geol 224:153–168. https://doi.org/10.1016/j.chemgeo.2005.07.019

Coca J, Ohde T, Redondo A, García-Weil L, Santana-Casiano M, González-Dávila M, Arístegui J, Nuez EF, Ramos AG (2014) Remote sensing of the El Hierro submarine volcanic eruption plume. Int J Remote Sens 35:6573–6598. https://doi.org/10.1080/01431161.2014.960613

Crisp JA (1984) Rates of magma emplacement and volcanic output. J Volcanol Geoth Res 20:177–211. https://doi.org/10.1016/0377-0273(84)90039-8

Von Damm K (1990) Seafloor hydrothermal activity: black smoker chemistry and chimneys. Ann Rev Earth Planet Sci 18:173–204. https://doi.org/10.1146/annurev.earth.18.1.173

Dando PR, Stüben D, Varnavas SP (1999) Hydrothermalism in the Mediterranean Sea. Prog Oceanogr 44:333–367

Danovaro R, Canals M, Tangherlini M, Dell'Anno A, Gambi C, Lastras G, Amblas D, Sanchez-Vidal A, Frigola J, Calafat AM, Pedrosa R, Rivera J, Rayo X, Corinaldesi C (2017) A submarine volcanic eruption leads to a novel microbial habitat. Nat Ecol Evol 1:0144–0148. https://doi.org/10.1038/s41559-017-0144

Domínguez-Cerdeña ID, del Fresno C, Moreno AG (2014) Seismicity patterns prior to the 2011 El Hierro Eruption. Bull Seismol Soc Am 104(1):567–575. https://doi.org/10.1785/0120130200

Eugenio F, Martin J, Marcello J, Fraile-Nuez E (2014) Environmental monitoring of El Hierro Island submarine volcano, by combining low and high resolution satellite imagery. Int J Appl Earth Observations Geoinf 29:53–66. https://doi.org/10.1016/j.jag.2013.12.009

Fernández de Puelles ML, Gazá M, Cabanellas M, Herrera I, González-Vega A, Presas-Navarro C, Arrieta JM, Fraile-Nuez E (2021) Abundance and structure of the zooplankton community during a post-eruptive process: the case of the submarine volcano Tagoro (El Hierro; Canary islands) 2013–2018. Front Mar Sci 8:692885. https://doi.org/10.3389/fmars.2021.692885

Ferrera I, Arístegui J, González JM, Montero MF, Fraile-Nuez E, Gasol JM (2015) Transient changes in bacterioplankton communities induced by the submarine volcanic Eruption of El Hierro (Canary Islands). PLoS ONE 10:1–24. https://doi.org/10.1371/journal.pone.0118136

Fraile-Nuez E, González-Dávila M, Santana-Casiano JM, Arístegui J, Alonso-González IJ, Hernández-León S, Blanco MJ, Rodríguez-Santana A, Hernández-Guerra A, Gelado-Caballero MD, Eugenio F, Marcello J, De Armas D, Domínguez-Yanes JF, Montero MF, Laetsch DR, Vélez-Belchí P, Ramos A, Ariza AV, Comas-Rodríguez I, Benítez-Barrios VM (2012) The submarine volcano eruption at the island of El Hierro: physical-chemical perturbation and biological response. Sci Rep 2:486. https://doi.org/10.1038/srep00486

Fraile-Nuez E, Santana-Casiano J, González-Dávila M, Vázquez J, Fernández-Salas L, Sánchez-Guillamón O, Palomino D, Presas-Navarro C (2018) Cyclic behavior associated with the degassing process at the shallow submarine volcano Tagoro, canary Islands, Spain. Geosciences 8:457. https://doi.org/10.3390/geosciences8120457

Fricke H, Giere O, Stetter K, Alfredsson GA, Kristjansson JK, Stoffers P, Svavarsson J (1989) Hydrothermal vent communities at the shallow subpolar Mid-Atlantic ridge. Mar Biol 102:425–429. https://doi.org/10.1007/BF00428495

Furushima Y, Nagao M, Suzuki A, Yamamoto H, Maruyama T (2009) Periodic behavior of the bubble jet (Geyser) in the taketomi submarine hot springs of the southern part of yaeyama archipelago Japan. Mar Technol Soc J 43:13–22. https://doi.org/10.4031/MTSJ.43.3.1

Gaines S, Roughgarden J (1985) Larval settlement rate: a leading determinant of structure in an ecological community of the marine intertidal zone. Proc Natl Acad Sci 82:3707–3711. https://doi.org/10.1073/pnas.82.11.3707

García-Davis S, Reyes CP, Lagunes I, Padrón JM, Fraile-Nuez E, Fernández JJ, Díaz-Marrero AR (2021) Bioprospecting antiproliferative marine microbiota from submarine volcano Tagoro. Front Mar Sci 8:1–18. https://doi.org/10.3389/fmars.2021.687701

Gasol JM, Li Zweifel U, Peters F, Fuhrman JA, Hagström Å (1999) Significance of size and nucleic acid content heterogeneity as measured by flow cytometry in natural planktonic bacteria. Appl Environ Microbiol 65:4475–4483. https://doi.org/10.1128/aem.65.10.4475-4483.1999

Gómez-Letona M, Arístegui J, Ramos AG, Montero MF, Coca J (2018) Lack of impact of the El Hierro (Canary Islands) submarine volcanic eruption on the local phytoplankton community. Sci Rep 8(4667):1–12. https://doi.org/10.1038/s41598-018-22967-6

González-Dávila M, Santana-Casiano JM, Millero FJ (2005) Oxidation of iron (II) nanomolar with H_2O_2 in seawater. Geochim Cosmochim Acta 69:83–93. https://doi.org/10.1016/j.gca.2004.05.043

González-Dávila M, Santana-Casiano JM, Millero FJ (2006) Competition between O_2 and H_2O_2 in the oxidation of Fe(II) in natural waters. J Solution Chem 35:95–111. https://doi.org/10.1007/s10953-006-8942-3

González-Vega A, Fraile-Nuez E, Santana-Casiano JM, González-Dávila M, Escánez-Pérez J, Gómez-Ballesteros M, Tello O, Arrieta JM (2020) Significant release of dissolved inorganic nutrients from the shallow submarine volcano Tagoro (Canary Islands) based on seven-year monitoring. Front Mar Sci 6:2465. https://doi.org/10.3389/fmars.2019.00829

González-Vega A, Callery I, Arrieta JM, Santana-Casiano JM, Domínguez-Yanes JF, Fraile-Nuez E (2022) Severe Deoxygenation Event Caused by the 2011 Eruption of the Submarine Volcano Tagoro (El Hierro Canary Islands). Front Mar Sci 9:834691. https://doi.org/10.3389/fmars.2022.834691

Guieu C, Bonnet S, Petrenko A, Menkes C, Chavagnac V, Desboeufs K, Maes C, Moutin T (2018) Iron from a submarine source impacts the productive layer of the Western Tropical South Pacific (WTSP). Sci Rep 8:9075. https://doi.org/10.1038/s41598-018-27407-z

Hernández-León S, Gómez M, Arístegui J (2007) Mesozooplankton in the Canary current system: the coastal-ocean transition zone. Prog Oceanogr 74:397–421. https://doi.org/10.1016/j.pocean.2007.04.010

Longpré MA, Stix J, Klügel A, Shimizu N (2016) Mantle to surface degassing of carbon-and sulphur-rich alkaline magma at El Hierro, Canary Islands. Earth Planet Sci Lett 460:268–280. https://doi.org/10.1016/j.epsl.2016.11.043

Lupton JE, Delaney JR, Johnson HP, Tivey MK (1985) Entrainment and vertical transport of deep-ocean water by buoyant hydrothermal plumes. Nature 316:621–623. https://doi.org/10.1038/316621a0

Marcello J, Eugenio F, Estrada-Allis S, Sangrà P (2015) Segmentation and tracking of anticyclonic eddies during a submarine volcanic eruption using ocean colour imagery. Sensors 15:8732–8748. https://doi.org/10.3390/s150408732

Olivé-Abelló A, Vinha B, Machín F, Zerbetto F, Bakalis E, Fraile-Nuez E (2021) Analysis of volcanic thermohaline fluctuations of tagoro submarine volcano (El Hierro Island, canary islands, Spain). Geosciences (Switzerland) 11:374. https://doi.org/10.3390/geosciences11090374

Rivera J, Lastras G, Canals M, Acosta J, Arrese B, Hermida N, Micallef A, Tello O, Amblas D (2013) Construction of an oceanic island: Insights from the El Hierro (Canary Islands) 2011–2012 submarine volcanic eruption. Geology 41:355–358. https://doi.org/10.1130/G33863.1

Rizzo AL, Caracausi A, Chavagnac V, Nomikou P, Polymenakou PN, Mandalakis M, Kotoulas G, Magoulas A, Castillo A, Lampridou D, Marusczak N, Sonke JE (2019) Geochemistry of CO2-rich gases venting from submarine volcanoes: the case of Kolumbo (Hellenic Volcanic Arc, Greece). Front Earth Sci 7:60. https://doi.org/10.3389/feart.2019.00060

Rona PA, Bemis KG, Jones CD, Jackson DR, Mitsuzawa K, Silver D (2006) Entrainment and bending in a major hydrothermal plume, Main Endeavour Field, Juan de Fuca Ridge. Geophys Res Lett 33:L19313. https://doi.org/10.1029/2006GL027211

Santana-Casiano JM, González-Dávila M, Millero FJ (2005) Oxidation of nanomolar level of Fe(II) with oxygen in natural waters. Environ Sci Technol 39:2073–2079. https://doi.org/10.1021/es049748y

Santana-Casiano JM, Fraile-Nuez E, González-Dávila M, Baker ET, Resing JA, Walker SL (2016) Significant discharge of CO_2 from hydrothermalism associated with the submarine volcano of El Hierro Island. Sci Rep 6:1–9. https://doi.org/10.1038/srep25686

Santana-Casiano JM, González-Dávila M, Fraile-Nuez E, De Armas D, González AG, Domínguez-Yanes JF, Escanez J (2013) The natural ocean acidification and fertilization event caused by the submarine eruption of El Hierro. Sci Rep 3:1140. https://doi.org/10.1038/srep01140

Santana-Casiano JM, González-Dávila M, Fraile-Nuez E (2018) The emissions of the Tagoro submarine volcano (Canary Islands, Atlantic Ocean): effects on the physical and chemical properties of the seawater. In: Volcanoes—geological and geophysical setting, theoretical aspects and numerical modeling, applications to industry and their impact on the human health. InTech, pp 1–21. https://doi.org/10.5772/intechopen.70422

Santana-González C, Santana-Casiano JM, González-Dávila M, Fraile-Nuez E (2017) Emissions of Fe(II) and its kinetic of oxidation at Tagoro submarine volcano, El Hierro. Mar Chem 195:129–137. https://doi.org/10.1016/j.marchem.2017.02.001

Shanks AL (2009) Pelagic larval duration and dispersal distance revisited. Biol Bull 216:373–385. https://doi.org/10.1086/BBLv216n3p373

Somoza L, González FJ, Barker SJ, Madureira P, Medialdea T, de Ignacio C, Lourenço N, León R, Vázquez JT, Palomino D (2017) Evolution of submarine eruptive activity during the 2011–2012 El Hierro event as documented by hydroacoustic images and remotely operated vehicle observations. Geochem Geophys Geosyst 18:3109–3137. https://doi.org/10.1002/2016GC006733

Sotomayor-García A, Rueda JL, Sánchez-Guillamón O, Urra J, Vázquez JT, Palomino D, Fernández-Salas LM, López-González N, González-Porto M, Santana-Casiano JM, González-Dávila M, Fraile-Nuez E (2019) First macro-colonizers and survivors around tagoro submarine Volcano, Canary Islands, Spain. Geosci 9(1):52

Sotomayor-Garcia A, Rueda JL, Sánchez-Guillamón O, Vázquez JT, Palomino D, Fernández-Salas LM, López-González N, González-Porto M, Urra J, Santana-Casiano JM, González-Dávila M, Fraile-Nuez E (2020) Geomorphic features, main habitats and associated biota on and around the newly formed Tagoro submarine volcano, Canary Islands, Seafloor Geomorphology as Benthic Habitat. (pp 835–846) Elsevier

Speer KG (1989) a forced baroclinic vortex around a hydrothermal plume. Geophys Res Lett 16:461–464. https://doi.org/10.1029/GL016i005p00461

Tassi F, Capaccioni B, Caramanna G, Cinti D, Montegrossi G, Pizzino L, Quattrocchi F, Vaselli O (2009) Low-pH waters discharging from submarine vents at Panarea Island (Aeolian Islands, southern Italy) after the 2002 gas blast: origin of hydrothermal fluids and implications for volcanic surveillance. Appl Geochem 24:246–254. https://doi.org/10.1016/j.apgeochem.2008.11.015

Thompson G (1983) Hydrothermal Fluxes in the Ocean. In: chemical Oceanography. Elsevier, pp 271–337. https://doi.org/10.1016/B978-0-12-588608-6.50011-0

Treml EA, Roberts JJ, Chao Y, Halpin PN, Possingham HP, Riginos C (2012) Reproductive output and duration of the pelagic larval stage determine seascape-wide connectivity of marine populations. In: Presented at the integrative and comparative biology, pp 525–537. https://doi.org/10.1093/icb/ics101

Uiblein F, Bordes F (1999) Complex trophic interactions around ocean islands. Ocean Challenge

Wilson ST, Hawco NJ, Armbrust EV, Barone B, Björkman KM, Boysen AK, Burgos M, Burrell TJ, Casey JR, DeLong EF, Dugenne M, Dutkiewicz S, Dyhrman ST, Ferrón S, Follows MJ, Foreman RK, Funkey CP, Harke MJ, Henke BA, Hill CN, Hynes AM, Ingalls AE, Jahn O, Kelly RL, Knapp AN, Letelier RM, Ribalet F, Shimabukuro EM, Tabata RKS, Turk-Kubo KA, White AE, Zehr JP, John S, Karl DM (2019) Kīlauea lava fuels phytoplankton bloom in the North Pacific Ocean. Science 365:1040–1044. https://doi.org/10.1126/science.aax4767

Witt V, Ayris PM, Damby DE, Cimarelli C, Kueppers U, Dingwell DB, Wörheide G (2017) Volcanic ash supports a diverse bacterial community in a marine mesocosm. Geobiology 15:453–463. https://doi.org/10.1111/gbi.12231

9 Tagoro Submarine Volcano as a Natural Source of Significant Dissolved Inorganic Nutrients

Alba González-Vega, Jesús M. Arrieta, Magdalena Santana-Casiano, Melchor González-Dávila, Carolina Santana-González, Jesús M. Mercado, José Escánez-Pérez, Carmen Presas-Navarro, and Eugenio Fraile-Nuez

Abstract

The shallow submarine volcano Tagoro releases high amounts of inorganic nutrients (N, P, Si, Fe) into the surrounding waters. These emissions have been intensely monitored during both the eruptive stage (October 2011–March 2012) and the post-eruptive hydrothermal stage (March 2012—ongoing). The obtained seven-years dataset comprises over 3300 water samples analysed for concentrations of silicate ($Si(OH)_4$), phosphate (PO_4), nitrate + nitrate($NO_3^- + NO_2^-$), ammonium (NH_4^+), and iron ($Fe(II)$) in the area affected by the volcanic emissions (> 250 km^2) as well as outside of this area for reference. This chapter provides an overview on the main results obtained from this comprehensive dataset. as well as contextualizing these results by comparison with other volcanic and hydrothermal nutrient sources in the world. The results show that the hydrothermal emissions from Tagoro volcano cause nutrient enrichments both in the vicinities of the vents (88–130 m depth) and the water column up to 50 m above the seabed. The emissions can occasionally also be injected into the mixed layer and might even reach the surface. The transport of these emissions presents a flux of comparable magnitude to other important nutrient fluxes in the region, such as those related to the NW-African coastal upwelling, when compared per unit of area. We highlight the importance of accounting for shallow hydrothermal and volcanic inputs worldwide as significant nutrient sources that can exert an important influence on the ecosystems of superficial nutrient-poor waters.

A. González-Vega
Universidad de Las Palmas de Gran Canaria (ULPGC), 35017 Las Palmas de Gran Canaria, Spain

A. González-Vega (✉) · J. M. Arrieta · J. Escánez-Pérez · C. Presas-Navarro · E. Fraile-Nuez (✉)
Centro Oceanográfico de Canarias, Instituto Español de Oceanografía (IEO), Consejo Superior de Investigaciones Científicas (CSIC), 38180 Santa Cruz de Tenerife, Spain
e-mail: alba.gonzalez@ieo.csic.es

E. Fraile-Nuez
e-mail: eugenio.fraile@ieo.csic.es

M. Santana-Casiano · M. González-Dávila · C. Santana-González
Instituto de Oceanografía y Cambio Global (IOCAG), Universidad de Las Palmas de Gran Canaria, 35017 Las Palmas de Gran Canaria, Spain

J. M. Mercado
Centro Oceanográfico de Málaga, Instituto Español de Oceanografía (IEO), Consejo Superior de Investigaciones Científicas (CSIC), 29640 Málaga, Spain

P. J. González (ed.), *El Hierro Island*, Active Volcanoes of the World,
https://doi.org/10.1007/978-3-031-35135-8_9

Keywords

Physical oceanography · Marine ecosystems · Dissolve inorganic nutrients · El Hierro submarine volcanic eruption · Canary Islands · Atlantic Ocean

9.1 Introduction

Inorganic nutrients are molecules required for the growth of phytoplankton. These nutrients have long been recognized to play a key role in controlling ocean productivity, which accounts for about half of the carbon fixed by photosynthesis on Earth (Field et al. 1998). Inorganic nutrients tend to be present in relatively small concentrations in the euphotic zone (Chavez et al. 2011). Thus, studying their inputs and cycles is key to understand the patterns of growth limitations of marine phytoplankton.

Two of the most important nutrients in seawater are nitrogen (N) and phosphate (P), essential for cellular structures and processes (Paytan and McLaughlin 2007; Zehr and Kudela 2011). Additionally, silicon (Si) is required for the growth of diatoms (Tréguer and De La Rocha 2013). But even in zones where these three elements are in abundance, it has been observed that phytoplankton growth can still be limited by other elements, mainly iron (Fe) (Martin et al. 1994).

The most important supply of nutrients to surface waters comes from remineralization of organic matter in subsurface layers, which generates pools of regenerated nutrients that are then injected to the surface through physical processes such as vertical mixing or upwelling (Ducklow and Plank 2019). Additional inputs can also be significant and influence the patterns of nutrient limitation (Moore et al. 2013). Terrestrial inputs such as atmospheric and fluvial fluxes can be important, but not all inputs come from land. Nutrients can also be directly injected into the ocean through volcanic and hydrothermal activity (Karl et al. 1988a).

It is well known that submarine volcanic activity releases nutrients into seawater, including Fe, Si, P, and N (González-Vega et al. 2020; Karl et al. 1989, 1988b; Santana-Casiano et al. 2013; Sarradin et al. 1999; Tunnicliffe et al. 1986). However, most of the existing studies describing these emissions come from deep hydrothermal settings, at depths where the ambient nutrient concentrations are generally high. Very few studies have described volcanic or hydrothermal emissions of inorganic nutrients to shallow, nutrient-poor waters (González-Vega et al. 2020; Kilias et al. 2013; Santana-Casiano et al. 2013; Sedwick and Stuben 1996).

Tagoro submarine volcano releases significant amounts of inorganic nutrients (N, P, Si, Fe) to the surrounding waters since its eruption in 2011 (González-Vega et al. 2020; Santana-Casiano et al. 2013; Santana-González et al. 2017). During the eruptive stage (October 2011–March 2012), these emissions reached an extensive area around El Hierro island, particularly to the south and south-west (see Chap. 8). Since March 2012, Tagoro volcano entered a post-eruptive or degassing stage, with diffuse-flow hydrothermal discharge that affects a smaller area of roughly 500 m around the main cone and close to the seafloor (< 50 m) (Santana-Casiano et al. 2016). In this stage, the combined effect of the physical anomalies can generate a buoyant plume that returns downward forming vortex features (Speer 1989). Theoretically, this vortex could be of sufficient magnitude to trap water in a local recirculation that could have a crucial effect on the expected dispersal of properties, including minerals, tracers, and organisms. In that sense, enrichments of inorganic nutrients that are still found in the area (González-Vega et al. 2020) could both be transported into the surrounding waters or/and recirculated around the volcano.

The submarine eruption of Tagoro volcano and the subsequent hydrothermal activity represent a unique opportunity to observe and monitor the release of inorganic nutrients from an active shallow submarine volcano. This chapter constitutes a review on inorganic nutrient emissions from active submarine volcanoes and hydrothermal systems. We will focus on the 7-year dataset obtained from Tagoro volcano, providing the magnitude, spatial distribution and transport, and relative ratios of the main inorganic nutrients (N, P, Si, Fe). In parallel, we will put all this information in context by reviewing the existing studies on nutrient emissions from other active submarine volcanoes and hydrothermal systems worldwide.

9.2 Sampling Strategies

The emissions of inorganic nutrients (N, P, Si, Fe) from Tagoro submarine volcano were studied throughout 7 years of monitoring (2011–2018) comprising 15 oceanographic cruises carried out at an average of twice a year, mostly around March and October. Throughout these years of monitoring, the different sampling strategies applied played a central role in understanding the system.

The first sampling strategy consisted on vertical hydrographic stations distributed around El Hierro island (Fig. 9.1a), collecting water samples for nutrient analysis in the whole water column (sampling depths were generally consistent, with some variations depending on the vertical distribution of the volcanic plume at the time). During the eruptive stage, a wide area up to hundreds of km^2 was affected (see Sect. 9.3.1) and the distribution of the volcanic plume was very variable depending on local dynamics, therefore the spatial planning of the hydrographic stations varied accordingly following the plume distribution. For the degassing stage, a more consistent grid of stations was established around the island of El Hierro (Fig. 9.1a), which were consistently sampled in different oceanographic cruises. These stations mainly served as reference stations, since area affected by the emissions during the post-eruptive stage was drastically reduced to roughly 500 m around the main cone of the volcano. In order to adapt to these new circumstances, a high-resolution transect of vertical stations separated by tens of meters (20–140) was established across the volcanic edifice (Fig. 9.1b) following the alignment of the main and secondary cones (Santana-Casiano et al. 2016). This transect was consistently sampled in several oceanographic cruises during the post-eruptive stage.

Furthermore, seeing as the emissions in this stage were found mostly in the first 50 m above the seafloor, a methodology called tow-yo was also applied, consisting of a sawtooth-shaped sampling, lowering and raising the rosette between 1 and 40 m above the seabed while the vessel moves slowly (0.2–0.4 kn) (Santana-Casiano et al. 2016). These transects were carried out throughout the whole volcanic edifice and particularly centered over the main and secondary craters (squared area in Fig. 9.1b).

Finally, eight samples were also obtained directly from the hydrothermal vents (Fig. 9.1b) through the use of the remotely operated vehicle (ROV) Liropus 2000 from the Spanish Institute of Oceanography (IEO), using a syringe-like system specifically designed for this kind of sampling.

In total, about 3300 water samples were analyzed for the seven-years study of inorganic nutrients from Tagoro volcano, collected from 221 vertical hydrographic stations, 43 tow-yo transects, and 8 vent samples.

9.3 Nutrient Emissions from Tagoro Volcano

Significant anomalies of increased concentrations in inorganic nutrients were found in both the eruptive and the post-eruptive stages, underlining a nutrient discharge from Tagoro volcano (González-Vega et al. 2020; Santana-Casiano et al. 2013; Santana-González et al. 2017). In this section we will discuss separately the data of each inorganic nutrient: silicate ($Si(OH)_4$), phosphate (PO_4), nitrogen species (NO_3^- + NO_2^-, NH_4^+), and iron (TdFe(II)); while also contextualizing these data with a compilation of studies from other volcanic and hydrothermal emissions in the world, presented in Table 9.1.

The variety of sampling strategies discussed previously (see Sect. 9.2) is reflected in Fig. 9.2. Here, the concentrations of inorganic nutrients (Si, P, N, Fe) are presented as boxplots divided into five categories: (i) ambient nutrient

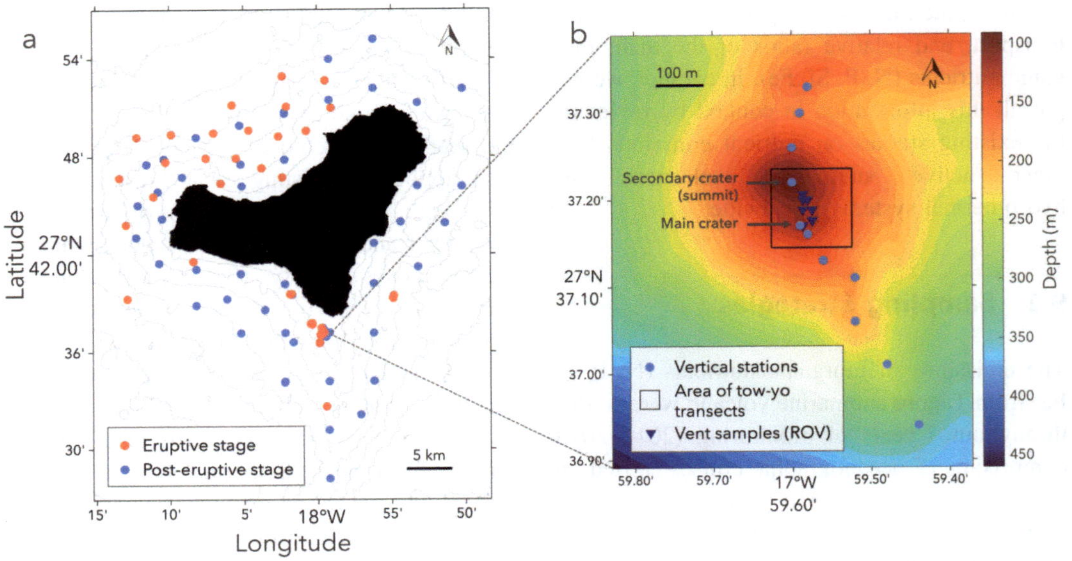

Fig. 9.1 **a** Map of vertical profiles carried out around the island of El Hierro during both the eruptive and post-eruptive stages. **b** Map of sampling during the post-eruptive stage in the Tagoro volcanic edifice, including a high-resolution transect of vertical profiles, an area around the main cone where the tow-yo transects were carried out, and discrete vent samples taken with ROV

concentrations from reference stations, unaffected by the volcano; (ii) concentrations during the eruptive stage in the water column; and concentrations during the post-eruptive stage: (iii) in the water-column (first 150 m); (iv) near the seafloor (tow-yo methodology); and (v) in the hydrothermal vents (collected with ROV).

9.3.1 Silicate

Volcanic and hydrothermal emissions constitute one of the main inputs of silica to the ocean (Tréguer and De La Rocha 2013). In fact, silicate is often used as a mixing tracer for hydrothermal fluids, since it presents a linear trend with temperature (Edmond et al. 1979; Sansone and Resing 1995; Sedwick and Stuben 1996). Significant emissions of Si have been widely described for hydrothermal vents and submarine volcanoes, with enrichments of one to three orders of magnitude with respect to ambient seawater (Karl et al. 1988a, 1989; Sedwick and Stuben 1996; Wilson et al. 2019).

In Tagoro volcano, silicate was found to be the most enriched nutrient by the volcanic emissions (Fig. 9.2a). Silicate concentrations under normal conditions in this area are typically very low (0.5 µmol kg^{-1}). During the eruption, concentrations as high as 22 µmol kg^{-1} were measured in the water column. In the post-eruptive stage, measurements from 16 samples taken directly from the vents with a ROV showed concentrations up to 189 µmol kg^{-1}, which represents an increase of three orders of magnitude with respect to normal conditions. The data show a decrease with the distance to the source, but remarkable concentrations up to 19 µmol kg^{-1} were still found in the water column (90–130 m depth).

9.3.2 Phosphate

Hydrothermal processes have often been described as sink within the phosphorus cycle, due to a net loss by formation of suspended iron oxyhydroxides and by direct uptake of phosphate

Table 9.1 Concentrations of inorganic nutrients in Tagoro submarine volcano and in other volcanic and hydrothermal settings. For each nutrient, reference data from unaffected waters (Ref.) and data from the area affected by the emissions (Aff.) are provided. Data may be provided as a single number, as mean ± standard deviation, or as a range (minimum–maximum). Depth of sampling and description of the sampling strategy are also specified

Study	Location	Depth	Sampling	$Si(OH)_4$		PO_4		$NO_3^- + NO_2^-$		NH_4^+		Fe	
		(m)		Ref	Aff	Ref	Aff	Ref	Aff	Ref	Aff	Ref	Aff
Santana-Casiano et al. 2013; González-Vega et al. 2020	**Tagoro submarine volcano (Canary Islands), eruptive stage**	10–100	Water column (vertical profiles)	0.45 µmol kg^{-1}	0.31–21.76 µmol kg^{-1}	0.04 µmol kg^{-1}	0.02–0.79 µmol kg^{-1}	0.41 µmol kg^{-1}	0.04–8.06 µmol kg^{-1}	–	–	–	–
González-Vega et al. 2020; Santana-González et al. 2017	**Tagoro submarine volcano (Canary Islands), post-eruptive stage**	90–130	Water column (vertical profiles)	0.81 µmol kg^{-1}	0.53–18.99 µmol kg^{-1}	0.08 µmol kg^{-1}	0.02–0.29 µmol kg^{-1}	1.53 µmol kg^{-1}	0.32–6.21 µmol kg^{-1}	–	–	0.3 nmol l^{-1}	0–44.6 nmol l^{-1}
González-Vega et al. 2020; Santana-González et al. 2017	**Tagoro submarine volcano (Canary Islands), post-eruptive stage**	123	Near seafloor (tow-yo)	0.31 µmol kg^{-1}	0.20–28.10 µmol kg^{-1}	0.11 µmol kg^{-1}	0.1–0.98 µmol kg^{-1}	1.47 µmol kg^{-1}	0.16–3.62 µmol kg^{-1}	–	–	0.3 nmol l^{-1}	2.2–1278.9 nmol l^{-1}
González-Vega et al. 2020	**Tagoro submarine volcano (Canary Islands), post-eruptive stage**	103–129	Vents (ROV)	0.53 µmol kg^{-1}	59.7–189.40 µmol kg^{-1}	0.12 µmol kg^{-1}	0.30–1.60 µmol kg^{-1}	1.44 µmol kg^{-1}	1.44–3.26 µmol kg^{-1}	0.02 µM	0.51–1.97 µM	–	–
Tunnicliffe et al. (1986)	Explorer Ridge (Juan de Fuca Ridge)	1900	Vents (submersible)	27.2 µg at l^{-1}	26.7–70.7 µg at l^{-1}	2.62 µg at l^{-1}	2.00–3.07 µg at l^{-1}	34.51 µg at l^{-1}	26.25–38.11 µg at l^{-1}	5.9 µg at l^{-1}	3.6–20.3 µg at l^{-1}	–	19,900 nmol
Sarradin et al. (1999)	Lucky Strike and Menez Gwen (Mid-Atlantic Ridge)	1700	Vents (submersible)	–	–	1.22 µmol l^{-1}	0.10–0.87 µmol l^{-1}	18.2 µmol l^{-1}	7.79–18.10 µmol l^{-1}	0.49 µmol l^{-1}	0.35–6.75 µmol l^{-1}	–	–
Karl et al. (1988a), (1989)	Pele's Vents (Loihi Seamount)	1000	Vents (submersible)	102.1 µM	108.6–1285 µM	2.93 µM	1.68–3.91 µM	42.10 µM	11.9–42.73 µM	0.72 µM	0.18–5.56 µM	0.05 µM	1010 µM
Sedwick and Stuben (1996)	Beach fumaroles at Vulcano Island (Aeolian Archipelago)	0.3	Fumaroles (funnel)	87.9 µmol kg^{-1}	631–1630 µmol kg^{-1}	0.04 µmol kg^{-1}	0.13–1.23 µmol kg^{-1}	0.13 µmol kg^{-1}	0.02–0.09 µmol kg^{-1}	19.2 µmol kg^{-1}	247–643 µmol kg^{-1}	1.25 µmol kg^{-1}	2.66–8.92 µmol kg^{-1}
Wilson et al. (2019)	Lava from Kilauea reaching seawater (Hawai'i)	5–45	Water column (vertical profiles)	2.9 µmol l^{-1}	23.3 ± 5.3 µmol l^{-1}	0.14 µmol l^{-1}	0.17 ± 0.04 µmol l^{-1}	0.03 µmol l^{-1}	3.1 ± 1.4 µmol l^{-1}	3.3 nmol l^{-1}	47.6 ± 4.9 nmol l^{-1}	–	–
Karl et al. (1988b)	Guaymas Basin (Gulf of California)	2000	Vents (submersible)	–	–	3.22 µM	0.8–3.37 µM	38.68 µM	2.82–35.80 µM	0.84 µM	505–11,559 µM	–	–

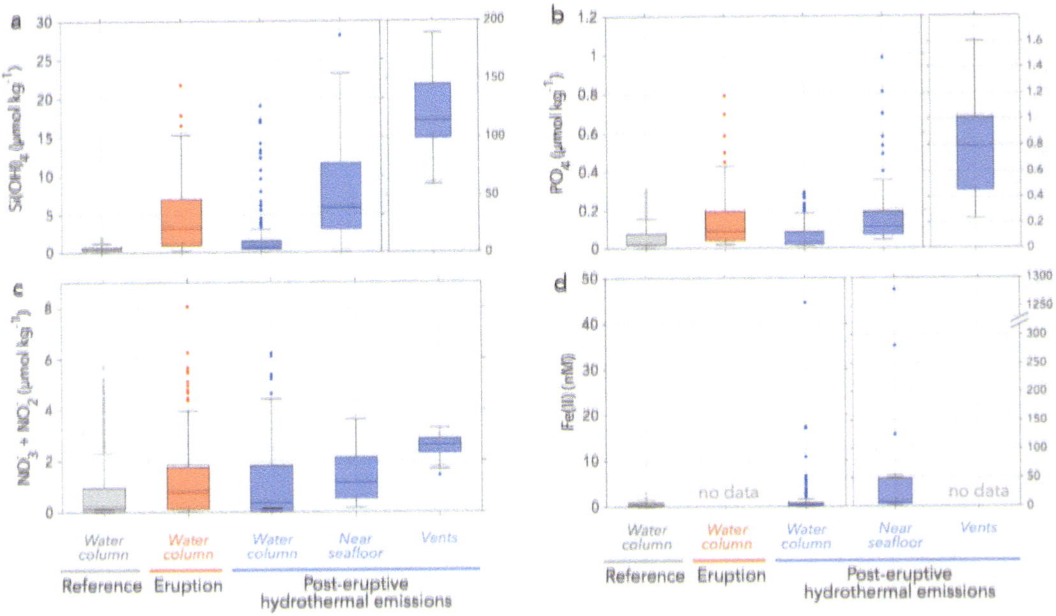

Fig. 9.2 Boxplots of nutrient concentrations in reference stations (grey), eruptive stage (orange), and post-eruptive stage (blue). The boxes are also classified according to the sampling strategy followed: vertical profiles for water-column samples (data from the first 150 m), tow-yos for near-seafloor samples, and ROV for vent samples

through reaction with basalt (Föllmi 1996; Paytan and McLaughlin 2007). This is typically true for deep hydrothermal systems (Karl et al. 1988b; Sarradin et al. 1999). However, in shallow systems, even with this loss, a net enrichment can occur given the extremely low concentrations in the ambient seawater (Sedwick and Stuben 1996; Wilson et al. 2019).

This is also the case for Tagoro volcano (Fig. 9.2b). In comparison to the ambient concentrations of around 0.1 μmol kg^{-1}, concentrations of up to 1.60 μmol kg^{-1} were found in the hydrothermal vents. The concentrations decrease as the distance to the source increases. In the water column, the maximums were found during the eruptive stage, with concentrations up to 0.79 μmol kg^{-1}. Smaller amounts of up to 0.29 μmol kg^{-1} reach the water column in the post-eruptive stage, but note that these values are still an enrichment with concentrations more than double of the ambient level.

9.3.3 Nitrogen Species (NO_3^-, NO_2^-, NH_4^+)

Inorganic nitrogen available for phytoplankton uptake can be found in the forms of nitrate (NO_3^-), nitrite (NO_2^-), and ammonium (NH_4^+). Ammonium is the preferred source as it is more energetically favorable, but the oxidized forms (nitrate and nitrite) are also commonly used (Zehr and Kudela 2011). Ammonium has been often described to be highly enriched in hydrothermal settings, with an enrichment of typically one order of magnitude related to background (Karl et al. 1989; Sarradin et al. 1999; Sedwick and Stuben 1996; Tunnicliffe et al. 1986; Wilson et al. 2019), although concentrations as high as 11,559 μM (ambient levels: 0.84 μM) have been described by Karl et al. (1988b) at the Guaymas Basin. Oxidized forms of nitrogen have also been frequently studied in hydrothermal sites, often jointly as nitrate + nitrite (NO_3^- + NO_2^-), since NO_2^-

concentrations are usually extremely low. $NO_3^- + NO_2^-$ typically do not show such big increments, and its concentrations in the emitted fluids can in fact represent either an enrichment or a net loss depending on the ambient concentrations (Karl et al. 1988b, 1989; Sarradin et al. 1999; Sedwick and Stuben 1996; Tunnicliffe et al. 1986). In that sense, shallow volcanoes and vents are bound to be more enriched in all forms of nitrogen due to the low ambient levels of this nutrient (Wilson et al. 2019).

At Tagoro volcano, nitrogen was mostly measured as $NO_3^- + NO_2^-$ (Fig. 9.2c). During the eruption, concentrations as high as 8.06 μmol kg^{-1} were found (reference: 0.41 μmol kg^{-1}). In the post-eruptive stage, concentrations of up to 6.21 μmol kg^{-1} are found in the water column; however, unlike the case of phosphate and silicate, nitrate does not show an increase when sampled closer to the source, exhibiting concentrations roughly in the same range as the ambient seawater in the samples collected from the vents with ROV. It has been suggested that nitrifying processes occur in hydrothermal sites, oxidizing NH_4^+ into NO_2^- and NO_3^- (Baker et al. 2012; Kilias et al. 2013; Lam et al. 2004; 2008). Although ammonium data from Tagoro submarine volcano is scarce, some samples taken from the vents with ROV support this hypothesis, showing NH_4^+ increases of almost 100-fold with respect to normal conditions. The occurrence of nitrification would explain the presence of oxidized nitrogen ($NO_3^- + NO_2^-$) in the water column.

9.3.4 Iron

During the eruptive stage, Tagoro volcano emitted large amounts of gases and reduced chemical species, changing the physical–chemical properties of the seawater (Santana-Casiano et al. 2013; see Chap. 8). Due to gas emissions, the seawater carbonate species were strongly affected and the pH decreased from 8.1 to values as low as 5.1. The concentration of Fe(II) and reduced sulfur increased to 50 μmol l^{-1} and 200 μmol l^{-1}, respectively, contributing to the reduction of the redox potential of the system (Santana-Casiano et al. 2013). Just after the molten eruptive phase, changes in the composition of the gas and reduced species emissions were observed, and the O_2 concentrations decreased due to the oxidation of these reduced species. Moreover, the Fe(II) emitted was oxidized to Fe(III) under the oxic seawater conditions (Santana-Casiano et al. 2016). The formed Fe(III) precipitated into various mineral species, mainly oxyhydroxide (De Baar 2001), resulting in the formation of massive deposits of iron. Iron mineralization within the seafloor sediments was described in the area by González et al. (2020) and related to hydrothermal iron enrichment.

It is calculated that around 4% of the total emitted iron could be stabilized against loss from solution due to complexation by dissolved organic ligands (Bennett et al. 2008; Resing et al. 2015), and also by incorporation into inorganic or organic colloids, which reside within the dissolved size fraction (Resing et al. 2015). Moreover, reduced species of Fe and S form a FeS colloidal complex, which remains suspended in the water and is modified as the pH changes (Luther et al. 2001). These nanoparticles can remain suspended in the deep sea for years with slower settling rates (Yücel et al. 2011), solubilizing and releasing the Fe(II).

During the post-eruptive stage, the high-resolution transects carried out along the volcanic edifice revealed positive anomalies of TdFe(II) concentration, which was also accompanied by important negative anomalies in the pH signal (−0.17 units), and which were limited to an area of 0.1 km^2 and depths between 90 and 160 m, close to the main cone (Santana-González et al. 2017). The ambient TdFe(II) concentrations in non-affected areas and in surface waters were lower than 0.3 nmol l^{-1}, with some exceptions at the depth of maximum chlorophylls that presented an increase in TdFe(II) concentration in the order of 1–2 nmol l^{-1} (Santana-González et al. 2017). In the area affected by the hydrothermal emissions, higher Fe(II) concentrations were observed in an intermittent mode at around 100 m depth and in a limited area of around 500 m around the main cone (Santana-

González et al. 2017). In the water column (first 150 m) of this affected area, the concentrations were between 0 and 20 nmol l^{-1}, while the samples collected near the seafloor were within a range of 2–50 nmolL^{-1}, with even higher punctual values. The maximum TdFe(II) concentration was recorded in march 2014 at around 100 m depth with a concentration of 1278.9 nmol l^{-1}, which was reflected in the pH values ranging from 7.76 to 7.84, and which decreased to 281 nmol l^{-1} in only 1.5 h. All these changes were associated with variable CO_2 emissions from hydrothermal vents located in the volcano area (Santana-Casiano et al. 2016). The low pH values enabled to maintain Fe(II) in solution for a longer period due to a decrease in its oxidation rate.

Oxidation kinetic studies were carried out to estimate the Fe(II) oxidation rate constant. These studies were done in samples close to the main cone (5 m above the seafoor) affected by the hydrothermal emissions and showed that Fe(II) oxidated faster than those expected in oligotrophic seawater. This can be explained by the higher amount of macronutrients, in particular silicates, which increased in concentrations to 16.3 fold in the water column and 325.4 fold in the vent (González-Vega et al. 2020). The macronutrients accelerate the oxidation of Fe(II) in the area (Santana-González et al. 2017), as indicated for the Fe(II) oxidation kinetics in high nutrient media (González et al. 2010; Samperio-Ramos et al. 2016).

TdFe(II) emitted in the area may act as a fertilization event (Ardyna et al. 2019; Tagliabue et al. 2017) in the seawater around the Tagoro submarine volcano near the island of El Hierro.

9.4 Extent and Transport of the Emissions

9.4.1 Eruptive Stage

During the eruption of the submarine volcano Tagoro, a wide area around El Hierro island was drastically affected, particularly south and southwest of the island. Figure 9.3a shows the areal distribution of the nutrient plume, defined as the area where nutrient concentrations were above the normal conditions as measured in background samples. This distribution was observed in November 2011 during the eruptive phase. The silicate ($Si(OH)_4$) plume is the most extent, with an area of 255 km^2, which is almost the size of El Hierro island (268 km^2). The phosphate (PO_4) and nitrate ($NO_3^- + NO_2^-$) plumes reached smaller areas of 118 and 95 km^2, respectively.

The nutrient plumes, however, reached even further regions episodically depending on the local current dynamics, as shown in Fig. 9.3b in late November 2011. The nutrient plumes were drifted to the north of the island, due to changes in the local current conditions. This represents an additional area of roughly 70–120 km^2, but note that the southern and western limitations of the plume were not determined, meaning that the affected area was even larger.

The distributions of both of these plumes are in agreement with the discoloured areas in the water as observed in satellite images from the same periods (Fig. 9.3c, d; see Chap. 8, Sect. 8.3.5). Note that the observed light-green color of the water in these images was not caused by an increase in chlorophyll-*a*, but rather as an effect of the chemicals injected by the volcanic emissions (Eugenio et al. 2014).

9.4.2 Post-eruptive Stage

From March 2012, once the eruption ended, Tagoro volcano entered a new phase, known as post-eruptive or degassing stage, which still remains active. In this new stage, the volcanic activity consists of diffuse-flow, low-temperature hydrothermal emissions with release of heat, gases, metals, and inorganic nutrients (Santana-Casiano et al. 2013). These emissions are limited to a much smaller area within the volcanic edifice and close to the seabed, rarely reaching the surface. Figure 9.4 shows the average distribution of the nutrient plumes where their concentrations are above normal conditions. This plume typically remains within 400 m around the main

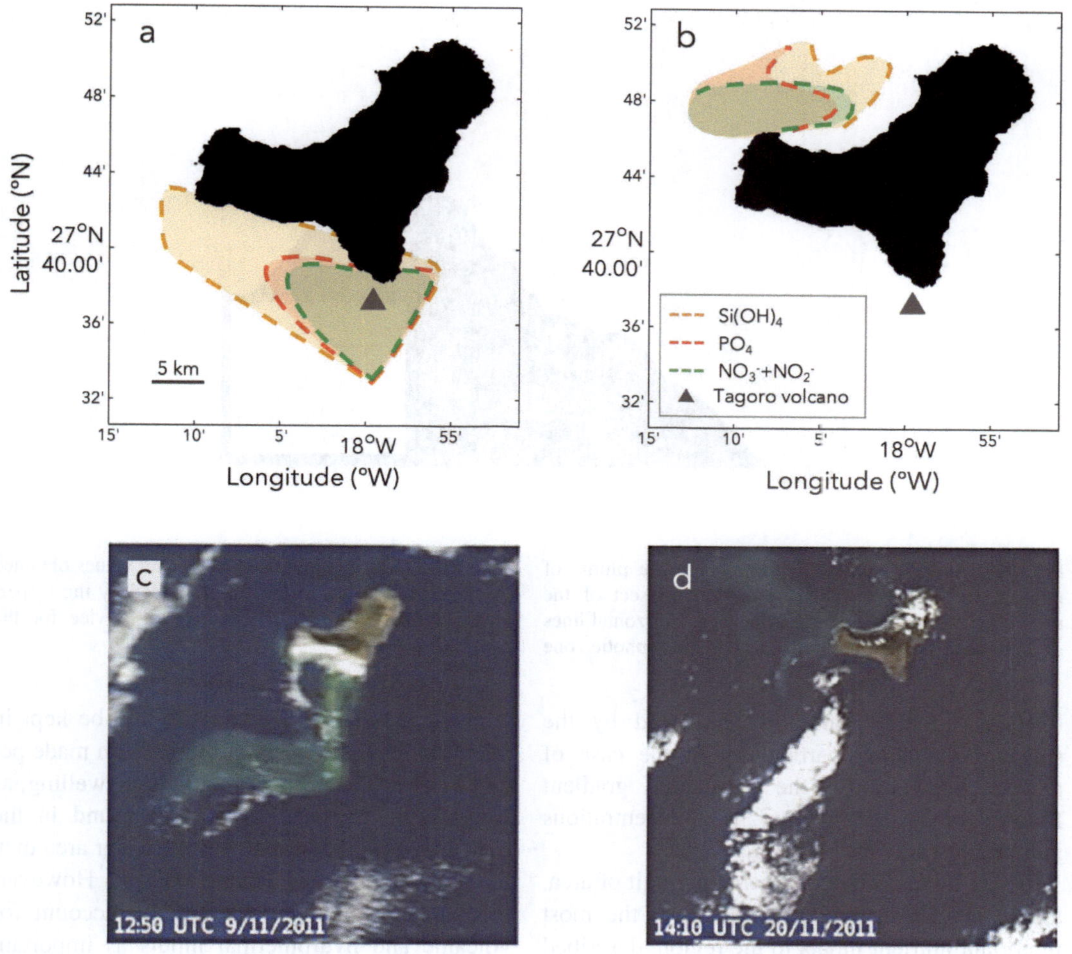

Fig. 9.3 **a** Areal distribution of the nutrient plumes during early November 2011 at 70 m depth. The plume is defined as the area where nutrient concentrations are above normal conditions, and is delimited by a dashed line. **b** A plume transported north of El Hierro island during late November 2011. Note that the south and west limits of the plume were not determined. **c**, **d** NASA MODIS RGB images of El Hierro island for the corresponding periods (early and late November 2011, respectively). Note that the color of the plume is not related to chlorophyll-*a*, but to the chemicals injected by the volcano (Eugenio et al. 2014)

crater, with an asymmetry to the south-west (due to the direction of the predominant current), and about 50 m above the seabed, although occasionally the plume has been observed to reach the surface.

Figure 9.4 also shows the depth of the euphotic zone as well as the mixed layer in the region, shown as minimums and maximums. These depths vary seasonally and depend on environmental conditions. Therefore, the nutrient plumes might occasionally reach the mixed layer and the euphotic zone, where they have the potential to exert an impact on photosynthetic organisms.

In order to estimate the transport of inorganic nutrients provided by Tagoro volcano in the post-eruptive stage, nutrient fluxes were calculated from a vertical section perpendicular to the predominant current (González-Vega et al. 2020). The results of these fluxes, in mol m^{-2} $year^{-1}$, were: 3.19 ± 1.17 for $NO_2^- + NO_3^-$, 0.02 ± 0.01 for PO_4, and 0.60 ± 1.35 for Si

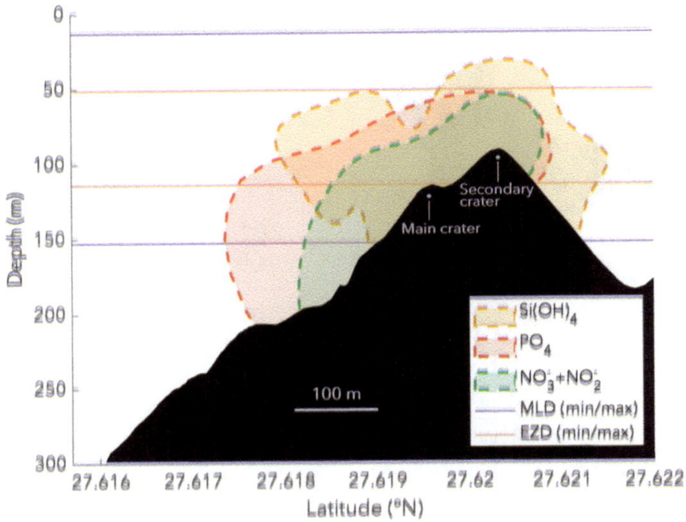

Fig. 9.4 Average vertical distribution of the plume of nutrient emissions in a high-resolution transect of the main cone during the post-eruptive stage. Horizontal lines show the mixed layer depth (MLD) and the euphotic zone depth (EZD), as minimum and maximum values obtained from monthly averaged datasets provided by the Copernicus Marine Environment Monitoring Service for the period of 1993–2019

$(OH)_4$. The high variability indicated by the standard deviation, particularly in the case of silicate, accounts for the substantial gradient found within the plume, as concentrations increase towards the seafloor.

These fluxes, when compared per unit of area, are of similar magnitude to some of the most important nutrient inputs to the region, described in detail by Barton et al. (1998). In this study, the northwest African upwelling was estimated to present a nitrogen flux of 5.87 mol m^{-2} $year^{-1}$. The Canary archipelago is located in the transition zone between this coastal upwelling and the open-ocean oligotrophic waters, with presence of various mesoscale processes that also supply nutrients to the area. For example, filaments from the coastal upwelling reaching the Canary region present an estimated nitrogen flux of 0.29 mol m^{-2} $year^{-1}$ (Barton et al. 1998). Additionally, a permanent cyclonic eddy originated in the lee region south of the island of Gran Canaria also supplies nutrients at a rate of 1.36 mol m^{-2} $year^{-1}$ for nitrogen (Barton et al. 1998).

Therefore, the estimated nitrate flux from Tagoro volcano of 3.19 mol m^{-2} $year^{-1}$ is of similar or even higher magnitude than these regional inputs. Of course, it should be kept in mind that these comparisons have been made per unit of area. The northwest African upwelling, as well as the mesoscale processes found in the transition zone, represent a much wider area than the plume emitted by Tagoro volcano. However, these results highlight the need to account for volcanic and hydrothermal inputs as important sources of nutrients to the ocean. Here we have accounted for one single monogenetic volcano with post-eruptive diffuse flow; however, its contribution has been proven to be non-neglectable. Therefore, it is crucial to work on assessing volcanogenic nutrient inputs worldwide, especially from shallow emissions, which have the potential of impacting the photosynthetic primary production.

9.5 Local Effects by the Nutrient Emissions of Tagoro Volcano

9.5.1 Nutrient Ratios

In 1934, Alfred C. Redfield made an observation that would forever impact the study of elemental stoichiometry in the ocean: he suggested that nitrogen and phosphorus were found in a constant ratio of 16:1 both in seawater and in phytoplankton biomass (Redfield 1934, 1960). This tight relationship between organisms and their environment is caused by a mutual feedback: phytoplankton consume and incorporate these elements in a certain proportion, and with their death and decomposition, the elements return to the water in the same proportion. Thus, life and its environment are locked in a state of mutual influence (Gruber and Deutsch 2014).

Even though Redfield's proportions are true as a global average and particularly consistent in the deep ocean, the reality is more complex. It was later revealed that optimal N:P ratios in phytoplankton can vary between 8.2:1 and 45.0:1 depending on the growth conditions, and that the Redfield ratio of 16:1 represents an average of species-specific N:P ratios (Klausmeier et al. 2004). Furthermore, changes in the regional or even global nutrient sources can lead to changes in the elemental ratios of the seawater (Pahlow and Riebesell 2000), which would provide some selective advantage for those organisms whose requirements are more adapted to the ambient stoichiometry (Zehr and Kudela 2011).

Nonetheless, the study of the relationships among nutrients in the ocean has evolved beyond Redfield's observations for N and P. Hydrothermal and volcanic settings are a good example, as they emit fluids that expectedly present a different composition than the ambient seawater. More specifically, silicate has been commonly described as a mixing tracer, assessed by a linear dependence with temperature in hydrothermal fluids (Edmond et al. 1979; Sansone and Resing 1995; Sansone et al. 1991). This means that if other chemical species are plotted against Si and show a linear behavior, conservative mixing is assumed; whereas a non-linear relation indicates non-conservative mixing (Sedwick and Stuben 1996).

For Tagoro submarine volcano, the relationships among N, P and Si were studied during the eruptive and post-eruptive stages (González-Vega et al. 2020). Firstly, in order to assess whether silicate could be used as a mixing tracer, the relationship between silicate (Si) and temperature (T) was studied from the near-seafloor and vent samples during the post-eruptive stage (Fig. 9.5a). The near-seafloor data are distributed around a small range of T and Si, but the vent samples are more evenly distributed and do indeed present a linear relationship. The correlation is statistically significant ($p < 0.001$) and presents an $r^2 = 0.878$. Thus, it can be assumed that silicate can be used as a mixing tracer for Tagoro volcano.

Next, both N and P were plotted against Si in order to assess their mixing behavior. It should be kept in mind that these are not elemental stoichiometries, as N is only represented by the species NO_3^- and NO_2^- (NH_4^+ is not included due to the reduced number of samples available), P only by PO_4, and Si by $Si(OH)_4$. However, we will name them as N, P, and Si for simplicity. Figure 9.5b, c show the N:Si and P:Si ratios, respectively, during the eruptive and post-eruptive stages. The post-eruptive stage is split into three datasets based on the grade of dilution (water column, near-seafloor, and vent samples). The water column samples show the most similar stoichiometry to the ambient seawater, followed by near-seafloor samples, and finally the vent samples, indicating the differential composition of the hydrothermal fluids and the effect of mixing with ambient seawater.

Even though all linear correlations were statistically significant ($p < 0.001$), the data are notably scattered (r^2 between 0.5 and 0.7), except for the case of N:Si in the vent samples ($r^2 = 0.925$). This wide distribution of the data is most likely due to a variety of processes that may be affecting the concentration of these species, e.g., (a) the possible use of NO_3^- as an electron acceptor in the absence of oxygen during the eruptive stage (Kristensen and Holmer 2001); (b) the withdrawal of phosphate by formation of suspended iron oxyhydroxides (Föllmi 1996; Paytan and McLaughlin 2007); (c) processes related to the nitrogen cycle, such as oxidation of NH_4^+ to NO_2^- and NO_3^- (Baker et al. 2012; Kilias et al. 2013; Lam et al. 2008; 2004); or

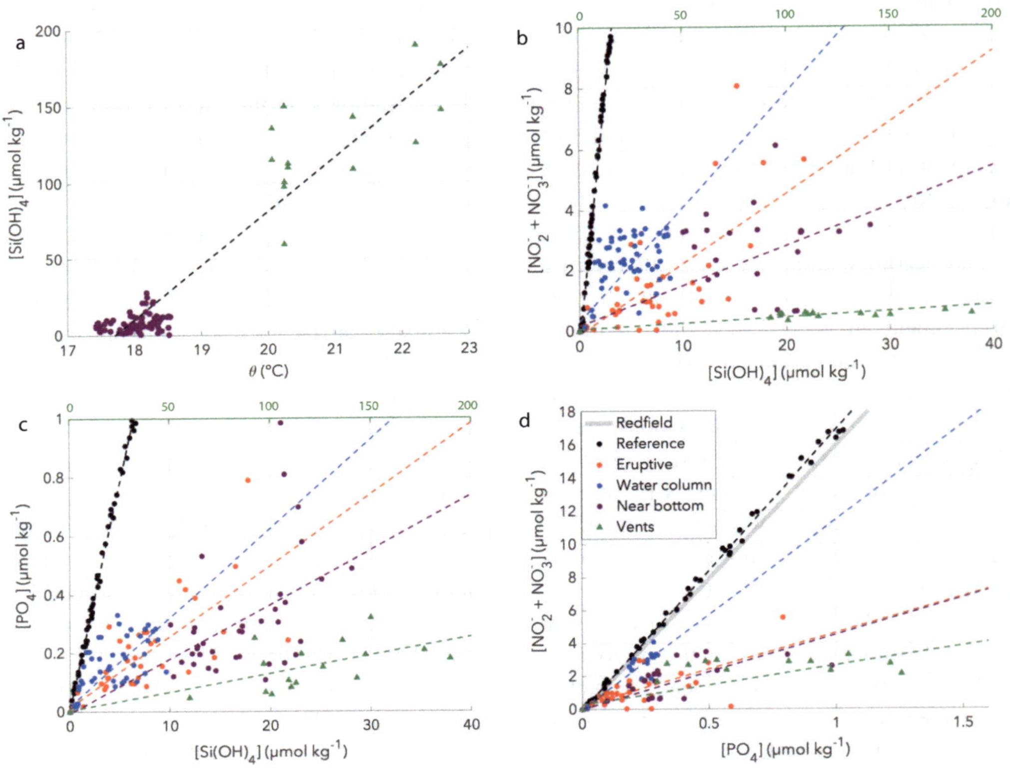

Fig. 9.5 Nutrient ratios calculated as linear correlations, **a** (NO_3^- + NO_2^-):PO_4 ratio, abbreviated as N:P (Redfield's 16:1 ratio is shown), **b** (NO_3^- + NO_2^-):$Si(OH)_4$ as N:Si, and **c** PO_4:$Si(OH)_4$ as P:Si, for the reference stations, the eruptive stage, and the degassing stage classified according to the grade of dilution of the samples (water-column/near-bottom/vents). Upper x-axis in panels **b**, **c** corresponds to silicate concentration in the vent samples (green). **d** Relation between silicate and potential temperature from the closest samples to the source

(d) nutrient uptake by primary producers. Thus, even though N:Si and P:Si were linearly correlating, the notable scattering of the data and the complexity of processes involved indicate a non-linear mixing behavior.

Finally, the N:P ratio (Redfield's legacy) was also studied at samples from Tagoro volcano (Fig. 9.5d). The ambient seawater, unaffected by the volcano, presented a ratio of 16.95:1, remarkably similar to Redfield's 16:1 ratio. The samples taken within the volcanic plume showed ratios varying from 4.38:1 in the vent samples to 11.44:1 in the water column samples. This highlights the potential of changes in the regional nutrient inputs to modify the elemental ratios of seawater at a local scale.

9.5.2 Effects on Local Communities

Local enrichments of microbial biomass and enhanced activity in deep-sea hydrothermal vents has long been recognised (Baker et al. 2012; Karl et al. 1988b, 1989; Lam et al. 2008), but the effects of shallow hydrothermal and volcanic emissions on surface communities have received attention only in recent years. In the Southern Ocean, hydrothermal vents emitting large amounts of bioavailable iron to surface waters have been observed to trigger massive phytoplankton blooms (Ardyna et al. 2019). A similar bloom in the North Pacific Ocean was stimulated by lava reaching the seawater from subaerial Kilauea volcano, characterized by high concentrations of metals and nutrients (Wilson et al. 2019). Atmospheric deposition of volcanic ash in the North-East Pacific has also been observed to act as a fertilizer for the surface ocean (Langmann et al. 2010), as well as support diverse bacterial colonisation in mesocosm experiments (Witt et al. 2017). High rates of planktonic and benthic production have been described in the hydrothermally influenced waters of the crater of the Ushishir Volcano in the Pacific Ocean (Tarasov et al. 1990). In the Western Tropical South Pacific, patches of high chlorophyll concentrations in the vicinity of volcanic islands were suggested to be sustained by a submarine source of hydrothermal fluids (Guieu et al. 2018). Buck et al. (2018) also suggested that the injection of iron-rich hydrothermal fluids from Ahyi Seamount (Mariana arc) into the mixed layer could potentially have an impact on oceanic productivity in the region. At slightly deeper settings (400–500 m), enhanced chemosynthesis and nitrification has been described in a Mediterranean arc-volcano (Kilias et al. 2013) as well as in the Western Pacific (Sorokin et al. 1998).

Benthic communities have been well characterised in the newly formed substrate of Tagoro volcano (Sotomayor-García et al. 2019; see Chap. 8). Two newly discovered benthic habitats were described around the summit and main craters: (1) hydrothermal vents with microbial mats, and (2) sulfurous-like fields mainly colonized by small hydrozoan colonies. The microbial mats covering the hydrothermal fields are formed by filaments made of bacterial trichomes enveloped within a sheath and colonized by euphotic bacteria (Danovaro et al. 2017). A newly recorded species (*Thiolava veneris*) was found in these filaments, with the potential to utilize organic and inorganic carbon as well as the ability to uptake sulfur and nitrogen compounds (Danovaro et al. 2017). Downslope of the volcanic complex, further highly diverse habitats have been observed, including annelids, arthropods, cnidarians, and mollusks as the first colonizers. A recent study has also revealed an increase in zooplankton abundance in waters affected by the volcanic emissions, with a higher presence of non-calanoid copepods and a decline in the diversity of the copepods community, indicating that the volcanic inputs have a significant effect on these organisms (Fernández de Puelles et al. 2021).

At the phytoplankton level, the effect of the emissions is yet unclear. Assessing a potential effect is a difficult task because of temporal variations related to the cyclicity of the emissions (Fraile-Nuez et al. 2018). A study carried out in vertical stations near the volcano (but outside of the area reached by the nutrient plume) showed no significant difference in chlorophyll-*a* and community structure respect to background stations (Gómez-Letona et al. 2018). However, positive chlorophyll-*a* anomalies have been observed above the main crater (González-Vega 2018).

9.6 Conclusions and Future Works

From the information gathered in the present chapter regarding the release of inorganic nutrients from Tagoro volcano, we conclude that: (i) inorganic nutrients (N, P, Si, Fe) have been emitted in large amounts and in bioavailable forms during both the eruptive and post-eruptive stages; (ii) these nutrient emissions (which still remain active) significantly enrich the water column and can be transported into the surrounding waters, and (iii) the emissions can occasionally be injected into the mixed layer and reach the euphotic zone.

However, there are still some gaps of information that can be assessed in future studies. For example, a more specific study on the dynamics of the inorganic nitrogen species emitted from Tagoro volcano would shed light on the knowledge of the system. We have concluded that nitrate (NO_3^-) and nitrite (NO_2^-) are not the main forms of nitrogen found in the hydrothermal source, and that ammonium (NH_4^+) may be the primary form; however, we still lack information about the relative concentrations of all these species in the whole water column and how they evolve as the hydrothermal plume spreads.

The question of whether these emissions have a direct impact on phytoplankton communities also remains to be answered. A more targeted study is required to assess this hypothesis, taking into account the horizontal and vertical distributions of the hydrothermal plumes, the temporal variations related to cyclicity (Fraile-Nuez et al. 2018), and the local dynamics that could be occurring as an effect of the physical anomalies, such as buoyant plumes and vortex features (Speer 1989).

Lastly, the study of nutrient release at Tagoro volcano comprises one of the largest datasets of inorganic nutrients from an active submarine volcano in the world. This time series needs to continue in order to be able to observe long-term changes in the activity and behavior of the hydrothermal system.

Acknowledgements The acquisition of the seven-years dataset of inorganic nutrients from Tagoro volcano was possible thanks to the funding of several institutions and organizations: (i) Ministerio de Economía y Competitividad del Gobierno de España (MINECO) and FEDER through VULCANO-I (CTM2012-36317), VULCANO-II (CTM2014-51837-R), ECOFEMA (CTM2010-19517), EACFe (CTM2014-52342-P) projects. (ii) The Spanish Institute of Oceanography (IEO) through BIMBACHE (IEO-2011-2012), RAPROCAN-III (IEO-2010-2012), VULCANA-I (IEO-2015-2017), VULCANA-II (IEO-2018-2020) and VULCANA-III (IEO-2021-2023) projects. (iii) Special funding from IEO for the use of the ROV Liropus 2000 in several cruises. (iv) Fundación Caja-Canarias through CYCLOVENT (2019SP18) project, which also provided financial support for A.G-V. (v) The European project CARBOCHANGE-264879. We would also like to thank the officers and crew of the R/V *Ramón Margalef* and *Ángeles Alvariño* from the IEO and the technician team of ACSM (ROV) for their help at sea. Finally, the authors are grateful to Pablo J. González, editor of this book, for his support during the writing process.

References

Ardyna M, Lacour L, Sergi S, d'Ovidio F, Sallée J-B, Rembauville M, Blain S, Tagliabue A, Schlitzer R, Jeandel C, Arrigo KR, Claustre H (2019) Hydrothermal vents trigger massive phytoplankton blooms in the Southern Ocean. Nat Commun 1–8. https://doi.org/10.1038/s41467-019-09973-6

Baker BJ, Lesniewski RA, Dick GJ (2012) Genome-enabled transcriptomics reveals archaeal populations that drive nitrification in a deep-sea hydrothermal plume. ISME J 6:2269–2279. https://doi.org/10.1038/ismej.2012.64

Barton ED, Arístegui J, Tett P, Canton M, García-Braun J, Hernández-León S, Nykjaer L, Almeida C, Almunia J, Ballesteros S, Basterretxea G, Escánez J, García-Weill L, Hernández-Guerra A, López-Laatzen F, Molina R, Montero MF, Navarro-Peréz E, Rodríguez JM, Van Lenning K, Vélez H, Wild K (1998) The transition zone of the Canary Current upwelling region. Prog Oceanogr 41:455–504. https://doi.org/10.1016/S0079-6611(98)00023-8

Bennett SA, Achterberg EP, Connelly DP, Statham PJ, Fones GR, German CR (2008) The distribution and stabilisation of dissolved Fe in deep-sea hydrothermal plumes. Earth Planet Sci Lett 157–167. https://doi.org/10.1016/j.epsl.2008.01.048

Buck NJ, Resing JA, Baker ET, Lupton JE (2018) Chemical Fluxes From a Recently Erupted Shallow Submarine Volcano on the Mariana Arc. Geochem Geophys Geosyst 19:1660–1673. https://doi.org/10.1029/2018GC007470

Chavez FP, Messié M, Pennington JT (2011) Marine primary production in relation to climate variability and change. Annu Rev. Mar Sci 3:227–260. https://doi.org/10.1146/annurev.marine.010908.163917

Danovaro R, Canals M, Tangherlini M, Dell'Anno A, Gambi C, Lastras G, Amblas D, Sanchez-Vidal A, Frigola J, Calafat AM, Pedrosa-Pàmies R, Rivera J, Rayo X, Corinaldesi C (2017) A submarine volcanic eruption leads to a novel microbial habitat. Nat Ecol Evol 1:1645–1648. https://doi.org/10.1038/s41559-017-0144

De Baar HJW (2001) Distributions, sources and sinks of iron in seawater. Biogeochem Iron Seawater 123–253

Ducklow H, Plank T (2019) Volcano-stimulated marine photosynthesis. Science 365:978–979. https://doi.org/10.1126/science.aay8088

Edmond JM, Measures C, McDuff RE, Chan LH, Collier R, Grant B, Gordon LI, Corliss JB (1979) Ridge crest hydrothermal activity and the balances of the major and minor elements in the ocean: the Galapagos data. Earth Planet Sci Lett 46:1–18. https://doi.org/10.1016/0012-821x(79)90061-x

Eugenio F, Martin J, Marcello J, Fraile-Nuez E (2014) Environmental monitoring of El Hierro Island submarine volcano, by combining low and high resolution satellite imagery. Int J Appl Earth Obs Geoinf 29:53–66. https://doi.org/10.1016/j.jag.2013.12.009

Fernández de Puelles ML, Gazá M, Cabanellas M, González-Vega A, Herrera I, Presas-Navarro C, Arrieta JM, Fraile-Nuez E (2021) Abundance and structure of the zooplankton community during a post-eruptive process: the case of the submarine volcano Tagoro (El Hierro; Canary islands) 2013–2018. Front Mar Sci 8:692885. https://doi.org/10.3389/fmars.2021.692885

Field CB, Behrenfeld MJ, Randerson JT, Falkowski P (1998) Primary production of the biosphere: integrating terrestrial and oceanic components. Science

281:237–240. https://doi.org/10.1126/science.281.5374.237

Föllmi KB (1996) The phosphorus cycle, phosphogenesis and marine phosphate-rich deposits. Earth-Sci Rev 40:55–124. https://doi.org/10.1016/0012-8252(95)00049-6

Fraile-Nuez E, Santana-Casiano J, González-Dávila M, Vázquez J, Fernández-Salas L, Sánchez-Guillamón O, Palomino D, Presas-Navarro C (2018) Cyclic behavior associated with the degassing process at the shallow submarine volcano Tagoro, Canary Islands, Spain. Geosciences 8:457–515. https://doi.org/10.3390/geosciences8120457

Gómez-Letona M, Arístegui J, Ramos AG, Montero MF, Coca J (2018) Lack of impact of the El Hierro (Canary Islands) submarine volcanic eruption on the local phytoplankton community. Sci Rep 8:4667. https://doi.org/10.1038/s41598-018-22967-6

González AG, Santana-Casiano JM, Pérez N, González-Dávila M (2010) Oxidation of Fe(II) in natural waters at high nutrient concentrations. Environ Sci Technol 44:8095–8101. https://doi.org/10.1021/es1009218

González FJ, Rincón-Tomás B, Somoza L, Santofimia E, Medialdea T, Madureira P, López-Pamo E, Hein JR, Marino E, de Ignacio C, Reyes J, Hoppert M, Reitner J (2020) Low-temperature, shallow-water hydrothermal vent mineralization following the recent submarine eruption of Tagoro volcano (El Hierro, Canary Islands). Mar Geol 430:106333. https://doi.org/10.1016/j.margeo.2020.106333

González-Vega A, Fraile-Nuez E, Santana-Casiano JM, González-Dávila M, Escánez-Pérez J, Gómez-Ballesteros M, Tello O, Arrieta JM (2020) Significant release of dissolved inorganic nutrients from the shallow submarine volcano Tagoro (Canary Islands) based on seven-year monitoring. Front Mar Sci 6:2465. https://doi.org/10.3389/fmars.2019.00829

González-Vega A (2018) Study of the significant release of inorganic nutrients during eruptive and degasification stages by the submarine volcano Tagoro, El Hierro island. Universidad de Las Palmas de Gran Canaria. http://hdl.handle.net/10553/74654

Gruber N, Deutsch CA (2014) Redfield's evolving legacy. Nat Geosci 7:853–855. https://doi.org/10.1038/ngeo2308

Guieu C, Bonnet S, Petrenko A, Menkes C, Chavagnac V, Desboeufs K, Maes C, Moutin T (2018) Iron from a submarine source impacts the productive layer of the Western Tropical South Pacific (WTSP). Sci Rep 8:9075. https://doi.org/10.1038/s41598-018-27407-z

Karl DM, McMurtry GM, Malahoff A, Garcia MO (1988a) Loihi Seamount, Hawaii: a mid-plate volcano with a distinctive hydrothermal system. Nature 335:532–535. https://doi.org/10.1038/335532a0

Karl DM, Taylor GT, Novitsky JA, Jannasch HW, Wirsen CO, Pace NR, Lane DJ, Olsen GJ, Giovannoni SJ (1988b) A microbiological study of Guaymas Basin high temperature hydrothermal vents. Deep Sea Res Part 1 Oceanogr Res Pap 35:777–791. https://doi.org/10.1016/0198-0149(88)90030-1

Karl DM, Brittain AM, Tilbrook BD (1989) Hydrothermal and microbial processes at Loihi Seamount, a mid-plate hot-spot volcano. Deep Sea Res Part 1 Oceanogr Res Pap 36:1655–1673. https://doi.org/10.1016/0198-0149(89)90065-4

Kilias SP, Nomikou P, Papanikolaou D, Polymenakou PN, Godelitsas A, Argyraki A, Carey S, Gamaletsos P, Mertzimekis TJ, Stathopoulou E, Goettlicher J, Steininger R, Betzelou K, Livanos I, Christakis C, Bell KC, Scoullos M (2013) New insights into hydrothermal vent processes in the unique shallow-submarine arc-volcano, Kolumbo (Santorini). Greece Sci Rep 3:13–33. https://doi.org/10.1038/srep02421

Klausmeier CA, Litchman E, Daufresne T, Levin SA (2004) Optimal nitrogen-to-phosphorus stoichiometry of phytoplankton. Nat Commun. https://doi.org/10.1038/nature02508

Kristensen E, Holmer M (2001) Decomposition of plant materials in marine sediment exposed to different electron acceptors (O_2, NO_3^- and SO_4^{2-}), with emphasis on substrate origin, degradation kinetics, and the role of bioturbation. Geochim Cosmochim Acta 65:419–433. https://doi.org/10.1016/S0016-7037(00)00532-9

Lam P, Cowen JP, Jones RD (2004) Autotrophic ammonia oxidation in a deep-sea hydrothermal plume. FEMS Microbiol Ecol 47:191–206. https://doi.org/10.1016/S0168-6496(03)00256-3

Lam P, Cowen JP, Popp BN, Jones RD (2008) Microbial ammonia oxidation and enhanced nitrogen cycling in the Endeavour hydrothermal plume. Geochim Cosmochim Acta 72:2268–2286. https://doi.org/10.1016/j.gca.2008.01.033

Langmann B, Zakšek K, Hort M, Duggen S (2010) Volcanic ash as fertiliser for the surface ocean. Atmos Chem Phys 10:3891–3899. https://doi.org/10.5194/acp-10-3891-2010

Luther GW, Rozan TF, Taillefert M, Nuzzio DB, Di Meo C, Shank TM, Lutz RA, Cary SC (2001) Chemical speciation drives hydrothermal vent ecology. Nature 410:813–816. https://doi.org/10.1038/35071069

Martin JH, Coale KH, Johnson KS, Fitzwater SE, Gordon RM, Tanner SJ, Hunter CN, Elrod VA, Nowicki JL, Coley TL, Barber RT, Lindley S, Watson AJ, Van Scoy K, Law CS, Liddicoat MI, Ling R, Stanton T, Stockel J, Collins C, Anderson A, Bidigare R, Ondrusek M, Latasa M, Millero FJ, Lee K, Yao W, Zhang JZ, Friederich G, Sakamoto C, Chavez F, Buck K, Kolber Z, Greene R, Falkowski P, Chisholm SW, Hoge F, Swift R, Yungel J, Turner S, Nightingale P, Hatton A, Liss P, Tindale NW (1994)

Testing the iron hypothesis in ecosystems of the equatorial Pacific Ocean. Nature 371:123–129. https://doi.org/10.1038/371123a0

Moore CM, Mills MM, Arrigo KR, Berman-Frank I, Bopp L, Boyd PW, Galbraith ED, Geider RJ, Guieu C, Jaccard SL, Jickells TD, La Roche J, Lenton TM, Mahowald NM, Marañón E, Marinov I, Moore JK, Nakatsuka T, Oschlies A, Saito MA, Thingstad TF, Tsuda A, Ulloa O (2013) Processes and patterns of oceanic nutrient limitation. Nat Geosci 6:701–710. https://doi.org/10.1038/ngeo1765

Pahlow M, Riebesell U (2000) Temporal trends in deep ocean Redfield ratios. Science 287:831–833. https://doi.org/10.1126/science.287.5454.831

Paytan A, McLaughlin K (2007) The oceanic phosphorus cycle. Chem Rev 107:563–576. https://doi.org/10.1021/cr0503613

Redfield AC (1960) The biological control of chemical factors in the environment. Sci Prog 11:150–170

Redfield AC (1934) On the proportions of organic derivatives in sea water and their relation to the composition of plankton. James Johnstone Memorial Volume, pp 176–192

Resing JA, Sedwick PN, German CR, Jenkins WJ, Moffett JW, Sohst BM, Tagliabue A (2015) Basin-scale transport of hydrothermal dissolved metals across the South Pacific Ocean. Nature 523:200–203. https://doi.org/10.1038/nature14577

Samperio-Ramos G, Casiano JMS, Dávila MG (2016) Effect of ocean warming and acidification on the Fe(II) oxidation rate in oligotrophic and eutrophic natural waters. Biogeochemistry 128:19–34. https://doi.org/10.1007/s10533-016-0192-x

Sansone FJ, Resing JA (1995) Hydrography and geochemistry of sea surface hydrothermal plumes resulting from Hawaiian coastal volcanism. J Geophys Res 100:13555. https://doi.org/10.1029/95jc01120

Sansone FJ, Resing JA, Tribble GW, Sedwick PN, Kelly KM, Hon K (1991) Lava-seawater interactions at shallow-water submarine lava flows. Geophys Res Lett 18:1731–1734. https://doi.org/10.1029/91GL01279

Santana-Casiano JM, González-Dávila M, Fraile-Nuez E, de Armas D, González AG, Domínguez-Yanes JF, Escánez J (2013) The natural ocean acidification and fertilization event caused by the submarine eruption of El Hierro. Sci Rep 3:1140. https://doi.org/10.1038/srep01140

Santana-Casiano JM, Fraile-Nuez E, González-Dávila M, Baker ET, Resing JA, Walker SL (2016) Significant discharge of CO2 from hydrothermalism associated with the submarine volcano of El Hierro Island. Sci Rep 6:25686. https://doi.org/10.1038/srep25686

Santana-González C, Santana-Casiano JM, González-Dávila M, Fraile-Nuez E (2017) Emissions of Fe(II) and its kinetic of oxidation at Tagoro submarine volcano, El Hierro. Mar Chem 195:129–137. https://doi.org/10.1016/j.marchem.2017.02.001

Sarradin P-M, Caprais J-C, Riso R, Kerouel R, Aminot A (1999) Chemical environment of the hydrothermal mussel communities in the Lucky Strike and Menez Gwen vent fields, Mid Atlantic ridge. Cah Biol Mar 40:93–104

Sedwick P, Stuben D (1996) Chemistry of shallow submarine warm springs in an arc-volcanic setting: Vulcano Island, Aeolian Archipelago, Italy. Mar Chem 53:147–161. https://doi.org/10.1016/0304-4203(96)00020-5

Sorokin YI, Sorokin PY, Zakuskina OY (1998) Microplankton and its functional activity in zones of shallow hydrotherms in the Western Pacific. J Plankton Res 20:1015–1031. https://doi.org/10.1093/plankt/20.6.1015

Sotomayor-García A, Rueda JL, Sánchez-Guillamón O, Urra J, Vázquez JT, Palomino D, Fernández-Salas LM, López-González N, González-Porto M, Santana-Casiano JM, González-Dávila M, Presas-Navarro C, Fraile-Nuez E (2019) First macro-colonizers and survivors around tagoro submarine Volcano, Canary Islands, Spain. Geosciences 9:52. https://doi.org/10.3390/geosciences9010052

Speer KG (1989) A forced baroclinic vortex around a hydrothermal plume. Geophys Res Lett 16:461–464. https://doi.org/10.1029/GL016i005p00461

Tagliabue A, Bowie AR, Boyd PW, Buck KN, Johnson KS, Saito MA (2017) The integral role of iron in ocean biogeochemistry. Nature 543:51–59. https://doi.org/10.1038/nature21058

Tarasov VG, Propp MV, Propp LN, Zhirmunsky AV, Namsakakv BB, Gorlenko VM, Starynin DA (1990) Shallow-Water Gasohydrothermal Vents of Ushishir Volcano and the Ecosystem of Kraternaya Bight (The Kurile Islands). Mar Ecol 11:1–23. https://doi.org/10.1111/j.1439-0485.1990.tb00225.x

Tréguer PJ, De La Rocha CL (2013) The world ocean silica cycle. Annu Rev Mar Sci 5:477–501. https://doi.org/10.1146/annurev-marine-121211-172346

Tunnicliffe V, Botros M, De Burgh ME, Dinet A, Johnson HP, Juniper SK, McDuff RE (1986) Hydrothermal vents of Explorer Ridge, northeast Pacific. Deep Sea Res Part 1 Oceanogr Res Pap 33:401–412. https://doi.org/10.1016/0198-0149(86)90100-7

Wilson ST, Hawco NJ, Armbrust EV, Barone B, Björkman KM, Boysen AK, Burgos M, Burrell TJ, Casey JR, DeLong EF, Dugenne M, Dutkiewicz S, Dyhrman ST, Ferrón S, Follows MJ, Foreman RK, Funkey CP, Harke MJ, Henke BA, Hill CN, Hynes AM, Ingalls AE, Jahn O, Kelly RL, Knapp AN, Letelier RM, Ribalet F, Shimabukuro EM, Tabata RKS, Turk-Kubo KA, White AE, Zehr JP, John S, Karl DM (2019) Kīlauea lava fuels phytoplankton bloom in the North Pacific Ocean. Science 365:1040–1044. https://doi.org/10.1126/science.aax4767

Witt V, Ayris PM, Damby DE, Cimarelli C, Kueppers U, Dingwell DB, Wörheide G (2017) Volcanic ash supports a diverse bacterial community in a marine mesocosm. Geobiology 15:453–463. https://doi.org/10.1111/gbi.12231

Yücel M, Gartman A, Chan CS, Luther GW (2011) Hydrothermal vents as a kinetically stable source of iron-sulphide-bearing nanoparticles to the ocean. Nat Geosci 4:367–371. https://doi.org/10.1038/ngeo1148

Zehr JP, Kudela RM (2011) Nitrogen cycle of the open ocean: from genes to ecosystems. Annu Rev Mar Sci 3:197–225. https://doi.org/10.1146/annurev-marine-120709-142819

Microbial Communities Surrounding an Underwater Volcano Near the Island of El Hierro (Canary Islands)

10

Isabel Ferrera, Jesús M. Arrieta, Marta Sebastián, and Eugenio Fraile-Nuez

Abstract

Underwater hydrothermal systems release nutrient-rich fluids that are favorable for microbial activity. This allows the growth of a plethora of microorganisms that can be found inhabiting hydrothermal plumes, metalliferous sediments, or forming dense microbial biofilms or mats on the surrounding of the vents. Most of the current knowledge on the microbiology associated with underwater volcanic activity comes from the study of deep-sea hydrothermal vents. However, much less is known about shallow underwater hydrothermal systems. The submarine volcanic eruption that took place near El Hierro (Canary Islands) in October 2011 and gave rise to the Tagoro volcano, located only 1.8 km from land and approximately 89 m below the sea surface, provided a rare occasion to study almost in real time the microbiology associated with an underwater eruptive process, from its birth to the current degassing state, allowing investigations of its effects on the surrounding pelagic and benthic communities. In this chapter, we summarize what is known 10 years after the formation of the Tagoro submarine volcano, and we discuss what questions should be pursued in order to foster our knowledge of the microbiology associated with this volcano and its surrounding waters.

Keywords

Hydrothermal systems · Shallow underwater volcanoes · Tagoro · Microbial diversity · Microbial activity · Venus' hair

10.1 Introduction

Microbes inhabit every ecosystem on Earth and play key ecological roles; they are the engines that drive nutrient cycles and energy flow and represent, by far, the largest share of the metabolic diversity of the biosphere. Underwater volcanoes are no exception to that rule and can harbor many types of microorganisms. Active submarine volcanoes provide nutrient-rich fluids (González-Vega et al. 2020; Chap. 9) that are favorable for microbial activity, allowing the

I. Ferrera (✉)
Centro Oceanográfico de Málaga, Instituto Español de Oceanografía, IEO-CSIC, 29640 Fuengirola, Málaga, Spain
e-mail: isabel.ferrera@ieo.csic.es

J. M. Arrieta · E. Fraile-Nuez
Centro Oceanográfico de Canarias, Instituto Español de Oceanografía, IEO-CSIC, C/ Farola del Mar, nº22, 38180 Santa Cruz de Tenerife, Spain

M. Sebastián
Instituto de Ciencias del Mar, ICM-CSIC, Pg. Marítim de La Barceloneta, 37-49, 08003 Barcelona, Spain

P. J. González (ed.), *El Hierro Island*, Active Volcanoes of the World,
https://doi.org/10.1007/978-3-031-35135-8_10

growth of microbes on or near hydrothermal systems (Emerson and Moyer 2010). Microbes can be found inhabiting hydrothermal plumes, metalliferous sediments, forming dense microbial biofilms or mats on the surrounding of the vents, or thriving directly on the vents, along the gradient from superheated waters in the vicinity of some vents to the cold temperatures typical of deep oceanic waters. The heterogeneity of physical and geochemical conditions in these underwater ecosystems makes active volcanoes potential hotspots of microbial diversity (Emerson and Moyer 2010). Yet, our knowledge of the microbiology of submarine volcanoes is still in its infancy, which is not just due to the technical challenges associated with sampling underwater volcanic areas. In fact, it is also because of the historical limitations of studying microorganisms in the wild, constrained by our inability to grow the majority of them in the laboratory (Rappé and Giovannoni 2003). Three decades ago, the advent of molecular methods allowed us to explore, for the first time, microbes in nature circumventing the need for culturing, an approach that revolutionized the field of environmental microbiology. PCR amplification and sequencing of ribosomal RNA genes allowed the initial identification of the dominant microorganisms in the environment (Pace et al. 1986; Stahl et al. 1984; Takai and Horikoshi 1999; Ward et al. 1992), and the development of high-throughput sequencing enabled uncovering microbial diversity at an unprecedented scale (Flores et al. 2011, 2012; Meyer et al. 2013; Sogin et al. 2006; Zinger et al. 2011). Further, the sequencing of protein-coding genes was used to identify and infer phylogenetic relationships among microorganisms involved in biogeochemical processes (Hügler et al. 2010; Nercessian et al. 2005; Roussel et al. 2011). The rise of the -omics era, in which metagenomics, metatranscriptomics, and metaproteomics can reveal the function of the whole community without PCR biases, has provided massive amounts of information on the genetic and functional diversity of microorganisms (Fortunato and Huber 2016; Olins et al. 2017; Reysenbach et al. 2020). Today, genome-centric approaches allow the access to individual environmental genomes (Danovaro et al. 2017; Merino et al. 2020; Reysenbach et al. 2020), greatly expanding the catalog of uncultivated microorganisms from these ecosystems. Yet, isolating bacteria and archaea remains a crucial step in complementing culture-independent methodologies, and allows to obtain complete genomes and the experimental study of microbial physiology in the laboratory. Altogether, the various approaches now available in the field of environmental microbiology contribute to advancing our knowledge of the microorganisms inhabiting these unique ecosystems. Nevertheless, despite the significant gains achieved in recent years, the microbiology of underwater volcanoes is understudied compared to other ecosystems, given the additional challenges of sample collection. Together, we are only at the beginning of understanding the diversity and metabolic potential of microbes in these submarine ecosystems.

10.2 Microbial Communities Around the Tagoro Underwater Volcano

The submarine volcanic eruption that took place near El Hierro (Canary Islands) in October 2011 provided a unique opportunity to study the microbiology associated with an underwater eruption. The birth of the Tagoro submarine volcano, located only 1.8 km from land and approximately 89 m below the sea surface, offered a rare occasion to observe almost in real time how a devastating event disrupting and destroying life under the sea opened at the same time new windows for life to begin again. Initial geophysical surveys of the volcanic eruption were followed by a series of hydrographic cruises, which allowed the study of the changes in seawater physical and chemical parameters as well as their effects on the marine ecosystem (see Chap. 8). The discharge of high temperature hydrothermal fluids, magmatic gases and volcanic particles from October 2011 to March 2012 warmed the water column and changed the seawater chemistry, including a significant decrease in pH and oxygen and an increase in iron and nutrients near the volcano (Fraile-Nuez et al. 2012; Santana-Casiano et al. 2013). Effects on the pelagic communities included the death of fish, which appeared floating on the surface while no fish schools could be acoustically detected within the volcano-affected area (Ariza et al. 2014; Fraile-Nuez et al. 2012). Moreover, the deep scattering layer produced by plankton and nekton echoes was shallower than usual in affected waters, and diel vertical migration of micronekton was disrupted (Ariza et al. 2014; Fraile-Nuez et al. 2012). A clear deoxygenation near the volcano was found from 75 to 175 m, yet oxygen profiles were normal in the post-eruptive stages (Fraile-Nuez et al. 2012). Once the eruptive stage was over, the Tagoro submarine volcano began an active hydrothermal phase, involving the discharge of heat, gases, and metals from multiple vents dispersed around the main cone by percolation of vent fluids through the highly permeable volcanic edifice (Fraile-Nuez et al. 2018; González-Vega et al. 2020; Santana-Casiano et al. 2016). In this new phase, the oceanographic conditions in the large affected area north and south of the island returned to normal, but cyclic emissions producing important physical–chemical anomalies in the interior of the main and secondary craters were documented (Fraile-Nuez et al. 2018). Today, the Tagoro submarine volcano is still active, in a degassing phase, but physical–chemical anomalies are only detectable in a $\sim$ 400 m radius around the main crater and close to the seabed. Recent studies have shown that the Tagoro system currently exhibits physical and chemical anomalies, with an increase in potential temperature of 2.55 °C, a decrease in salinity of 1.02, a decrease in density of 1.43 kg m^{-3}, and a decrease in pH of 1.25 units (Fraile-Nuez et al. 2018; Chap. 8 of this book).

Bacterioplankton (i.e., bacteria and archaea) around the Tagoro submarine volcano was also monitored. These organisms constitute the smallest but most abundant organisms of marine plankton and may be crucial in understanding the magnitude of sudden disturbances, particularly because of their fast response to environmental changes (i.e., Ferrera et al. 2020; Mock et al. 2016; Nogales et al. 2011). The effects that the volcanic disturbance had on prokaryote abundance, activity, diversity, and community structure were studied through the eruptive and post-eruptive phases (November 2011–April 2012) (Ferrera et al. 2015; Fraile-Nuez et al. 2012). Subsequently, multidisciplinary surveys were continued in the area but, unfortunately, there was little microbiological research. Still, some studies provided evidence that the Tagoro submarine volcano, after evolving from the eruptive to a degassing phase, gave rise to a new ecosystem in which microorganisms involved in iron, sulfur, and methane cycles were able to grow (González et al. 2020), including previously undescribed microbes (Danovaro et al. 2017). Sampling for microbiological studies was resumed with some regularity in 2018, taking advantage of the hydrographic expeditions led by the Spanish Institute of Oceanography that monitors the area twice a year. Current efforts focus on the effect of volcanic emissions and the associated nutrients on photosynthetic productivity of the water column and on the isolation and culturing of microbes associated with the diffusive flows. These expeditions are an opportunity to study the effects of a volcano in a degassing stage on the photosynthesis-based microbial food web of the water column and also to explore whether the volcano continues to provide the reduced materials and conditions that sustain chemosynthetic primary production at the venting sites. Below, we summarize what is known about the microbial communities around the Tagoro submarine volcano, from its birth to the current state, and we discuss what questions should be pursued in order to foster our knowledge of the microbiology associated with this volcano, both in the pelagic and benthic realms.

10.2.1 Abundance and Activity of Microbial Communities in Waters Surrounding the Tagoro Submarine Volcano

The abundance of photosynthetic picoplankton and heterotrophic prokaryotes in the surrounding waters of the volcano was assessed during various cruises that took place shortly after the onset of the eruption until it ceased. Cells with the typical characteristics of *Prochlorococcus* and *Synechococcus* cyanobacteria were enumerated using flow-cytometry in samples collected from November 2011 to February 2012. The data indicated that the abundance of these members of the picophytoplankton were not affected in surface waters but both cyanobacterial groups declined threefold in deeper waters closer to the volcano, compared with stations far from the volcano (Fraile-Nuez et al. 2012). Conversely, in these initial surveys, heterotrophic prokaryote's abundance increased with depth at stations affected by the volcanic emissions (Fraile-Nuez et al. 2012). Further analyses comparing samples collected from November 2011 to April 2012 at subsurface waters and oxygen depleted waters in various areas around El Hierro island revealed that: (i) the number of prokaryotes was higher during the submarine eruption than in the post-eruption stages, (ii) the differences were more evident when comparing the control station to the stations in the vicinity of the volcano, and (iii) the oxygen-depleted waters ($\sim$70–200 m) were in general more affected than subsurface waters (Ferrera et al. 2015).

The study of Ferrera et al. (2015) also assessed the nucleic acid content of bacterioplankton as a single cell-based proxy of cell activity. Previous investigations have shown that the delineation of bacterioplankton into two populations based on their nucleic acid content (high nucleic acid content—HNA—or low nucleic acid content—LNA) can be attributed partly to a difference in their physiological state and that HNA cells tend to be more active (Gasol et al. 1999; Lebaron et al. 2001; Servais et al. 2003), although it is known that nucleic acid content may also partially reflect the genome size of the cells (Vila-Costa et al. 2012). The percentage of presumably more active cells was significantly different between sampling periods (eruption and post-eruption) and between depths (subsurface, 0–70 m; oxygen depleted waters, 70–200 m) (Ferrera et al. 2015). The proportion of HNA cells was greater in the affected zone and near the volcano during the eruptive phase, particularly in oxygen-depleted waters (Fig. 10.1) (Ferrera et al. 2015). Flow cytometric analyses further revealed the presence of two types of particles distinct from the typical HNA or LNA populations. Based on the cytograms, one of them could represented particles and was observed associated with the discharge of vent material appearing mainly in stations closer to the volcano. The second population could represent cells attached to these particles and was detected in significantly higher amounts in the volcano area (Ferrera et al. 2015).

Overall, this study showed that the eruption promoted an increase in the abundance and activity of heterotrophic prokaryotes during the eruption, notably in depths close to the volcano but values returned to normal levels in the post-eruptive period. This increase could be a consequence of the enrichment of Fe(II) and macronutrients over the volcano promoted by the eruption as reported by Santana-Casiano et al. (2013) and González-Vega et al. (2020). Nonetheless, the distinct response of cyanobacteria and heterotrophic prokaryotes to the volcano's influence suggests that there was selection under the rapidly changing conditions.

In subsequent studies, bioorthogonal non-canonical amino acid tagging (BONCAT) (Dieterich et al. 2006; Hatzenpichler et al. 2014; Leizeaga et al. 2017) was used to evaluate the effect of the volcano in a degassing state on microbial communities. This approach enables the visualization of protein synthesis and coupled with catalyzed reporter deposition fluorescent in situ hybridization (CARDFISH), it allows the phylogenetic identification of the active cells (Fig. 10.2). Samples for BONCAT were collected in October 2016 when the degasification of the volcano was driven by periodic pulses

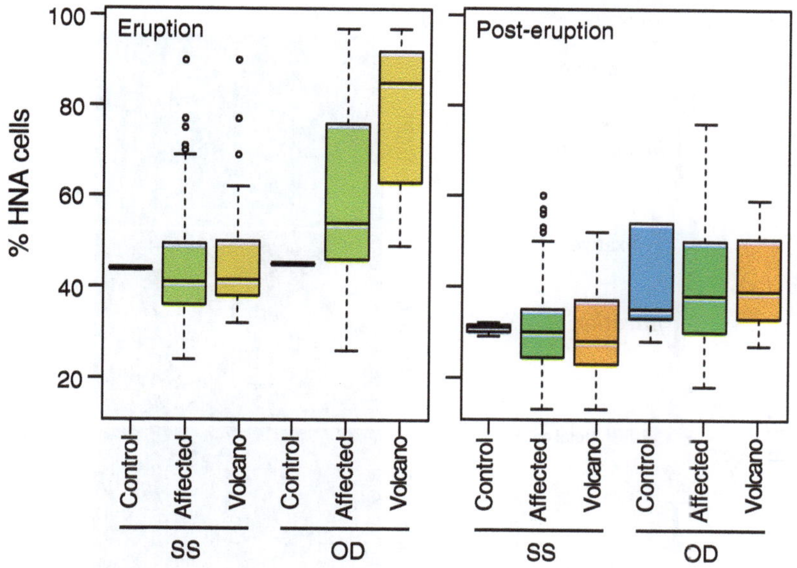

Fig. 10.1 Distribution of the percentage of high-nucleic acid content (% HNA) cells in samples collected during the eruptive phase (top panel) and the post-eruptive phase (bottom panel) grouped by depth (SS: subsurface, 0–70 m; OD: oxygen depleted waters, 70–200 m) and location (Control: stations in the control zone; Affected: stations in all affected areas, Volcano: affected stations in the vicinity of the volcano). Modified from Ferrera et al. (2015)

(Fraile-Nuez et al. 2018). Single cell activity of prokaryotes increased in waters affected by the volcanic emissions, which were detected by pH anomalies (Martínez-Miralles 2017), providing evidence that underwater volcanos have an impact on communities in surrounding waters long after the eruption is over.

It is well-known that underwater eruptions can release nutrients that may stimulate the growth of phytoplankton (Buck et al. 2018; Guieu et al. 2018; Santana-Casiano et al. 2013). A clear example was recently documented during the last eruption of the Kilauea volcano (Hawaii), when a unique perturbation of lava-impacted seawater produced a large-scale ocean fertilization event, which revealed how marine ecosystems respond to exogenous inputs of nutrients (Wilson et al. 2019). There, the extensive plume of chlorophyll *a* was even detectable in satellite images. In contrast, Gómez-Letona et al. (2018) did not find evidence of significant phytoplankton blooms occurring with the Tagoro eruption. They hypothesized that a bloom did not occur because nutrients were diluted by efficient renewal of ambient water in the area. However, new studies around the Tagoro have revealed an increase in zooplankton abundance in waters affected by the volcano, with a higher presence of non-calanoid copepods and a decline in the diversity of the copepods' community, indicating that volcanic inputs have a significant effect on these organisms (Fernández de Puelles et al. 2021). Thus, the Tagoro submarine volcano can be considered a natural laboratory where to analyze the zooplankton community, in particular the factors that modulate its abundance and diversity (Fernández de Puelles et al. 2021).

10.2.2 Diversity and Structure of Microbial Communities in Waters, Sediments and Mineral Deposits Surrounding the Tagoro Submarine Volcano

A preliminary study investigating the effect of the volcanic eruption on the diversity of bacterioplankton used phylogenetic analyses of

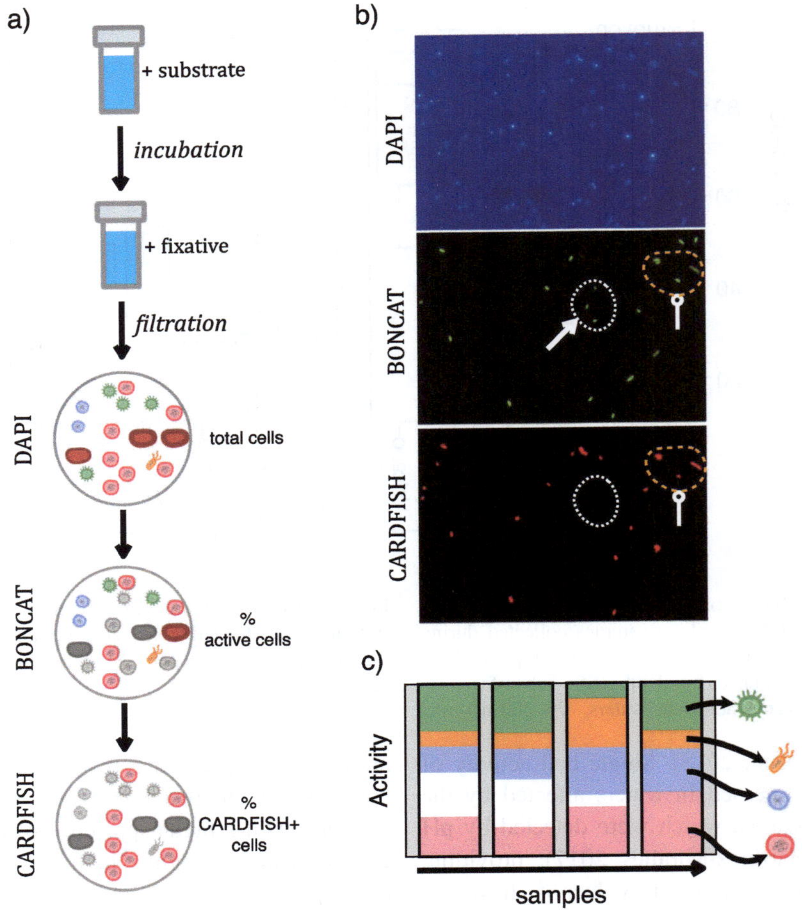

Fig. 10.2 Single cell analysis of prokaryotic activity. **a** Schematic flow of the bioorthogonal non-canonical amino acid tagging (BONCAT) technique. The sample is incubated with an amino acid surrogate and then the incubation is terminated with a fixative. The cells are then collected onto a 0.2 μm pore size filter which is divided into several portions. Each portion of the filter undergoes a CARDFISH hybridization with a group-specific probe (CARDFISH + cells), and a click-chemistry reaction through which the proteins synthesized during the incubation are fluorescently labelled (% of active cells). Samples are counterstained with DAPI (4′,6-diamidino-2-phenylindole) to enumerate the total number of prokaryotic cells. **b** Examples of micrographs of DAPI, BONCAT and CARDFISH with a sample collected during the degassing stage of the Tagoro volcano (October 2016). In the BONCAT micrograph, an example of active cells not labelled with the Gammaproteobacteria probe is shown (arrow), and an example of active Gammaproteobacteria cells are signaled by the round tip arrow. **c** Schematic of the outcome of the results. With this analysis the contribution of different prokaryotic groups to the bulk activity of the prokaryotic community can be assessed

cyanobacterial 16S rRNA gene sequences (generated from clone libraries) and revealed a possible volcano-induced selection for certain *Prochlorococcus* ecotypes in affected waters (Fraile-Nuez et al. 2012). In a later, more comprehensive study, samples collected during four of the initial hydrographic cruises were used for amplicon sequencing of bacterial and archaeal 16S rRNA gene tags. This methodology is commonly used in microbial ecology studies and allows the detection of patterns of diversity, structure, and dynamics of microbial communities. The results indicated that bacterial communities during the eruption were overall less rich

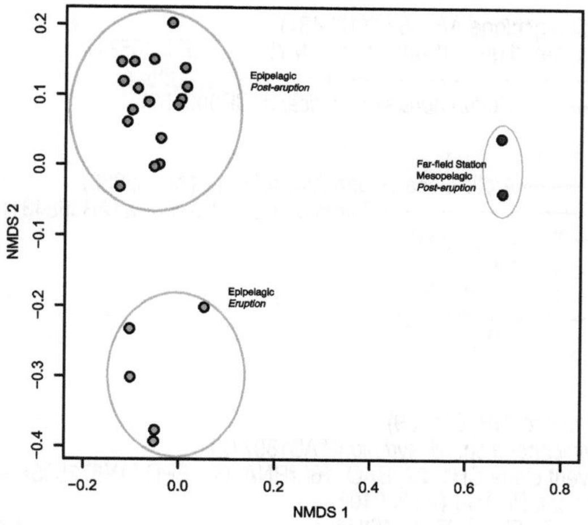

Fig. 10.3 Non-metrical multidimensional (nMDS) plot based on bacterial taxa in seawater samples collected from November 2011 to April 2012. The position of samples reflects how different bacterial assemblages are from each other based on their distance in a two-dimensional plot. The figure shows how bacterial communities collected during the eruption clustered together and were separate from the post-eruption communities. Modified from Ferrera et al. (2015)

(lower number of taxa) than those collected in the following months. In contrast, archaeal richness estimates in eruption and post-eruption phases were similar (Ferrera et al. 2015).

An index of differences in bacterial composition (beta-diversity) indicated that bacterial communities collected during the eruption grouped together and were separate from those collected after the eruption (Fig. 10.3; Ferrera et al. 2015). The presence of certain groups typically associated with hydrothermal systems (e.g., Epsilonproteobacteria and Zetaproteobacteria) supported the hypothesis that, as for abundance and activity, the volcano induced changes in the bacterial community structure. However, no clear clustering of eruptive or post-eruptive samples was observed for archaea, although the low number of sequences analyzed for the archaeal domain could explain the lack of pattern (see Ferrera et al. 2015).

Further, sequence tags revealed interesting differences between the erupting and post-erupting periods in the composition of the bacterial communities. For example, the proportion of Alphaproteobacteria, including the SAR11 group—a common member of marine bacterioplankton—was lower in samples collected in November 2011 than in samples collected months later. Moreover, despite the low abundance of Epsilonproteobacteria, recognized as an ecologically relevant group in hydrothermal environments for playing key roles in carbon and sulfur cycling (Nakagawa et al. 2005), they were significantly higher in November 2011. In fact, they were hardly detected in epi- or mesopelagic samples collected months after the eruption. Interestingly, the few tags found in the post-eruption period were different than those present during the eruption (see Fig. 10.4). Members of the candidate division ZB3, often associated with oxygen minimum zones and sulfidic environments, were also detected in the water column although at low abundances. Lastly, tags belonging to the Zetaproteobacteria, which are iron-oxidizers associated with seamounts (McAllister et al. 2019), were also detected in the water column near the Tagoro submarine volcano. Compounds released during the eruption included inorganic sulfur, reduced iron, manganese, and ammonium (Fraile-Nuez et al. 2012), all of which can serve as a source of energy for some of these organisms, which were likely accompanying the volcanic emissions and thrived in the water column during and after the eruption. As for archaea, the taxa that typically dominate hydrothermal systems (e.g., Nanoarchaea, Archeaoglobales, Thermococcales, Thermoplasmatales; e.g. Gugliandolo and Maugeri 2019; Huber et al. 2002; Kormas et al. 2006; Pagé et al. 2008; Reysenbach et al. 2020) were not detected near the Tagoro submarine volcano after its birth. Overall, the study by Ferrera et al. (2015) concluded that the geochemical changes in seawater resulted in changes in the bacterial community structure and composition, although these changes were transient. Microbial communities in the area were largely restored after the eruption.

In 2014, two years after the eruptive phase was over, a survey collected sediment and mineral deposits near degassing vents around the Tagoro and found the presence of various

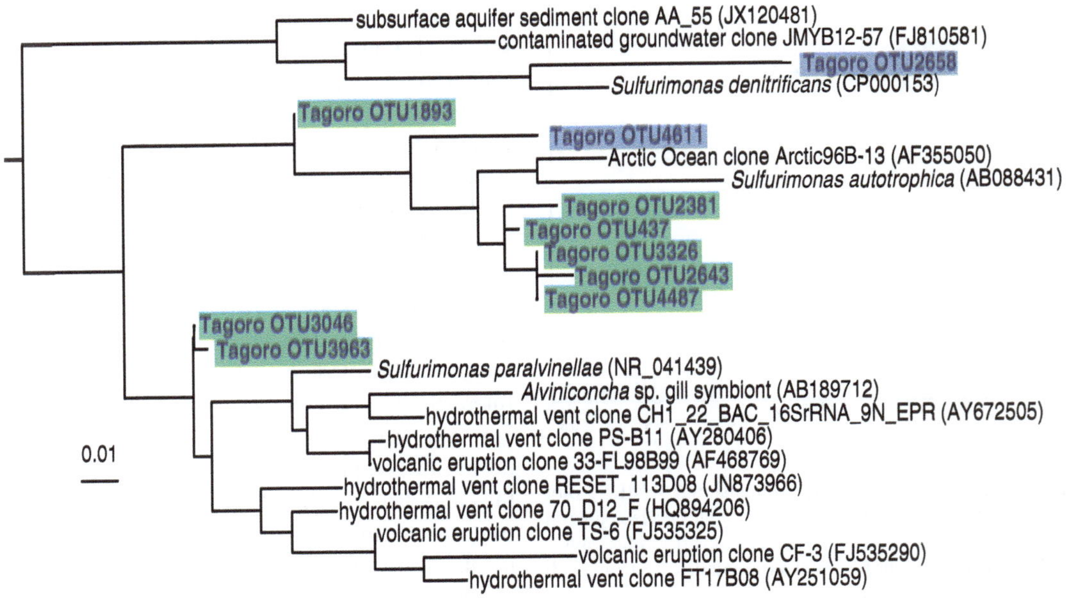

Fig. 10.4 Phylogenetic tree of epsilonprotebacterial 16S rRNA gene sequences retrieved from seawater samples collected around the Tagoro submarine volcano between November 2011 and April 2012. The color indicates the sampling period when sequences were detected (green: eruption, blue: post-eruption). Reference sequences from GenBank database are indicated by their accession number in parenthesis. The scale bar indicates substitutions per site. Modified from Ferrera et al. (2015)

organomineral structures, reflecting microbial iron-oxidation fueled by Fe-enriched hydrothermal fluids (González et al. 2020). Sequencing of 16S rRNA genes from a sediment sample unveiled that most bacterial and archaeal taxa recovered belonged to groups typically found in the seawater column, such as Acidobacteria, Oceanospirilalles, Bacteroidetes, and Thaumarchaeota. Moreover, organisms typically recovered from hydrothermal systems, such as iron-oxidizing bacteria from the Zetaproteobacteria class, sulfur-oxidizing bacteria from the Gammaproteobacteria (Thiotrichales and Chromatiales), sulfate-reducers (Deltaproteobacteria), and methanogenic and methane-oxidizing archaea were also present (González et al. 2020). Overall, this study provided evidence of the occurrence of diverse microorganisms involved in iron, sulfur and methane cycling in the Tagoro submarine volcano, some of which may participate in microbial biomineralization processes (i.e., the formation or accumulation of minerals by microorganisms). Further, sequence data generated from a hydrographic cruise that took place in October 2016 suggest that venting activity continued providing the conditions for the development of iron- and sulfur-oxidizing bacteria, since these microbes were relatively abundant in seawater samples with pH anomalies (Romano-Gude 2017).

10.2.3 New Forms of Microbial Life Emerging from the Tagoro Submarine Volcano

One of the most fascinating findings made from exploring the Tagoro submarine volcano was the discovery of what appears to be a new bacterium, named Venus' Hair (Danovaro et al. 2017). In 2014, 32 months after the eruption, an expedition found an area of about 2000 m^2 around the volcanic cone colonized by whitish microbial mats, which were made of filaments of up to a few centimeters long attached to the lava substrate.

Microscopic analysis of the filaments revealed that these were long trichomes of bacterial cells inside of 36–90 μm-wide sheaths. The filaments resembled those formed by members of the genus *Thioploca* within the Thiotrichales (Teske and Nelson 2006) but the organisms covering the Tagoro submarine volcano were firmly attached to the substrate whereas the *Thioploca* filaments typically do not. The lack of vacuoles in Venus' Hair, unlike other members of sulfur-oxidizing filamentous Thiotrichales (Jørgensen and Gallardo 1999), was another hint that Venus' Hair could be a new organism. The lack of a cultured isolate of the organism was circumvented by Danovaro and colleagues by powerful metagenomic analyses of DNA from the filaments, which enabled the reconstruction of various partial genomes, one of which (bin 11) dominated the sequence dataset. Phylogenomic analyses using multiple gene markers showed that bin 11 was affiliated with the family Thiotrichaceae (class Gammaproteobacteria). Bin 11 was closest to the genome of the filamentous *Thioploca araucae*, but these two genomes only shared about one third of predicted genes. These results indicated that the Venus' Hair filaments likely were not only a new species but a new genus of the order Thiotrichales. The proposed new name is *Candidatus* Thiolava veneris (Danovaro et al. 2017).

Genomic analyses of bin 11 revealed its unusual metabolic potential. Like other chemolithoautotrophs, *Candidatus* Thiolava veneris has the capability for CO_2 fixation, but it has not one but three mechanisms for carbon fixation: the Calvin–Benson–Bassham cycle, the reductive tricarboxylic acid cycle, and the C4-dicarboxylic acid cycle. Strikingly, it has also genes for the exploitation of organic substances, as heterotrophic organisms do, being thus a potential mixotroph. The genome of *Candidatus* Thiolava veneris also contains genes involved in nitrate reduction and different pathways for sulfur metabolism. In addition, its partial genome contains a gene similar to a known heavy metal translocating ATPase, which could be used for detoxification of metals leached from the solidified lava. Perhaps this unusual metabolic repertoire explains why this organism was able to massively colonize the seafloor after the destruction that followed the submarine volcanic eruption, leading to a novel microbial habitat. Further analyses of the organisms associated with the mat conducted by sequencing taxonomic ribosomal genes showed a diverse array of bacteria, archaea, and eukaryotes, many of which could not be assigned to organisms present in current databases, indicating that the new ecosystem could represent an important contribution to biodiversity. Overall, the data suggest that the filamentous bacteria act as the system's primary producer at the bottom of a food chain that supports higher trophic levels. Currently, the main crater area of the Tagoro is still covered with whitish mats likely formed by *Candidatus* Thiolava veneris (Fig. 10.5), which offers the possibility to further investigate this fascinating habitat.

Fig. 10.5 Image showing whitish filaments attached to mineral deposits in the Tagoro volcano (indicated by the white arrow), likely formed by *Candidatus* Thiolava veneris. Image captured by ROV Liropus 2000 from the Spanish Institute of Oceanography (IEO-CSIC) as part of the VULCANA-II project (April 2017)

10.3 Concluding Remarks and Future Perspectives

Shallow hydrothermal systems constitute an excellent natural laboratory for investigating the effects of high CO_2, temperature, and nutrients as well as lower pH on marine microorganisms. Moreover, eruptive events provide a unique opportunity to observe subseafloor microbial communities. Since 2011, the shallow Tagoro submarine volcano has been monitored by the Spanish Institute of Oceanography (IEO-CSIC) and other national and international marine research institutions and universities. The multidisciplinary monitoring carried out over the past 10 years has included a microbiological perspective, yet our knowledge of the microbial communities in this ecosystem is still sparse. Most of the focus of the surveys has been on evaluating the effects of the eruptive plume first and then the gaseous emissions on the surrounding pelagic communities (Ferrera et al. 2015; Fraile-Nuez et al. 2012; Gómez-Letona et al. 2018; Romano-Gude 2017; Martínez-Miralles 2017). These studies have concluded that the eruption significantly changed the abundance, activity, and diversity of prokaryotes. In particular, from monitoring the pelagic compartment during early stages of the Tagoro submarine volcano we have learnt that: (i) cyanobacterial cells suffered a decline at depths close to the volcano during the eruption; (ii) the number of heterotrophic prokaryotes in the surrounding waters of the volcano was generally higher during the eruption than in the post-eruption stage; (iii) the percentage of HNA potentially active cells was greater during the eruptive phase in the affected zone; (iv) the eruption caused a decrease in the diversity of bacterial communities; (v) certain typically hydrothermal taxa were detected in the water column during the eruption; and (vi) greater effects on diversity and community composition of archaea were not observed (Ferrera et al. 2015; Fraile-Nuez et al. 2012). Surveys in the degassing stage have shown that, despite most values returning to normal in the post-eruptive stages thanks to dilution with background seawater, there are physical (temperature, salinity, and density) anomalies still present in the area (see Chap. 8 of this book). Moreover, venting areas still have a certain influence in the area, as observed by the occurrence of hydrothermal microorganisms involved in iron cycling that are detected in the water column years later (Romano-Gude 2017) and the increase in the single-cell activity of marine prokaryotes in waters affected by the plume (Martínez-Miralles 2017). Conversely, other authors could not find any apparent impact on local phytoplankton communities (Gómez-Letona et al. 2018), while a significant change was detectable in the composition of the zooplankton community in waters affected by the volcanic emissions (Fernández de Puelles et al. 2021). These contrasting reports should motivate further investigation. A major caveat lies in the fact that any biological effect of volcanic inputs to the water column will occur after a certain time lag and, with the combined effect of uptake and dilution, it is difficult to distinguish between water masses affected by emissions and unaffected control waters. Current efforts led by the Spanish Institute of Oceanography (IEO-CSIC) focus on improving the design of the sampling strategy and finding appropriate tracers of volcanic affection in order to unequivocally identify the water parcels that receive volcanic inputs. This knowledge will be instrumental to properly assess the effect of diffuse volcanic emissions on the pelagic food web.

Regarding the benthic compartment, analyses of microbial mats, sediments, and mineral deposits in the post-eruptive stage revealed that years after its onset, the volcano was fueling the development of microbes involved in iron, sulfur, and methane cycling (Danovaro et al. 2017; González et al. 2020). One of the most remarkable findings was the discovery of *Candidatus* Thiolava veneris, a filamentous organism with the ability of massively colonizing the seafloor around Tagoro (Danovaro et al. 2017). The authors characterized this organism using microscopic and metagenomic approaches and, based on phylogenomic placement and structural differences, they proposed this organism as a new species. Unfortunately, the organism was not isolated in pure culture and the 16S rRNA gene sequence of *Candidatus* Thiolava veneris was never retrieved from the metagenomic assembled sequences. Due to the lack of ribosomal gene sequence data, the authors could not perform fluorescence in situ hybridization analysis on the filaments, which would have been the irrefutable proof that bin 11 belonged to the filamentous organism. Perhaps other single-cell molecular techniques like gene-FISH (Barrero-Canosa et al. 2017) or its variants (Guimarães et al. 2021) could be used to target *Candidatus* Thiolava veneris genes, which would help confirm the identity of this filamentous organism.

The assembly of single-amplified genomes (SAGs) and metagenome-assembled genomes (MAGs) have unveiled the existence of many new organisms and even phyla (Rinke et al. 2013; Parks et al. 2017). Yet, having a cultured 'type material' is still required by the International Code of Nomenclature of Prokaryotes (ICNP) to name a new species, despite the rising initiatives for naming uncultivated organisms (Konstantinidis et al. 2017; Murray et al. 2020), and the common practice in microbiology of defining species based on 16S rRNA gene sequences (Kim et al. 2014). Obtaining *Candidatus* Thiolava veneris in culture would help confirm its identity and move from the *Candidatus* category to a fully described organism. Most importantly, it would be possible to investigate its physiology under different, controlled scenarios and to test the metabolic versatility suggested so far only with metagenomic information.

With the exception of the study by Danovaro and colleagues which applied metagenomics, the analysis of prokaryotic diversity conducted on the Tagoro submarine volcano has been limited to sequencing ribosomal genes. By means of 16S rRNA gene sequencing, the presence of organisms typically associated with hydrothermal systems, such as members of the Epsilonproteobacteria, Zetaproteobacteria or Woesearchaeota, has been reported. However, no genomic information has been retrieved from these organisms. Thus, whether the ecotypes living in this volcano are identical to those reported from other hydrothermal environments is unknown. A long-standing question in microbial ecology is about the biogeography and microbial dispersion across very distant communities (Martiny et al. 2006; Mestre and Höfer 2020). The question is of particular interest in hydrothermal systems since these ecosystems are distantly distributed, and as a result can potentially harbor isolated microbial populations (Dick 2019; Whitaker et al. 2003). Some of the taxa in the vicinity of the volcano were not present in the surrounding waters and, intriguingly, *Candidatus* Thiolava veneris had not been previously reported elsewhere. Up to now, little effort has been put into comparing the diversity of the organisms inhabiting the Tagoro submarine volcano with other hydrothermal systems. Obtaining individual genomes from these organisms would help understand whether the Tagoro harbors communities similar to those of other geothermally active systems or it contains its own local diversity. Another issue that has not been fully investigated is the presence of symbiotic bacteria among the biota colonizing the newly formed ecosystem (Witt et al. 2017), as opposed to those from hydrothermal vents in mid-ocean ridges worldwide, which have been studied in detail. Approaching this subject would further contribute to that biogeography question, since symbiotic bacteria are selected by environmental conditions but also by their hosts, which themselves display strong biogeographical patterns (Bachraty et al. 2009; Dubilier et al. 2008; Petersen et al. 2010; Tunnicliffe et al. 1998).

Acknowledgements The authors would like to thank all people and institutions involved in the ten years spent monitoring the Tagoro submarine volcano, in particular, the Spanish Institute of Oceanography (IEO-CSIC), and the crew of the R/V *Ramon Margalef* and R/V *Angeles Alvariño*. Special thanks to the book editor, Pablo González, for his support during the writing process of this chapter, and to David L. Kirchman and another anonymous reviewer for critically reviewing it. We would also like to thank Ainslie Hone for proofreading this chapter. The authors received funding from projects BIMBACHE (IEO-2011-2012), RAPROCAN-III (IEO-2010-2012), VULCANA-I (IEO-2015-2017), VULCANA-II (IEO-2018-2020), and VULCANA-III (IEO-2021-2023) from the Spanish Institute of Oceanography (IEO-CSIC), funding from the Ministerio de Economía y Competitividad del Gobierno de España (MINECO) and FEDER through VULCANO-I (CTM2012-36317) and VULCANO-II (CTM2014-51837-R) projects, and from the Fundación Caja-Canarias through CYCLOVENT (2019SP18) project. I. Ferrera received the support of the Fundación BBVA through the 'Becas Leonardo a Investigadores y Creadores Culturales' 2019 Program and M. Sebastián had the institutional support of the 'Severo Ochoa Centre of Excellence' accreditation (CEX2019-000928-S). This publication reflects the views of only the authors, and their institutions and funding agencies cannot be held responsible for any use which may be made of the information contained therein.

References

Ariza A, Kaartvedt S, Røstad A, Garijo JC, Arístegui J, Fraile-Nuez E, Hernández-León S (2014) The submarine volcano eruption off El Hierro Island: effects on the scattering migrant biota and the evolution of the pelagic communities. PLoS ONE 9(7):e102354

Bachraty C, Legendre P, Desbruyeres D (2009) Biogeographic relationships among deep-sea hydrothermal vent faunas at global scale. Deep Sea Res Part I 56 (8):1371–1378

Barrero-Canosa J, Moraru C, Zeugner L, Fuchs BM, Amann R (2017) Direct-geneFISH: a simplified protocol for the simultaneous detection and quantification of genes and rRNA in microorganisms. Environ Microbiol 19(1):70–82

Buck NJ, Resing JA, Baker ET, Lupton JE (2018) Chemical fluxes from a recently erupted shallow submarine volcano on the Mariana Arc. Geochem Geophys Geosyst 19(5):1660–1673

Danovaro R, Canals M, Tangherlini M, Dell'Anno A et al (2017) A submarine volcanic eruption leads to a novel microbial habitat. Nat Ecol Evol 1(6):1–9

Dick GJ (2019) The microbiomes of deep-sea hydrothermal vents: distributed globally, shaped locally. Nat Rev Microbiol 17(5):271–283

Dieterich DC, Link AJ, Graumann J, Tirrell DA, Schuman EM (2006) Selective identification of newly synthesized proteins in mammalian cells using bioorthogonal noncanonical amino acid tagging (BONCAT). Proc Nat Acad Sci 103(25):9482–9487

Dubilier N, Bergin C, Lott C (2008) Symbiotic diversity in marine animals: the art of harnessing chemosynthesis. Nat Rev Microbiol 6(10):725–740

Emerson D, Moyer CL (2010) Microbiology of seamounts: common patterns observed in community structure. Oceanography 23(1):148–163

Fernández de Puelles ML, Gaza M, Cabanellas-Reboredo M, González-Vega A, Herrera Bravo De Laguna I, Presas Navarro C, Arrieta JM, Fraile Nuez E (2021) Abundance and structure of the zooplankton community during a post-eruptive process: the case of the submarine volcano Tagoro (El Hierro; Canary Islands), 2013–2018. Front Mar Sci 8

Ferrera I, Arístegui J, González JM, Montero MF, Fraile-Nuez E, Gasol JM (2015) Transient changes in bacterioplankton communities induced by the submarine volcanic eruption of El Hierro (Canary Islands). PLoS ONE 10(2):e0118136

Ferrera I, Reñé A, Funosas D, Camp J, Massana R, Gasol JM, Garcés E (2020) Assessment of microbial plankton diversity as an ecological indicator in the NW Mediterranean coast. Mar Pollut Bull 160:111691

Flores GE, Campbell JH, Kirshtein JD, Meneghin J et al (2011) Microbial community structure of hydrothermal deposits from geochemically different vent fields along the Mid-Atlantic Ridge. Environ Microbiol 13 (8):2158–2171

Flores GE, Shakya M, Meneghin J, Yang ZK et al (2012) Inter-field variability in the microbial communities of hydrothermal vent deposits from a back-arc basin. Geobiology 10(4):333–346

Fortunato CS, Huber JA (2016) Coupled RNA-SIP and metatranscriptomics of active chemolithoautotrophic communities at a deep-sea hydrothermal vent. ISME J 10(8):1925–1938

Fraile-Nuez E, González-Dávila M, Santana-Casiano JM, Arístegui J et al (2012) The submarine volcano eruption at the island of El Hierro: physical-chemical perturbation and biological response. Sci Rep 2:486

Fraile-Nuez E, Santana-Casiano JM, González-Dávila M, Vázquez JT et al (2018) Cyclic behavior associated with the degassing process at the shallow submarine volcano Tagoro, Canary Islands, Spain. Geosciences 8 (12):457

Gasol JM, Zweifel UL, Peters F, Fuhrman JA, Hagström Å (1999) Significance of size and nucleic acid content heterogeneity as measured by flow cytometry in natural planktonic bacteria. Appl Environ Microbiol 65(10):4475–4483

Gómez-Letona M, Arístegui J, Ramos AG, Montero MF, Coca J (2018) Lack of impact of the El Hierro (Canary Islands) submarine volcanic eruption on the local phytoplankton community. Sci Rep 8(1):1–12

González FJ, Rincón-Tomás B, Somoza L, Santofimia E et al (2020) Low-temperature, shallow-water hydrothermal vent mineralization following the recent submarine eruption of Tagoro volcano (El Hierro, Canary Islands). Mar Geol 430:106333

González-Vega A, Fraile-Nuez E, Santana-Casiano JM, González-Dávila M et al (2020) Significant release of dissolved inorganic nutrients from the shallow submarine volcano Tagoro (Canary Islands) based on seven-year monitoring. Front Mar Sci 6:829

Gugliandolo C, Maugeri TL (2019) Phylogenetic diversity of Archaea in shallow hydrothermal vents of Eolian Islands, Italy. Diversity 11(9):156

Guieu C, Bonnet S, Petrenko A, Menkes C, Chavagnac V, Desboeufs K, Maes C, Moutin T (2018) Iron from a submarine source impacts the productive layer of the Western Tropical South Pacific (WTSP). Sci Rep 8 (1):1–9

Guimarães NM, Azevedo NF, Almeida C (2021) FISH Variants. Methods Mol Biol (Clifton, NJ) 2246:17–33

Hatzenpichler R, Scheller S, Tavormina PL, Babin BM, Tirrell DA, Orphan VJ (2014) In situ visualization of newly synthesized proteins in environmental microbes using amino acid tagging and click chemistry. Environ Microbiol 16(8):2568–2590

Huber H, Hohn MJ, Rachel R, Fuchs T, Wimmer VC, Stetter KO (2002) A new phylum of Archaea represented by a nanosized hyperthermophilic symbiont. Nature 417(6884):63–67

Hügler M, Gärtner A, Imhoff JF (2010) Functional genes as markers for sulfur cycling and CO_2 fixation in microbial communities of hydrothermal vents of the Logatchev field. FEMS Microbiol Ecol 73(3):526–537

Jørgensen BB, Gallardo VA (1999) *Thioploca* spp.: filamentous sulfur bacteria with nitrate vacuoles. FEMS Microbiol Ecol 28(4):301–313

Kim M, Oh HS, Park SC, Chun J (2014) Towards a taxonomic coherence between average nucleotide identity and 16S rRNA gene sequence similarity for species demarcation of prokaryotes. Int J Syst Evol Microbiol 64(2):346–351

Konstantinidis KT, Rosselló-Móra R, Amann R (2017) Uncultivated microbes in need of their own taxonomy. ISME J 11(11):2399–2406

Kormas KA, Tivey MK, Von Damm K, Teske A (2006) Bacterial and archaeal phylotypes associated with distinct mineralogical layers of a white smoker spire from a deep-sea hydrothermal vent site (9° N, East Pacific Rise). Environ Microbiol 8(5):909–920

Lebaron P, Servais P, Agogué H, Courties C, Joux F (2001) Does the high nucleic acid content of individual bacterial cells allow us to discriminate between active cells and inactive cells in aquatic systems? Appl Environ Microbiol 67(4):1775–1782

Leizeaga A, Estrany M, Forn I, Sebastián M (2017) Using click-chemistry for visualizing in situ changes of translational activity in planktonic marine bacteria. Front Microbiol 8:2360

Martínez-Miralles S (2017) Effect of an underwater volcano in a degassing stage on the activity of marine prokaryotes. Master thesis, University of Barcelona

Martiny JBH, Bohannan BJ, Brown JH, Colwell RK et al (2006) Microbial biogeography: putting microorganisms on the map. Nat Rev Microbiol 4(2):102–112

McAllister SM, Moore RM, Gartman A, Luther III GW, Emerson D, Chan CS (2019) The Fe (II)-oxidizing Zetaproteobacteria: historical, ecological and genomic perspectives. FEMS Microbiol Ecol 95(4):fiz015

Merino N, Kawai M, Boyd ES, Colman DR, McGlynn SE, Nealson KH, Kurokawa K, Hongoh Y (2020) Single-cell genomics of novel actinobacteria with the Wood-Ljungdahl pathway discovered in a serpentinizing system. Front Microbiol 11:1031

Mestre M, Höfer J (2020) The microbial conveyor belt: connecting the globe through dispersion and dormancy. Trends Microbiol 29(6):482–492

Meyer JL, Akerman NH, Proskurowski G, Huber JA (2013) Microbiological characterization of post-eruption “snowblower” vents at Axial Seamount, Juan de Fuca Ridge. Front Microbiol 4:153

Mock T, Daines SJ, Geider R, Collins S et al (2016) Bridging the gap between omics and earth system science to better understand how environmental change impacts marine microbes. Glob Change Biol 22(1):61–75

Murray AE, Freudenstein J, Gribaldo S, Hatzenpichler R et al (2020) Roadmap for naming uncultivated Archaea and Bacteria. Nat Microbiol 5(8):987–994

Nakagawa S, Takai K, Inagaki F, Hirayama H, Nunoura T, Horikoshi K, Sako Y (2005) Distribution, phylogenetic diversity and physiological characteristics of epsilon-Proteobacteria in a deep-sea hydrothermal field. Environ Microbiol 7(10):1619–1632

Nercessian O, Bienvenu N, Moreira D, Prieur D, Jeanthon C (2005) Diversity of functional genes of methanogens, methanotrophs and sulfate reducers in deep-sea hydrothermal environments. Environ Microbiol 7(1):118–132

Nogales B, Lanfranconi MP, Piña-Villalonga JM, Bosch R (2011) Anthropogenic perturbations in marine microbial communities. FEMS Microbiol Rev 35(2):275–298

Olins HC, Rogers DR, Preston C, Ussler W III et al (2017) Co-registered geochemistry and metatranscriptomics reveal unexpected distributions of microbial activity within a hydrothermal vent field. Front Microbiol 8:1042

Pace NR, Stahl DA, Lane DJ, Olsen GJ (1986) The analysis of natural microbial populations by ribosomal RNA sequences. Adv Microbial Ecol 1–55

Pagé A, Tivey MK, Stakes DS, Reysenbach AL (2008) Temporal and spatial archaeal colonization of hydrothermal vent deposits. Environ Microbiol 10 (4):874–884

Parks DH, Rinke C, Chuvochina M, Chaumeil PA et al (2017) Recovery of nearly 8,000 metagenome-assembled genomes substantially expands the tree of life. Nat Microbiol 2(11):1533–1542

Petersen JM, Ramette A, Lott C, Cambon-Bonavita MA, Zbinden M, Dubilier N (2010) Dual symbiosis of the vent shrimp *Rimicaris exoculata* with filamentous gamma-and epsilonproteobacteria at four Mid-Atlantic Ridge hydrothermal vent fields. Environ Microbiol 12 (8):2204–2218

Rappé MS, Giovannoni SJ (2003) The uncultured microbial majority. Ann Rev Microbiol 57(1):369–394

Reysenbach AL, John ES, Meneghin J, Flores GE et al (2020) Complex subsurface hydrothermal fluid mixing at a submarine arc volcano supports distinct and highly diverse microbial communities. Proc Natl Acad Sci 117(51):32627–32638

Rinke C, Schwientek P, Sczyrba A, Ivanova NN et al (2013) Insights into the phylogeny and coding potential of microbial dark matter. Nature 499(7459):431–437

Romano-Gude D (2017) Microbial diversity study of an underwater volcano in a degassing stage: island of El Hierro (Canary Islands). Master thesis, University of Barcelona

Roussel EG, Konn C, Charlou JL, Donval JP et al (2011) Comparison of microbial communities associated with three Atlantic ultramafic hydrothermal systems. FEMS Microbiol Ecol 77(3):647–665

Santana-Casiano JM, González-Dávila M, Fraile-Nuez E, De Armas D, González AG, Domínguez-Yanes JF, Escánez J (2013) The natural ocean acidification and fertilization event caused by the submarine eruption of El Hierro. Sci Rep 3(1):1–8

Santana-Casiano JM, Fraile-Nuez E, González-Dávila M, Baker ET, Resing JA, Walker SL (2016) Significant discharge of CO_2 from hydrothermalism associated with the submarine volcano of El Hierro Island. Sci Rep 6(1):1–9

Servais P, Casamayor EO, Courties C, Catala P, Parthuisot N, Lebaron P (2003) Activity and diversity of bacterial cells with high and low nucleic acid content. Aquat Microb Ecol 33(1):41–51

Sogin ML, Morrison HG, Huber JA, Welch DM et al (2006) Microbial diversity in the deep sea and the underexplored "rare biosphere." Proc Natl Acad Sci 103(32):12115–12120

Stahl DA, Lane DJ, Olsen GJ, Pace NR (1984) Analysis of hydrothermal vent-associated symbionts by ribosomal RNA sequences. Science 224(4647):409–411

Takai K, Horikoshi K (1999) Genetic diversity of archaea in deep-sea hydrothermal vent environments. Genetics 152(4):1285–1297

Teske A, Nelson DC (2006) The Genera *Beggiatoa* and *Thioploca*. Prokaryotes 6:784–810

Tunnicliffe V, McArthur AG, McHugh D (1998) A biogeographical perspective of the deep-sea hydrothermal vent fauna. Adv Mar Biol 34:353–442

Vila-Costa M, Gasol JM, Sharma S, Moran MA (2012) Community analysis of high-and low-nucleic acid-containing bacteria in NW Mediterranean coastal waters using 16S rDNA pyrosequencing. Environ Microbiol 14(6):1390–1402

Ward DM, Bateson MM, Weller R, Ruff-Roberts AL (1992) Ribosomal RNA analysis of microorganisms as they occur in nature. Adv Microbial Ecol 219–286

Whitaker RJ, Grogan DW, Taylor JW (2003) Geographic barriers isolate endemic populations of hyperthermophilic archaea. Science 301(5635):976–978

Wilson ST, Hawco NJ, Armbrust EV, Barone B et al (2019) Kīlauea lava fuels phytoplankton bloom in the North Pacific Ocean. Science 365(6457):1040–1044

Witt V, Ayris PM, Damby DE, Cimarelli C, Kueppers U, Dingwell DB, Wörheide G (2017) Volcanic ash supports a diverse bacterial community in a marine mesocosm. Geobiology 15(3):453–463

Zinger L, Amaral-Zettler LA, Fuhrman JA, Horner-Devine MC et al (2011) Global patterns of bacterial beta-diversity in seafloor and seawater ecosystems. PLoS ONE 6(9):e24570

Impact of Tagoro Volcano Formation on Benthic Habitats and Associated Biota: A Review

11

Ana Sotomayor-García, José L. Rueda, Olga Sánchez-Guillamón, Javier Urra, Alejandro Martín-Arjona, Marcos González-Porto, Juan T. Vazquez, Desirée Palomino, Nieves López-González, Luis M. Fernández-Salas, J. Magdalena Santana-Casiano, Melchor González-Dávila, and Eugenio Fraile-Nuez

Abstract

Since Tagoro volcano erupted in 2011, several impacts have been associated to the volcano formation process, some of which are still present to date. This chapter is a review of the marine environmental perturbations caused by Tagoro volcano as a new geological structure, but thoroughly onto the partly annihilated benthic and demersal pre-existing biota, and the colonizing dynamics during the recovery process. Shallow recent volcanic activity in the NE Atlantic is uncommon, thus, Tagoro provides a unique opportunity to study, from the very beginning, the evolution of the unusual shallow hydrothermal systems, and the establishment of new marine habitats and associated biota. Distinct habitat types, with different associated deposit products (volcaniclastic aprons, lava balloons, lava ponds, etc.), have been described, and a description of the colonizing biota has been made from the available published works as well as from underwater imagery and benthic dredge samples taken during several field expeditions (Vulcano0313, 1013, 0314 and Vulcana0417). Those habitats included hard (rocky) and mixed (loose) substrate habitats, but also extreme habitats with hydrothermal vents and bacterial mats, accompanied by significant physical and chemical anomalies. Habitat preference by the observed taxa among the volcanic edifice has been explored through nMDS analyses, and a comparative analysis with published data of the typical fauna of the region (La Restinga, Mar de Las Calmas,

A. Sotomayor-García (✉)
Institut de Ciències del Mar, Consejo Superior de Investigaciones Científicas, Passeig Marítim de La Barceloneta 37, 08003 Barcelona, Spain
e-mail: anasg@icm.csic.es

J. L. Rueda · O. Sánchez-Guillamón · J. Urra · A. Martín-Arjona · J. T. Vazquez · D. Palomino · N. López-González
Instituto Español de Oceanografía (IEO), Consejo Superior de Investigaciones Científicas (CSIC), Centro Oceanográfico de Málaga, Puerto Pesquero, S/N, 29640 Fuengirola, Málaga, Spain

M. González-Porto · L. M. Fernández-Salas
Instituto Español de Oceanografía (IEO), Consejo Superior de Investigaciones Científicas (CSIC), Centro Oceanográfico de Cádiz, 11006 Cádiz, Spain

J. M. Santana-Casiano · M. González-Dávila
Instituto de Oceanografía y Cambio Global, Universidad de Las Palmas de Gran Canaria, 35017 Las Palmas, Spain

E. Fraile-Nuez (✉)
Instituto Español de Oceanografía (IEO), Consejo Superior de Investigaciones Científicas (CSIC), Centro Oceanográfico de Tenerife, Santa Cruz de Tenerife, C/ Farola del Mar, 22, 38180 Santa Cruz de Tenerife, Spain
e-mail: eugenio.fraile@ieo.csic.es

P. J. González (ed.), *El Hierro Island*, Active Volcanoes of the World,
https://doi.org/10.1007/978-3-031-35135-8_11

Marine Reserve, El Hierro Island), permitted to foresee the very first steps and direction of the recovery of the benthic and demersal communities. The impact caused by Tagoro onto the nearest littoral benthic communities, the ichthyofauna and the local fisheries have been described as well. Finally, some recommendations and further steps are given in order to adequately monitor the successional trend and environmental status of the benthic and demersal communities.

Keywords

Shallow underwater volcanoes · Tagoro · Benthic habitat diversity · Dynamics of marine ecology

11.1 Introduction

Underwater volcanic eruptions expose marine organisms to highly variable and extreme conditions in short time frames. Violent eruptions can destroy benthic-established communities by burying them with lavas and pyroclastic materials (Tolstoy et al. 2006; Sotomayor-García et al. 2019), but create at the same time a new and empty substrate, spotted with diffusive vents that can be colonized by organisms ranging from bacteria to metazoans (Tunnicliffe et al. 1997; Dunstan and Johnson 1998; Shank et al. 1998; Mullineaux et al. 2003; Cowen and Sponaugle 2009; Hernandez et al. 2011; Witt et al. 2017; Sotomayor-García et al. 2020). In 2011, Tagoro volcano (hereafter Tagoro) erupted south of El Hierro Island (Canary Islands) at 89 mbsl (meter below the sea level) (Fraile-Nuez et al. 2012; Rivera et al. 2013), annihilating most of the inhabiting benthic and demersal fauna in the area (Sotomayor-García et al. 2019). The immediate colonization and recovery process of those habitats has been described in Sotomayor-García et al. (2019, 2020). The former drastic conditions seemed to be advantageous to some organisms that have a wide range of tolerance or high capacity to adapt to new environments, high growth rates, high reproductive output and mobility, and also, to those that were common in the adjacent areas. Nevertheless, the eruption itself was responsible at the same time of a transitory fertilization event of the surrounding seawater (Fraile-Nuez et al. 2012; Santana-Casiano et al. 2013; González-Vega et al. 2020), latter on replaced by lateral fluxes of organic particles from distant active established habitats, or downwards from the ocean's surface waters as has been described in other environments by Smith et al. (2008) and Levin et al. (2016).

Magmatic activity at shallow depths (infralittoral-circalittoral) is often brief and punctual, followed by an extended-on-time degassing phase. In the case of Tagoro, the volcanic eruption lasted for 5 months (Rivera et al. 2013), and the venting phase remains active to date (Fraile-Nuez et al. 2018; González-Vega et al. 2020). The recovery of the habitats and associated biota is then subjected to a gentle fluid emission, i.e. visible shimmering water, and small but noticeable water temperature anomalies of around + 0.02 °C above the ambient, and large deposits of S and Fe mineral covering mantles on the nearby venting chimneys (González et al. 2020).

The first colonization steps of extreme environments in the ocean have been well documented for the deep-sea hydrothermal vents, at the spreading mid-ocean ridges (Gulmann et al. 2015; Alain et al. 2004; Marcus et al. 2009; O'Brien et al. 2015). The habitats of these hydrothermal vents may display repeated primary succession because community development is reset by continuous eruptive events on a decadal scale (Embley et al. 1998; Soule et al. 2007). Therefore, some knowledge has been achieved on the premier macrofauna that takes over (i.e., chemosymbiotic tubeworms and deep-sea mussels) within the associated productivity regimes, as well as on the chemosynthetic bacteria that make the community development viable (Van Dover 2000). Nevertheless, when it comes to eventual shallow-water eruptions, scarce information is available on the ecological succession for replenishing the eradicated habitat. The extent to which a benthic marine community is accessible is crucial to understand its resilience capacity. Colonization of new substrates by larvae of macrofaunal organisms from distant areas was found to be one of the major triggers for habitat recovery after an eruptive event at the East Pacific Rise (Mullineaux et al. 1998, 2010, 2012) and Juan de Fuca Ridge (Marcus et al. 2009). Moreover, it is important to take into account, that natural disasters only partly destroy the pre-existing communities, i.e., there are rarely no survivors to an eruption (Sousa 2001). Thus, these survivors may contribute to the reconstruction of the pre-disturbance communities (Platt and Connell 2003).

In this chapter, the impact caused by the eruption and further recovery of the habitats have been reviewed combining information published by different authors together with data gathered for a decade, from before (Oceana 2010) and after the eruption through Vulcano and Vulcana oceanographic expeditions from IEO (Instituto Español de Oceanografia, Sotomayor-García et al. 2019, 2020) (Fig. 11.1a). The impact of Tagoro formation onto benthic and demersal communities has been analyzed, but also onto pelagic and littoral systems, accounting for the ichtyofauna and flora of the surrounding ambient. Finally, some recommendations and further directions are given in order to maintain Tagoro's biological communities monitoring.

11.2 Physical–Chemical Anomalies of the Water Masses During the Eruptive Event and the Successive Recovery Phase

During the eruptive phase, where large quantities of mantle-derived gases, solutes and heat were released into the surrounding waters, Conductivity-Temperature and Depth (CTD) measurements of the waters affected by the volcanic emissions, revealed temperature and salinity anomalies up to + 3 °C and 0.3, respectively, at 80 mbsl. The major anomalies though were commonly detected around the principal and secondary cones. At the time of the eruption, maximum temperature anomalies of + 18.8 °C were observed close to the principal cone (around 210 mbsl) (Fraile-Nuez et al. 2012). Moreover, important anomalies in the carbonate system variables, pH_T (total pH), C_T (total dissolved inorganic carbon) and A_T (total alkalinity) were observed principally due to CO_2 emissions. At depths between 75 and 100 mbsl, pH values ranged from 5.8 to 6.4 (outside the affected area: pH = 8.0), which increased to values between 7.2 and 7.6 to the southwest of the volcano. The C_T at surface waters was as high as 7682 μmol kg^{-1} (outside the affected area, at the surface, C_T = 2100 μmol kg^{-1}). To the west, and following the direction of the volcanic plume, the C_T decreased gradually to 2780 μmol kg^{-1}. The emissions of reduced species such as iron and sulfur were also important. Between 75 and 125 mbsl, the concentration of Fe(II) reached values higher than 50 μmol kg^{-1} and the S(II) species presented concentrations of 476 μmol kg^{-1}. The Fe(II) emitted was oxidized to Fe(III) and precipitated into various mineral species, mainly oxy-hydroxides. The flux and oxidation of the reduced emitted species was so high that oxygen values were often below detection limit and patches of anoxia at 100 mbsl in the affected area were observed (Santana-Casiano et al. 2013; Ariza et al. 2014). Studies of volatiles and trace elements concentrations from the Tagoro lava balloons (Longpré et al. 2017) agreed with the changes in the CO_2 variables and redox species observed along the water column.

Five months after the beginning of the eruptive process, monitoring of the physical properties around Tagoro continued showing significant variations in the temperature and salinity, with small but notable differences of + 0.02 °C and − 0.018, respectively. During the degassing period, the area chemically affected by Tagoro and its hydrothermal emissions was concentrated 0.5 km around the principal cone. The composition of the gases emitted was mainly CO_2. These emissions generated a pH_T value of 6.1 (1.9 units below the normal values) and a C_T of 4.191 μmol kg^{-1} in the water column just over the volcanic edifice, around 90 mbsl (Santana-Casiano et al. 2013). Positive anomalies for

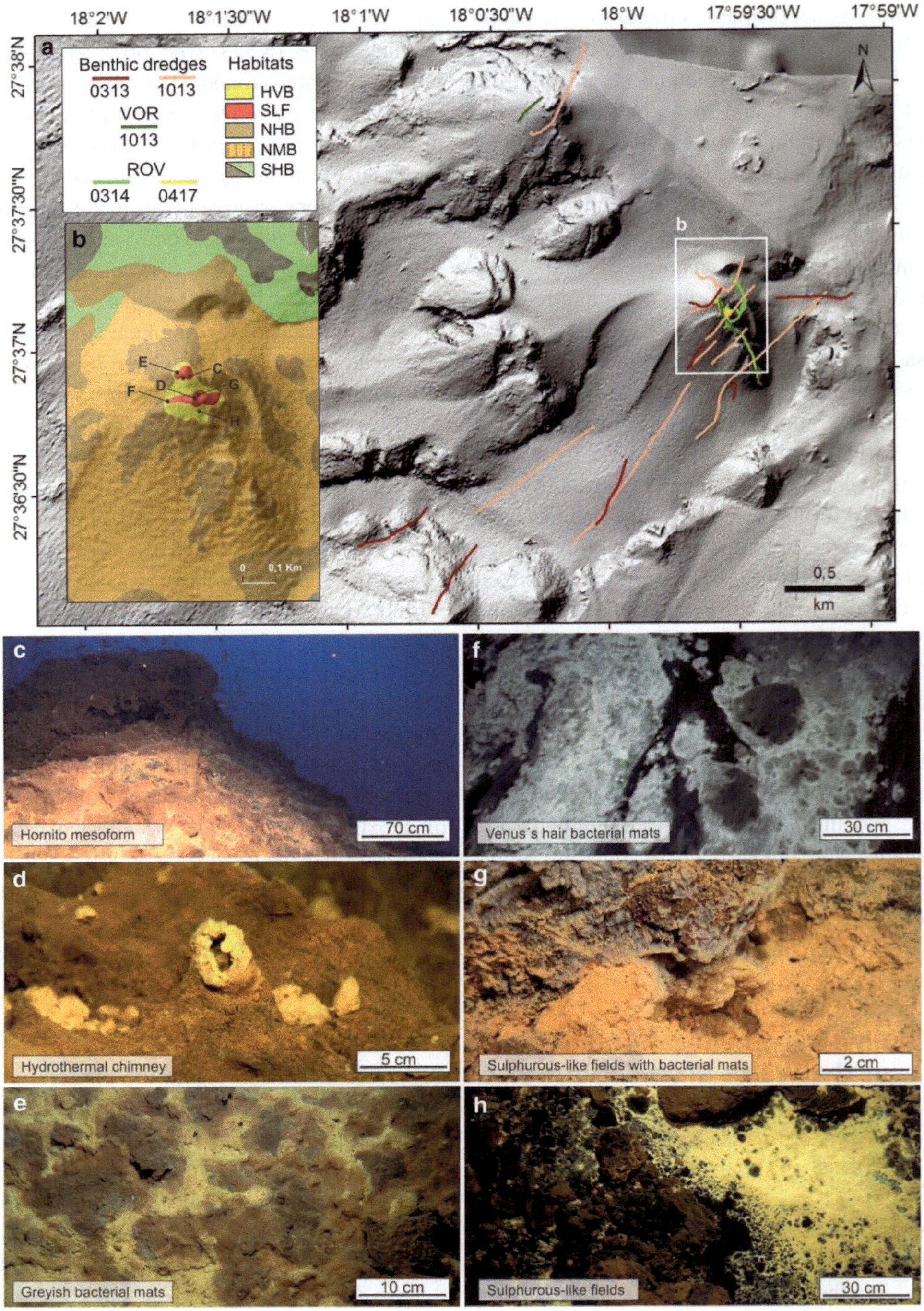

Fig. 11.1 **a** Location of benthic dredges and underwater imagery tracks recovered during Vulcano and Vulcana oceanographic expeditions. **b** Map displaying the main habitats of Tagoro volcanic complex: HVB, hydrothermal vent sites with bacterial mats; SLF, sulphurous-like fields, NHB and NMB, new hard and mixed bottoms and SHB, stable hard bottoms, and location of underwater images C to H. **c** Hornito mesoform, **d** hydrothermal chimney, **e** Greyish bacterial mats near the summit of Tagoro volcano, **f** Venus's hair bacterial mats, **g** sulphurous-like fields and small bacterial mats in the depression-like crater and **h** Sulphurous-like fields near to the depression

Fe(II) were still observed. A decrease in pH contributed to a decrease in the oxidation rate of Fe(II), and favored the persistence of this metal in the medium.

In October 2016, five years after the end of the eruptive phase, a physical mooring installed in the interior of a hydrothermal depressed area at 127 mbsl of the principal cone of Tagoro, registered important physical anomalies above the surface of the seabed in reference to the surrounding waters. In that sense, the hydrothermal system generated a micro-environment that were supporting continuously maximum potential temperature anomalies of up to + 2.55 °C, a decrease in salinity of − 1.02 and a decrease in density of − 1.43 kg m^{-3} (Fraile-Nuez et al. 2018). The emission of CO_2 was observed only above the principal cone reducing the pH by ~ 0.30 units in 2014 (Santana-Casiano et al. 2016) and 0.05 in 2019. The combined effect of temperature and salinity anomalies could generate a buoyant plume on the environment that returned shallower water downwards by diffusion, to close the circulation cells, which at the same time generated by advection a horizontal flow (Speer 1989). Theoretically, this vortex, is of sufficient magnitude to trap water in a local recirculation that could have a crucial effect on the expected dispersal of plume properties, including mineral particles, chemical tracers, and organisms (Speer 1989). Also, a general enrichment of macronutrients (nitrate, nitrite, phosphate and silicate) was observed during a period of eight years since the onset of the eruption (González-Vega et al. 2020), as well as an extended mineralization of the seafloor due to hydrothermal iron (Fe)-rich sediments (González et al. 2020).

11.3 Tagoro Formation and Establishment of Chemosynthesis-Based Habitats

The main geomorphic macro- (e.g., edifice, cones) and micro-features (e.g., hornitos) of Tagoro, as well as the newly formed habitats and associated biota, were previously characterized in different works (Somoza et al. 2017; Danovaro et al. 2017; Sotomayor-García et al. 2019, 2020). The eruptive activity, ranging from effusive to explosive with a wide range of eruption styles, combined with later hydrothermal venting, gave rise to the Tagoro volcanic complex (TVC) composed by the principal cone, various secondary cones and associated deposit products, including volcaniclastic aprons, lava balloons, lava ponds, bulbous pillow lavas, scoria and hydrothermal hornitos (Somoza et al. 2017).

One aspect that characterizes Tagoro is the absence of a previous volcanic-type structure to be reused for its growth, as mentioned in Chap. 3.2 (Vázquez et al., this volume). Therefore, the volcanic stacking resulting from these different eruptive styles, corresponds to new volcanic bottoms mainly generated on the previous physiography of the southern flank of El Hierro Island. The TVC is formed by this pile of volcanic material and the debris and volcanic flows that transported material to the southwest channeled through a pre-existing valley (Vázquez et al. 2016). In the adjacent areas, a pyroclastic deposit was produced, which was ejected towards the surface and later distributed by the water masses, especially towards low gradient areas of both the slope and the island shelf. This volcanic dynamic devastated large part of the bottoms where the stacking of Tagoro and debris-volcanic flows occur. Nevertheless, although the impact in adjacent areas was important, some species survived throughout the entire eruption, as discussed in the following section.

Once magmatic activity subsided at Tagoro, some of these newly formed seafloor deposits conformed two novel habitats with extreme characteristics: "Hydrothermal vents with the presence of bacterial mats" and "Sulphurous-like fields" (Sotomayor-García et al. 2020) (HVB and SLF in Fig. 11.1b). They both occur around the principal and secondary cones. Some authors indicated that these were the most extreme habitats in Tagoro (Danovaro et al. 2017; Sotomayor-García et al. 2020), basically colonized by bacteria that oxidize reduced sulphur and Fe emanating from vent sites. The study of these habitats represented a unique opportunity for improving the poor knowledge of these singular extreme habitats for the NE Atlantic (Sotomayor-García et al. 2019). Other habitats were formed by pyroclastic deposits consisting of new hard substrates and mixed bottoms (NHB and NMB, respectively, in Fig. 11.1b) and will be presented and discussed in the following section. Finally, stable hard bottoms that were not significantly covered by erupted material surrounded TVC and are also presented in the following section (SHB in Fig. 11.1b).

Hydrothermal vents occur over the hard substrates of the principal and secondary cones and depression-like craters, where significant physical–chemical anomalies associated with hydrothermal vents occurred during the post-eruption phase due to the degasification process of the TVC (Fraile-Nuez et al. 2018) (Fig. 11.1b, e, HVB). In underwater imagery numerous chimneys, siphons and holes were identified (Fig. 11.1d). The bacterial mats concentrated around venting areas and were characterized by a 10 cm thick band of purple to yellowish color in underwater photographs (Fig. 11.1e). Also, the occurrence of numerous hornitos at the summit of the principal cone could be indicative of the final eruptive activity with hydrothermal flows and alteration, where metric conical to cylindrical accumulations of scoriaceous lava spatters, and small eruptive vents with emissions of small amounts of lava fragments and degassing structures are still present (Fig. 11.1c). Hydrothermal venting beneath the hornitos resulted in the rising of water at high pressure through secondary conducts or fissures in the volcanic rocks promoting their alteration and the formation of Fe-oxyhydroxide precipitates and bacterial mats (Somoza et al. 2017; González et al. 2020). In addition, the hydrothermally altered lavas are represented by multicolor (white, black and red) seafloor patches, and the scoriaceous lava is composed of large brownish heterometric and subangular blocks showing coarse vesicular textures and degassing conduits. Pillow lavas are observed in the most distal deposits as flowing over the accumulation of lava balloons on the lower slopes of the secondary vents (Somoza et al. 2017).

Danovaro et al. (2017) investigated the massive colonization of this extreme new habitat ("Hydrothermal vents with bacterial mats"), and described a new chemosynthetic bacteria genus and species of the order Thiotrichales, commonly known as Venus's hair (*Thiolava veneris*) (Fig. 11.1f). These Venus`s hair bacterial mats are characterized by centimeter-long filaments of cells, each 3–6 µm in diameter, strung together inside a sheath with a width of 36–90 µm. Danovaro et al. (2017) indicated that Venus's hair displayed an unprecedented array of metabolic pathways, spanning from the exploitation of organic and inorganic carbon released by volcanic degassing to the uptake of sulfur and nitrogen compounds. This metabolic plasticity provided key competitive advantages for the rapid colonization of the new habitat created by the submarine eruption. Also, when analyzing Venus's hair DNA, DNA sequences of metazoan taxa belonging to meiofauna with different trophic strategies (e.g., arthropods, annelids and nematodes, –from grazers and epistrate-feeders to predators–), were detected as well (Danovaro et al. 2017). This could indicate that Venus's hair may also represents an important food source and sustain the entire life cycle of the smallest benthic metazoan organisms, which have been poorly studied at Tagoro.

Despite this first record of such bacteria at Tagoro, it is difficult to explain its origin and spreading mechanisms because Tagoro volcano is a clear example of an isolated vent system unconnected to any other known active hydrothermal site in the surroundings. According to metagenomic comparison of Venus's hair 16 s rDNA sequence, with other vent metagenomes (Danovaro et al. 2017), the prokaryotic assemblages detected in Venus's hair mats were completely different from those of other hydrothermal vent ecosystems investigated so far, even from those detected in the water masses around Tagoro. Only the prokaryotic assemblages from the Woody Crack Front vent site, located in the Mid-Atlantic Ridge showed ~ 15% similarity with those of the bacterial mats found at Tagoro (Danovaro et al. 2017). Typical hydrothermal vents are common in the Mid-Atlantic Ridge (Desbruyères et al. 2000), but examples of isolated and active vents at shallower locations of the Macaronesian region are still scarce and restricted to some volcanic islands and seamounts (Cardigos et al. 2005; Couto et al. 2015; Hernández et al. 2016). Recent and similar eruptive events formed newly seafloor substrates in the East Pacific Rise, Juan de Fuca Ridge, the Kermadec and the Mariana arcs (Kelley et al. 2002; de Ronde et al. 2005; Meyer et al. 2013; Gulmann et al. 2015) and they provided novel information on key eruptive and post-eruptive colonization processes. In general, deep-sea hydrothermal vents are characterized by high biomass but low biodiversity, including several endemic metazoans (Desbruyères et al. 2000). In contrast, shallow-water vents tend to have low biomass of a wider diverse fauna than deep-sea vents, but they generally display few or non-endemic species. Indeed, no endemic metazoans have yet been found strictly associated with vents of Tagoro and other shallow-water vents of the Macaronesian region (Cardigos et al. 2005; Sotomayor-García et al. 2019).

Finally, the "Sulphurous-like fields" of Tagoro are characterized by sulphur-looking light and sparse deposits, which can be easily re-suspended as seen in underwater images (Fig. 11.1g, h). These deposits cover large extensions surrounding the abovementioned hydrothermal vents and hornitos, covering the proximal extensions of the main depression-like crater and the shallower secondary cones, around 89–130 mbsl (Fig. 11.1b, SLF) (Sotomayor-García et al. 2020). No different species to those ones colonizing the bacterial mats could be detected in the underwater imagery. Further studies are needed to characterize in detail the biological communities (from bacteria to metazoans) occurring in this extreme habitat.

11.4 Impact of Tagoro Formation on Non-chemosynthesis-Based Habitats

As commented in the previous sections, the eruption promoted extreme physical–chemical anomalies of the water column in and around the main focal eruptive areas, during the first six months, the degasification phase. At the distanced areas from the principal and secondary cones, temperature anomalies were only observed close to the seabed (data extracted from tow-yo transects in the depth range of 60–250 mbsl). At depths between 60 and 120 mbsl, significant temperature anomalies of up to + 1.2 °C were only found during autumn, when the thermocline was located at around 60 mbsl. Conversely, during spring, the thermocline was found deeper than in autumn, at about 140 mbsl, and the temperature was significantly different from background water in the range of 140–250 mbsl of Tagoro, with a magnitude of up to + 0.8 °C (Fernández de Puelles et al. 2021). Chemical anomalies at the deeper waters were not observed and the variations occurred around the principal and secondary cones. During the post-eruptive phase in 2014, ΔpH values of − 0.25 at 100 mbsl were detected, and the same anomaly was observed in 2015 and 2016. In 2019, the ΔpH observed was only 0.005, below the detection limit (Santana-Casiano et al. 2013, 2016).

No specific information on the habitats and associated biota that occurred in the area where Tagoro was formed is available, but probably these communities contained components that are common at similar depths and bottoms of the south-eastern part of El Hierro Island. Oceana (2010) explored different parts of El Hierro Island and characterized the habitats and communities of circalittoral and bathyal hard bottoms down to 530 mbsl in areas close to Tagoro. This information could serve as a baseline for estimating the potential habitats and communities that once were present where this new volcano was formed. One of the explored areas was the

Marine Reserve "Mar de Las Calmas" (ES7020057), which is located a nautical mile north-westwards from Tagoro, and also the bathyal bottoms off La Restinga harbour, close-by to where Tagoro was later on formed. In the deep circalittoral and bathyal zones of those areas, Oceana (2010) documented the presence of dominant hard bottoms covered in some cases by a thick layer of sediments. In these areas, deep-sea sponge grounds were common and conformed by the hexactinellid sponge *Asconema setubalense* and several Lithistid sponges. Another common habitat was conformed by aggregations of gorgonians, including the common *Callogorgia verticillata, Ellisella flagellum* and *Acanthogorgia hirsuta*, as well as other gorgonians that are generally less abundant such as *Narella bellissima* and *Candidella imbricata*. Cold-water corals such as *Dendrophyllia cornigera* and *Anomocora fecunda*, and anthipatharians such as *Parantipathes hirondelle, Stichopathes* sp. and *Antipathes* sp., were also common components in those explored bottoms, and they were also observed years after in the adjacent areas of Tagoro by Sotomayor-García et al. (2019) (Fig. 11.2l–o). Some echinoderms were also abundant, such as the echinoids *Echinus* sp. and *Echinus melo*, as well as crinoids colonizing the rocks together with gorgonians and antipatharians (Oceana 2010). The most common crustaceans were *Plesionika edwardsii*, especially in crevices, and *Anamathia rissoana*, the latter sometimes found on large gorgonians (e.g., *Callogorgia verticillata*) (Oceana 2010). Molluscs were mainly represented by oysters assigned to *Neopycnodonte zibrowii* and the octopus *Eledone cirrhosa* (Oceana 2010).

Similar species were documented by Sotomayor-García et al. (2019) in deep areas of Tagoro including patches with large antipatharian colonies (mainly *Stichopathes* cf. *setacea* and *Anthipathes furcata*), dead colonies of unidentified scleractinians, hydrozoans (*Sertularella* sp., Plumulariide) and some sponges (*Axinella* sp., lithistid sponges). Some of the slow-growing species detected (e.g. black corals, scleractinians) were considered survivors from the eruption by those authors (Fig. 11.2i–k), mainly due to the fact that they were half buried by the erupted volcanic materials, when encountered in specific areas of Tagoro. These areas were located away from the main focal eruptive points of the volcanic edifice and lava debris apron. Later on, Sotomayor-García et al. (2019) indicated the presence of different habitats and communities, some of them similar to those reported by Oceana (2010), conforming the "Newly formed mixed bottoms" (NMB) and (Fig. 11.2a–c) and "Newly formed hard bottoms" (NHB) (Fig. 11.2d–h) habitats as well as the "Stable hard bottoms" (SHB) (Fig. 11.2l–o), outside the affected area.

Small (centimeters) to medium size (tens of centimeters) of basaltic pyroclasts characterize the new NHB habitat (Fig. 11.2d–h), surrounded by the NMB habitat (Fig. 11.2a–c), which is also composed by basaltic rocks of larger size (from tens of cm to meters), mainly represented by volcanic bombs and lava-balloons fragments (Sotomayor-García et al. 2020). The characteristics and natural instabilities of the NMB substrates make it to be less compact than the NHB habitat. The latter represents an ideal area for colonizers including small invertebrates such as the bivalve *Neopycnodonte cochlear* (Fig. 11.2 g), the annelid *Hyalopomatus* sp. and the decapod *Plesionika narval* (Fig. 11.2h), which was observed in high densities in the crevices of the volcanic bombs and lava-balloons. Also, the annelid *Hermodice carunculata* was observed in high abundance in the most recent underwater images (Fig. 11.2d–f).

The biodiversity reported by Sotomayor-García et al. (2019) in these newly formed habitats (NHB and NMB) is represented by seven different phyla, further described in the following section of this chapter. Some of the observed taxa were considered to be successful benthic colonizers of new substrates. Contrasting ecological parameters were observed in these habitats, with a higher number of taxa documented in the NHB but with higher abundance values in NMB habitats (Fig. 11.3b). Nevertheless, the most diverse phyla are similar in both habitats and include annelids such as the mobile taxa *Glycera* sp., *Chloeia* cf. *venusta* and

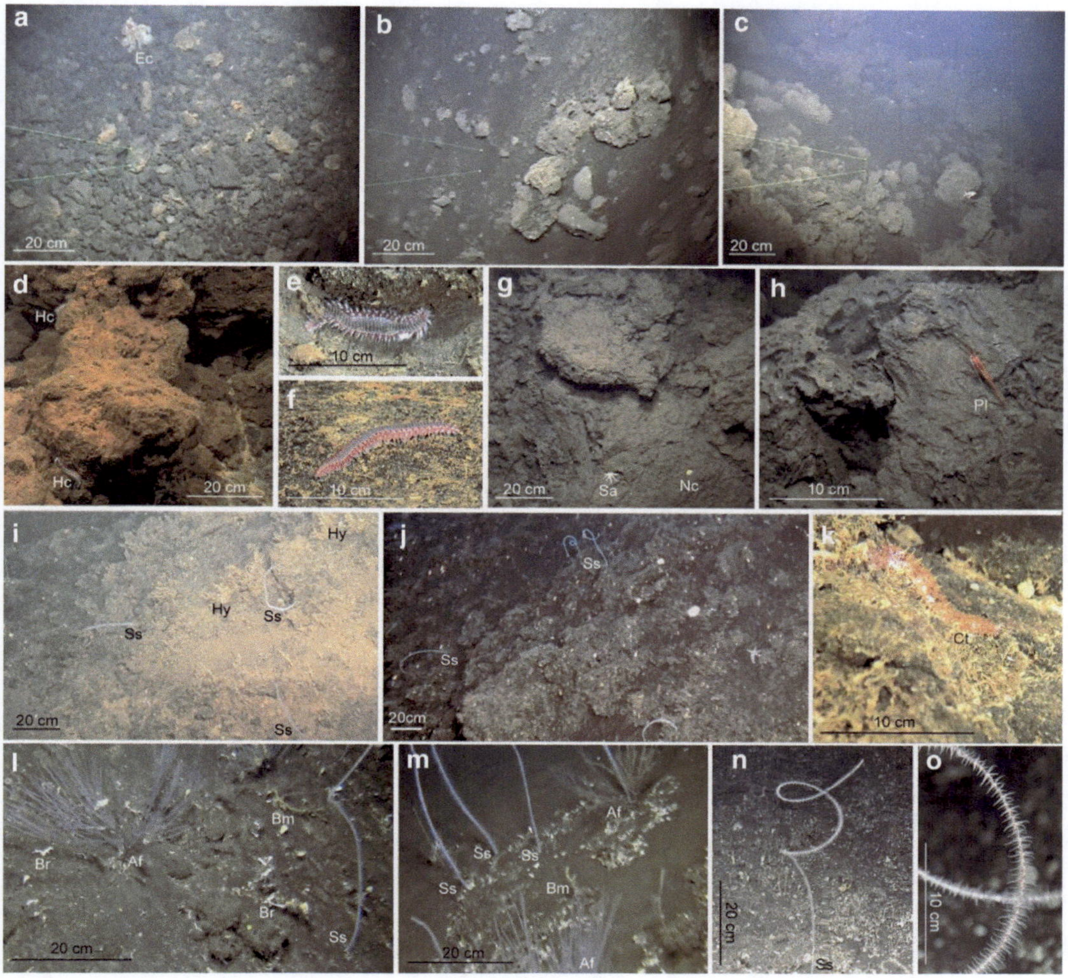

Fig. 11.2 Images of the different habitat types and species within Tagoro volcanic complex (TVC) (**a–k**) and at the undisturbed area (**l–o**). **a–c** Newly formed mixed substrates with small and large rocks (NMB), with an individual of *Eledone cirrhosa* (Ec) (Mollusca: Cephalopoda). **d–h** Newly formed pyroclastic rocks (NHB harboring *Hermodice carunculata* (Hc; see details in images **e**, **f** (Annelida: Polychaeta), *Neopycnodonte cochlear* (Nc) (Mollusca: Bivalvia), *Stylocidaris affinis* (Sa) (Echinodermata: Echinoidea), and *Plesionika* sp. (Pl) (Arthropoda: Decapoda) individuals. **i–k** Rocky bottoms within TVC (NHB) harbouring slow-growing twisted whip-shaped *Stichopathes* cf. *setacea* (Ss) (Cnidaria: Anthozoa) colonies, considered as survivors from the eruption, as well as other invertebrates such as *Coscinasterias tenuispina* (Ct) (Echinodermata: Asteroidea) and *Hyalopomatus* sp. (Hy) (Annelida: Polychaeta). **l, m** Stable undisturbed hard bottoms (SHB) colonized by fan-shaped *Antipathes furcata* (Af) (Cnidaria: Anthozoa), *S.* cf. *setacea* (Ss), small *Bebryce mollis* (Cnidaria: Anthozoa) and hydrozoans colonies. **n, o** Detail of a *S.* cf. *setacea* colony (Ss)

Onuphis sp.; arthropods including the dominant *Monadaeus couchii* and *Alpheus* sp.; cnidarians mostly represented by small colonies of the hydrozoan *Sertularella* cf. *tenella*; molluscs, mainly the ostreid bivalve *N. cochlear* (up to 100 ind. m^{-2}); and echinoderms such as the echinoid *Stylocidaris affinis* (Fig. 11.2g). The colonizing success of some of these taxa may have been magnified by a combined effect of (1) being dominant components of undisturbed hard bottoms located close by and the horizontal movements of individuals from these adjacent

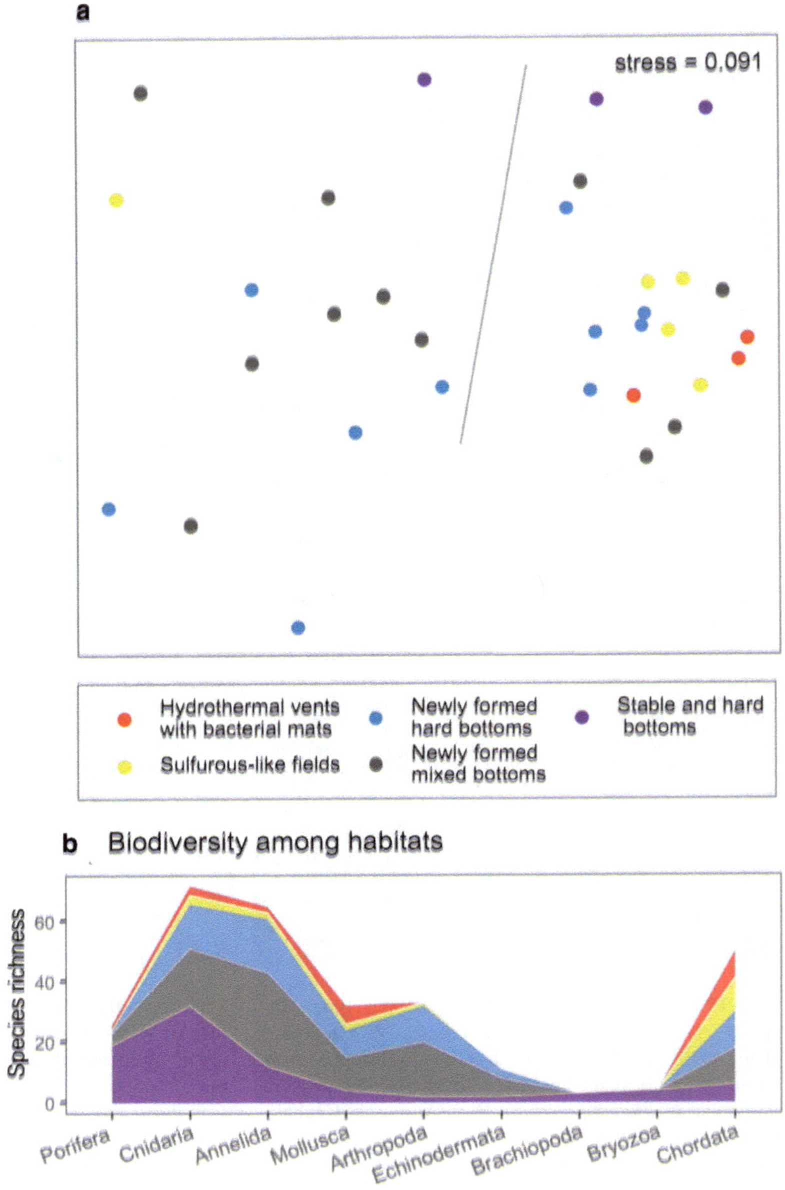

Fig. 11.3 **a** nMDS plot for samples and underwater images from sites of Tagoro volcanic complex (TVC) collected during Vulcano 0313, Vulcano 1013 and Vulcano 0314 oceanographic expeditions. The nMDS shows the samples distribution accordingly to presence/absence of the different taxa identified from both video images and benthic dredge. Color-coding corresponds to the habitats from which samples were collected. At the TVC: Hydrothermal vents with bacterial mats (red), Sulfurous-like fields (yellow), Newly formed hard bottoms (steel-blue), Newly formed mixed bottoms (pale-black). Ouside the TVC: Stable and hard bottoms (purlple). Vertical-ish gray line separates to the left benthic dredge samples and to the right video samples. **b** Area plot shows the distribution of the species richness (number of species) of each phyla identified at the different habitats. Same color-coding as Fig. 11.3a

habitats (Hernández et al. 2011); (2) the planktotrophic development of larvae during their life cycle that would permit their settlement in remote places (Cowen and Sponaugle 2009), and (3) some opportunistic organisms (e.g., annelids, molluscs) with high larval settlement success in free empty substrates (Dunstan and Johnson 1998) such as those conformed by newly formed volcanic rocks in Tagoro.

Despite recolonization after volcanic eruptions have been largely studied on mid-ocean ridges for vent chemosynthetic taxa, information on deep-sea community succession and recovery of meio and macrofaunal is scarce (Putts et al. 2019). Definitively, the eruption and subsequent formation of Tagoro caused drastic modifications. The extrusion of stacking layers of magmatic material, together with the exposure to hot, turbid, mineral-rich water, and changes in the hydrodynamic regime-caused by a change in the seafloor morphology, have seriously disturbed the biological communities inhabiting the original habitats (Fraile-Nuez et al. 2012; Santana-Casiano et al. 2013; Somoza et al. 2017; Sotomayor-García et al. 2019, 2020). These facts caused mortality either directly or indirectly, as observed in the submarine images.

In the case presented here, in a second stage following this devastating process, Tagoro provided a large amount of nutrients and elemental resources that promoted the recovery of the ecosystem (Santana-Casiano et al. 2013; Sotomayor-García et al. 2019; González-Vega et al. 2020). This recuperation process was progressive. Observations on the evolution of deep-sea coral communities on similar environments, have suggested a ~ 150 year time range for the development of a pioneer coral community (Putts et al. 2019), whereas the establishment of a 'mature' coral community supporting a more diverse array of tall, slower-growing taxa, may require far longer to fully develop (Putts et al. 2019). In Tagoro, fast-growing and small-size pioneering taxa such as some molluscs, arthropods and annelids will be followed by more complex and larger species of slow growth and long-lived nature. Initially, this young substrate will show a low species richness but a fast species accumulation, with the number of species increasing with time. The common taxa in El Hierro hard bottoms reported by Oceana (2010) and missing in these newly formed habitats include mostly slow-growing cnidarians such as colonial scleractinians and antipatharians, which may appear throughout the first century of community succession and colonization (Putts et al. 2019), if habitat and biological conditions are still adequate (i.e., species growth rate, dispersal capacity, circulation of the water masses, nutrients, etc.). This colonization process could be promoted in Tagoro by a potentially enhanced availability of food and larvae as observed in other high relief geomorphological features in other parts of the world (Mienis et al. 2007; Dolan et al. 2008). The colonization process of complex organisms such as the colonial scleractinians onto newly formed volcanic substrates seems to be partially linked to the nature of their internal skeletons and the rates of elements assimilation, which in turn can also limit growth rates (Roberts et al. 2006). In this line, pioneering scleractinian colonies with aragonite skeletons may be the first ones to colonize once the acidic conditions of the alteared water column permitted so, with such skeletons derived from ubiquitous elemental resources (e.g., calcium ions and dissolved inorganic carbon) whose concentrations increased enormously after the eruption (Santana-Casiano et al. 2013). Some slower-growing corals (e.g., Antipatharia) with proteinaceous skeletons may colonize in this succession stage, or later on, as they generally require more time for conforming colonies (Wagner et al. 2012). As time goes by, and excluding future climate-related or anthropic-related disturbances, the community inhabiting Tagoro will probably be characterized by branching skeletal coral colonies such as those observed in stable undisturbed hard bottoms close to Tagoro, which will slowly modify the benthic environment into a unique habitat, supporting increased biodiversity (Oceana 2010; Long and Baco 2014; Rossi et al. 2017).

11.5 Habitat Preference by Post-eruptive Benthic and Demersal Colonizing Fauna

Habitat preference by the different benthic and demersal species has been analyzed using multivariate similarity analyses (non-metric Multidimensional Scaling, nMDS) based on presence/absence (Jaccard index similarity matrix) information of the different taxa identified among the newly formed habitats and the non-buried habitats (Fig. 11.3a). The presence/absence data was retrieved from underwater videos and from benthic dredge samples obtained during Vulcano 0313, Vulcano 1013, and Vulcano 0314 oceanographic expeditions. Habitat preference of the most abundant taxa identified in the dredge samples (abundance was only counted for benthic dredge organisms) has been explored by overlaying their relative abundance onto the original nMDS. Presence/absence information and relative abundance data revealed the colonization dynamics across the different habitat types (inside the TVC affected area: "Hydrothermal vents with bacterial mats", HVB; "Sulfurous-like fields", SLF; "Newly formed hard bottom", NHB; and "Newly formed mixed bottoms", NMB; and outside the TVC affected area: "stable and hard bottoms" SHB, described in Sotomayor-García et al. 2020; Fig. 11.3a). A total of 6451 individuals, accounting for 116 taxa and nine phyla, were identified as inhabitants of Tagoro and its adjacent areas (Sotomayor-García et al. 2019) (Fig. 11.3b). In the multivariate analyses (nMDS), samples of organisms (ressembling assemblages) from the newly formed bottoms form a very disperse grouping (steel-blue and pale-black, Fig. 11.3a). Then, benthic communities of the extreme habitats: "Hydrothermal vents with bacterial mats" and "sulfurous-like fields", form a robust group (red and yellow symbols, Fig. 11.3a), and samples from less-impacted habitats located outside of Tagoro, also conformed a distinct group from those of the newly formed habitats (purple symbols, Fig. 11.3a).

Relative abundance of the most representative taxa identified in the benthic dredge samples has been plotted through modifying the dot size in the nMDS (from now on: bubble plots, Figs. 11.4. and 11.5). Bubble plots may provide information on the habitat preference for certain taxa, and emphasize the affinity of particular taxa to specific habitats. Macro-benthic taxa colonizing the "Hydrothermal vents with bacterial mats" and the "Sulfurous-like fields" (i.e. S and Fe deposits) are very scarce and displaying much lower abundances than in the other habitats (Fig. 11.3b). The dominant taxa included small-unidentified hydrozoans (size up to 2–3 cm), including some assigned to the *Sertularella* genus (e.g., *Sertularella* cf. *tenella*) (Fig. 11.4a). Small hydroids have been also detected in other shallow hydrothermal systems in the NE Atlantic (Fricke et al. 1989; Cardigos et al. 2005; Couto et al. 2015). Underwater images also detected small-unidentified serpulid polychaetes, and some fishes, especially *Seriola* sp.

Then, at the "Newly formed hard bottoms" and "Newly formed mixed bottoms", a higher biodiversity was detected when compared to the extreme habitats (Fig. 11.3b). A total of 64 taxa from different phyla were detected, except brachiopods and bryozoans. Composition of the community of these two newly formed habitats is somehow similar, however, the structure (abundance of the different taxa) differed significantly, being substantially higher at the mixed bottoms, as mentioned in a previous section. Mature complex communities have been observed where different types of substrates combine, with biological components typically associated to each type of substrate, resulting in high number of species (Snelgrove 1999). The most abundant species are cnidarians, annelids, molluscs and arthropods, and one species of echinoderm, the echinoid *Stylocidaris affinis*. Cnidarians were mainly represented by small colonies of hydrozoans also detected in the extreme habitats (*Sertularella* cf. *tenella* and other *Sertularella* species, Fig. 11.4a), and an unidentified species of Lafoeidae (Fig. 11.5a), which was also identified outside the TVC affected area. Annelids were represented by small sessile serpulids (*Hyalopomatus* sp., Fig. 11.4b) and mobile amphinomids, eunicids and glycerids (*Chloeia* cf. *venusta*, *Onuphis* sp., Fig. 11.4c, d, and *Glycera* sp.), considered the latter group as efficient predators and scavengers with high dispersal potential (Rouse and Pleijel 2001). Also in terms of abundance, the unidentified species of Lafoeidae, Syllidae, *Protula* sp., (Fig. 11.5a–c), and an unidentified Terebellidae species, were also important components of the new habitats, which were also identified outside Tagoro. Molluscs were essentially represented by vast amounts of the bivalve *Neopycnodonte cochlear* (locally reaching up to 100 individuals m^{-2}, Fig. 11.4e), and several individuals of unidentified opistobranchs detected in underwater images. Finally, arthropods were represented by the decapods *Monodaeus couchii*, especially in the mixed bottoms (Fig. 11.4f), *Alpheus* sp., *Munida* sp., *Plesionika narval* and other *Plesionika*

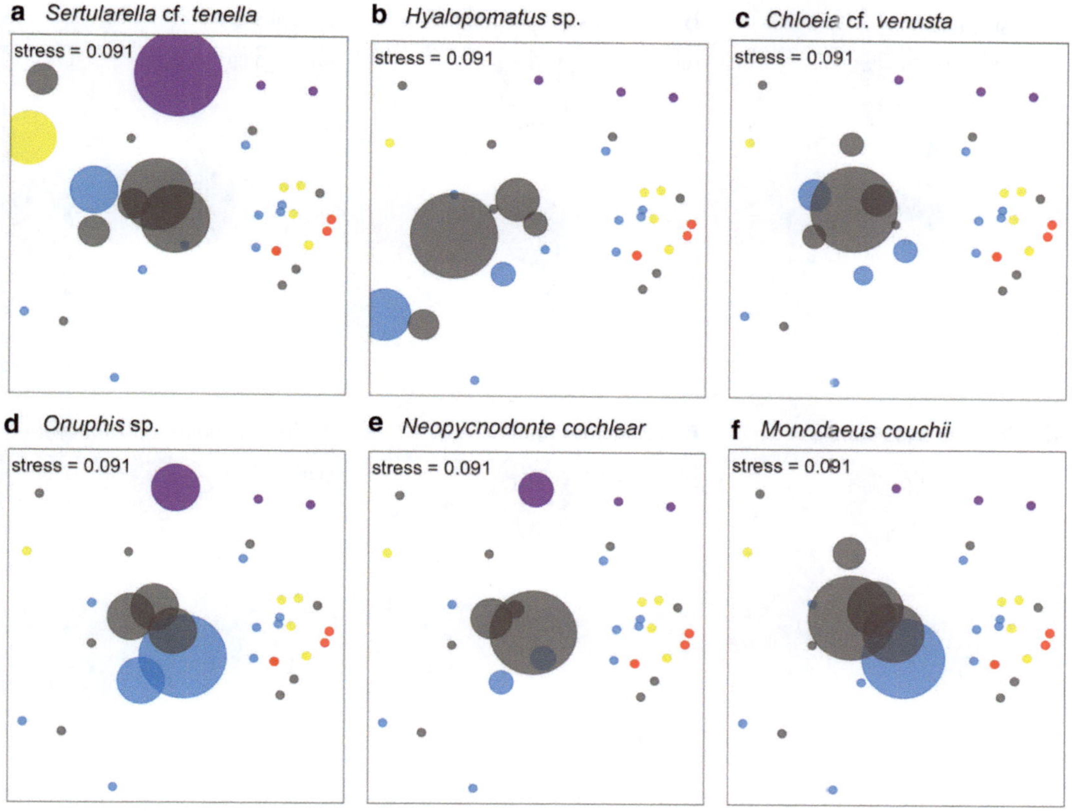

Fig. 11.4 nMDS with bubble plots illustrating the relative abundance of some of the most representative taxa identified in benthic dredge samples collected at Tagoro, among the different habitat types. Bubble size is indicative of relative abundance. *Sertularella* cf. *tenella* (**a**), *Hyalopomatus* sp. (**b**), *Chloeia* cf. *venusta* (**c**), *Onuphis* sp. (**d**), *Neopycnodonte cochlear* (**e**) and *Monodaeus couchii* (**f**)

species detected in underwater images. Some of these identified species are opportunistic with a planktotrophic development (Cowen and Spagnuole 2009), and with high settlement success rates that may favor them to be pioneer colonizers of these newly formed habitats (Gaines and Roughgarden 1985; Dunstan and Johnson 1998). Moreover, undisturbed areas surrounding Tagoro, may have acted as sources of larvae, recruits and adults of mobile taxa (Fig. 11.3a). Depending on a variety of factors influencing the viability of the transport, i.e., larvae mortality (Cowen et al. 2000), pelagic larvae duration (Shanks 2009), and larvae output according to new cohort sizes (Rouse and Pleijel 2001), the recovery of the disturbed area will be enhanced by the communities located closeby.

Outside Tagoro, benthic and demersal communities were highly diverse and some taxa were detected in high abundances. A total of 68 taxa, from nine phyla, were detected (Fig. 11.3b). Several species of colonial cnidarians (*Stichopathes* cf. *setacea, Antipathes furcata* and *Tanacetipathes cavernicola*, (Fig. 11.5d–f) co-occurs and provides a complex structure to a well-established circalittoral hard bottom community with dense antipatharian aggregations and large sponges (Jones et al. 1994). These slow growing and fragile organisms enhance the creation of three-dimensional habitats, providing settling substrates and shelter from predators (Collie et al. 1997; Kaiser et al. 1999). From those, only *Stichopathes* cf. *setacea* was detected in Tagoro, which when spotted in very specific

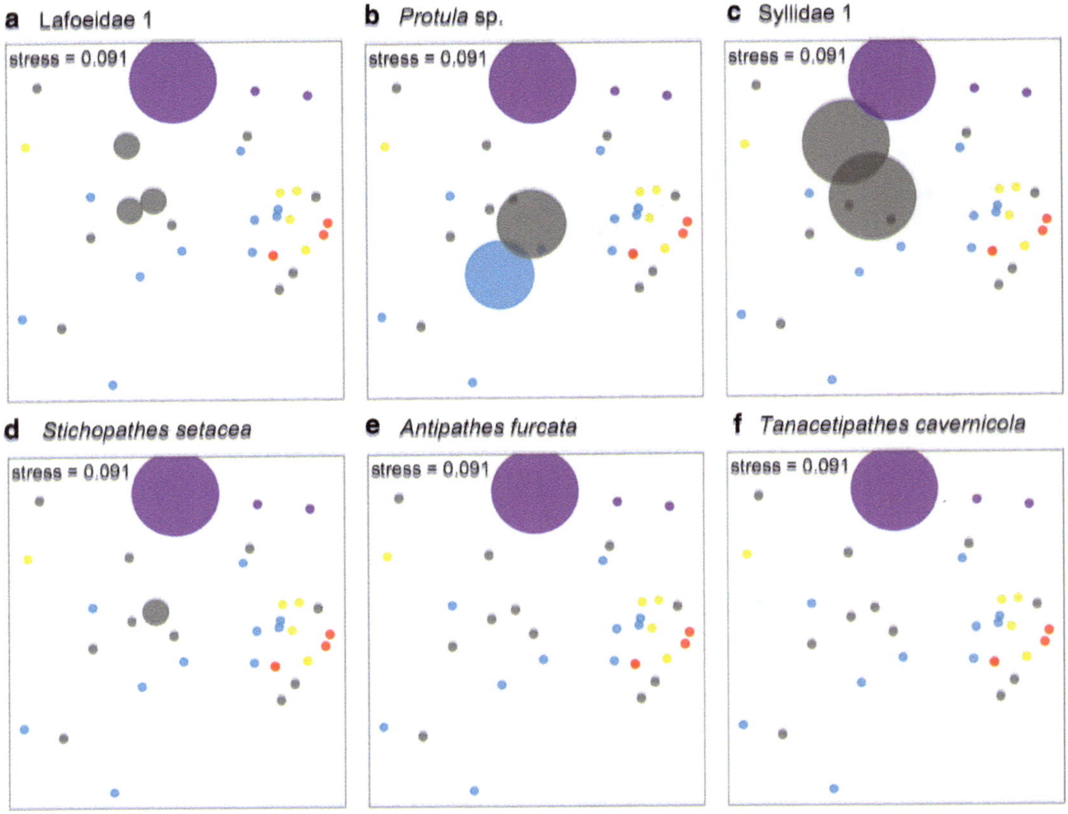

Fig. 11.5 Bubble plots illustrate the relative abundance of the most representative taxa identified in benthic dredge samples collected at Tagoro (NHB, steel-blue and NMB, pale-black dots), and outside the affected area (SHB, purple). Bubble size is indicative of relative abundance. Lafoeidae (**a**), *Protula* sp. (**b**), unid. Syllidae (**c**), *Stichopathes setacea* (**d**), *Antipathes furcata* (**e**), and *Tanacetipathes cavernicola* (**f**)

locations, was considered a survivor to the burial. Also, well-developed hydrozoan colonies were detected, mainly *Aglaophenia* sp.1 and 2, and *Sertularella* cf. *gayi*. A wide variety of epibionts visualized in underwater images, colonized some of the above-mentioned antipatharians, and mainly represented by vast populations of cirripeds (*Heteralepas cornuta*, also seen in Tagoros' new habitats), bryozoans (*Reteporella* sp.), and molluscs (*Neopycnodonte cochlear*, *Pteria hirundo*), being those, the most abundant phyla of the non-buried habitats. Some of these epibionts were identified as key colonizers of the new substrates. For example, the bivalve *N. cochlear* is a non-selective filter feeder, which grows attached to the substrate and adapts its shape to irregular surfaces. Moreover, it is a very common ostreid species of the circalittoral hard bottoms of the Macaronesia Region, including those of Canary Islands. The availability of uncolonized hard substrate and a potential event of natural fertilization that followed the eruption (Santana-Casiano et al. 2013), could probably favored the proliferation of phytoplankton and this common filter-feeding bivalve in Tagoro. Annelids were represented by serpulids, an unidentified Syllidae, *Protula* sp., and an unidentified species of Lafoeidae, as abovementioned (Fig. 11.5a–c). Moreover, different sponges were also detected in these complex and biodiverse habitats, including *Axinella* sp., *Ircina* sp., *Geodia* sp., and several other unidentified species of Demospongiae. Finally, non-previously detected phyla in Tagoro were observed in habitats located close-by, at the no-buried areas, such as brachiopods (*Megerlia truncata* and *Pajaudina atlantica*) and bryozoans (Frondiporidae, *Hornera* sp., and *Reteporella* sp.), hosted by habitat-structuring communities of black corals (essentially *Stichopathes* cf. *setacea*).

In general, there is a transition between communities with low number of taxa in extreme habitats to communities with small opportunistic colonizers in higher biodiversity at the newly formed habitats of Tagoro. Then, a well-established and complex benthic community is observed outside Tagoro, with some of its components also detected at Tagoro, based on the gradient of disturbances caused by the eruption. The former basically composed by small hydrozoans and serpulids, whereas the second is characterized by arthropods, annelids, cnidarians and hydrozoans. Finally, a three-dimensional structured community with complex ecological interactions, based on slow-growing cnidarians and large sponges, characterizes the habitats outside Tagoro.

11.6 Impact of Tagoro Formation on the Demersal and Pelagic Megafauna

The oceanographic characteristics of the southern part of El Hierro Island, sheltered from the trade winds and calmer waters, together with its steeply sloping seabed, promote the existence of a large fish biodiversity, especially given that the Canary Islands are considered to be oligotrophic (Bacallado et al. 1984). Two years before the volcanic eruption, Oceana (2010) documented different fish species in ROV high-resolution images recorded between 370 and 530 m depth, in areas close to the current location of Tagoro. The fish fauna included the seaperches *Anthias antias* and *Callanthias ruber,* the rockfishes *Helicolenus dactylopterus* and *Pontinus kuhlii*, the wrasse *Acantholabrus palloni,* the caproid *Antigonia capros,* the alfonsino *Beryx splendens,* the frogmouth *Chaunax pictus,* and the conger eel *Conger conger,* among other species. The coastal fish fauna of El Hierro is reasonable diverse and abundant (Bortone et al. 1991) with ca. 50 species observed in visual transects using scuba-diving, among which the ornate wrasse *Thalassoma pavo*, the canary damsel *Similiparma lurida* and the Azores chromis *Chromis limbata* represented dominant components.

This high fish biodiversity has enhanced an important local artisanal fishery that mainly has targeted the parrotfish *Sparisoma cretense,* the barred hogfish *Bodianus scrofa,* the triggerfish *Canthidermis sufflamen,* the groupers *Epinephelus marginatus* and *Mycteroperca fusca,* the morays *Muraena* spp., and the seabreams *Diplodus* spp., besides other pelagic fish species such as the wahoo *Acanthocybium solandri,* the european pilchard *Sardina pilchardus*, the blue jack mackerel *Trachurus picturatus,* the bogue *Boops boops* and tuna *Thunnus* spp., among others (Martín-Sosa and Cansado 2006).

Some of these fish species represent a food source for large predators such as cetaceans, being 16 the different species recorded in these waters (Aguilar de Soto et al. 2013). The bottlenose dolphin *Tursiops truncatus*, the short-fin pilot whale *Globicephala macrorhynchus,* the sperm whale *Physeter macrocephalus* or even big mysticetes, such as the humpback whale *Megaptera novaeangliae* and the rorquals *Balaenoptera* spp. are common cetaceans in this area. Besides, El Hierro holds important resident populations of Blainville's and Cuvier's beaked whales, *Mesoplodon densirostris* and *Ziphius cavirostris*, in deep waters close to the shore.

During the eruption, large quantities of mantle-derived gases, solutes and heat were released into the surrounding waters of Tagoro and caused important physical–chemical anomalies, as mentioned before. A combination of these perturbations mainly driven by the release of high concentration of reduced compounds, derived into an extreme deoxygenation events and caused high mortalities of several demersal species and some pelagic ones (mainly myctophids and hatchetfishes) in the area during the most acute episodes of the eruption (Fig. 11.6). Many local and regional media reported this, however, no detailed information has been found on reports or scientific papers. Ariza et al. (2014) suggested that the

Fig. 11.6 Images of dead fishes floating on the surface after different eruption episode of Tagoro. **a–c** Numerous dead fishes close to the shore, including a high number of *Beryx splendens* (Bs) and *Diplodus* spp (Ds), among others. **d–f** injuries associated with fast depressurization and high temperature events were observed in dead individuals such as *Balistes capriscus* (Bc), *Kyphosus sectatrix* (Ks) or *Conger conger* (Cc), among others

magnitude of the eruption did not leave any scope for adaptation, and only those species with capacity of fast vertical and horizontal mobility could avoid massive mortality events. Some of the species killed by the eruption were reported in publications by the local media including the black belly rosefish *Helicolenus dactylopterus,* the slimehead *Hoplostethus mediterraneus,* the beardfish *Polimixia nobilis,* as well as *Balistes capriscus* (Fig. 11.6e), *B. explendens* (Fig. 11.6 b), *C. conger* (Fig. 11.6f), *S. lurida* and different species of genus *Diplodus* spp. (Fig. 11.6b), among others (Fig. 11.6). They all showed injuries associated with fast depressurization (López 2013). The eruptions also caused an impact on the artisanal fishing fleet at La Restinga harbour, because fishing was banned for several months due to a potential chemical pollution of commercial species in these waters due to the erupted materials and substances (Acosta et al. 2011).

The first fish colonizers of TVC observed during the Vulcano and Vulcana oceanographic expeditions (from 2013 to 2017) were *A. anthias*, and the black-tail comber *Serranus atricauda*, also reported in Sotomayor-García et al. (2019) (Fig. 11.7e), followed by others species such as *Muraena helena, C. conger,* rockfish *Scorpaena* sp, the common stingray *Dasyatis pastinaca* and the alfonsino *Beryx splendens* (Fig. 11.7d, f–i). All of them are generally common species in the adjacent unaffected areas, which could have favored the recolonization process of Tagoro. Moreover, some of these species (e.g. *M. helena, C. conger*) take advantage of the cavities and crevices existing in hard substrates as an ideal habitat for establishing their populations. Furthermore, pelagic species such as the jacks *Seriola dumerili and S. rivoliana* or the barracuda *Sphyraena virindensis* were also observed after the eruption (Fig. 11.7a–c), although its pelagic and migratory habits do not allow drawing conclusion as a permanent resident. This was also the case for large shoals of small pelagics (e.g., *Sardina pilchardus*) reported by different media.

The post-eruptive stage has showed the restoration capacity of pelagic ecosystems after volcanic perturbations (Ariza et al. 2014). The evaluation of mesopelagic fauna a year after the eruption, as well as some of their predators such as the beaked whales that are monitored by University of La Laguna, showed no significant differences in their populations before and after the eruption (Aguilar de Soto et al. 2013).

Fig. 11.7 ROV Images from Instituto Español de Oceanografía—ROV of ichthyofauna colonizers within Tagoro: **a** *Seriola rivoliana*. **b** *Seriola dumerili*. **c** School of *Anthias antias* with some *Sphyraena viridensis* (Sv) individuals on the background. **d** A *Muraena helena* individual sheltering in a crevice of newly formed hard habitats. **e** School of *Anthias antias* with a *Serranus atricauda* (Sa) individual centered in the image. **f** A *Dasyatis pastinaca* individual moving over newly formed hard substrates. **g** School cf *Anthias antias* with a *Scorpaena* sp. (Ss) individual on newly formed bottoms. **h** A *Conger conger* individual spotted inside a newly formed volcanic rock cavity. **i** A *Beryx splendens* individual close to a newly formed volcanic rocky substrate

Nevertheless, further studies about the re-colonization process by the pelagic and demersal megafauna are required to have a clear picture on the impact caused on them by the eruptions.

11.7 Impact of Tagoro Formation on the Littoral Benthic Communities

Although the eruption took place some kilometers away from the coast, some studies have demonstrated that the same volcano which was responsible for the creation of a highly corrosive environment, affecting marine biota, also injected to the water column a huge concentration of inorganic nutrients and Fe(II), that could contribute positively to the recuperation of the marine ecosystem (Santana-Casiano et al. 2013; González-Vega et al. 2020). Nevertheless, other phytoplankton monitoring studies could not find significant evidences on that sense (Gómez-Letona et al. 2018).

Some authors detected changes in the littoral benthic communities before and after the volcanic eruption, which were probably linked to changes of the water column. Sangil (2013) studied the marine vegetation of the Marine

Reserve "Mar de Las Calmas" before and after the eruption and detected a decline of the macroalgae *Lobophora variegata*, especially in those littoral areas that were closer to Tagoro. The same trend was detected with some coralline algae and this seemed linked to the pH decrease in the water caused by the eruptions, and to the lower concentrations of oxygen and the presence of sulfurous components in the water column. Nevertheless, the environmental conditions during the eruption and the decline of those dominant macroalgae on the rocky shores provided a fast increase of other non-incrusting macroalgae for few months (e.g., *Cladophora liebethrutii, Pseudochlorodesmis furcellata, Ceramium* spp., *Polysiphonia* spp.). Finally, a recovery of *Lobophora variegata* populations was recorded in the area once the water masses recovered the original characteristics. Gil-Díaz et al. (2014) indicated that the calcified brown alga *Padina pavonica* was sensitive to acute and chronic environmental pH changes as the ones detected during the eruption. Indeed, a high decalcification of this macroalgae was detected in November 2011 at the inner harbor of La Restinga (close to the eruption), in contrast to the low decalcification of other *P. pavonica* from areas far away from the eruption. The authors concluded that these brown algae could be use as a bio-indicator of ocean acidification.

The effects of the eruption on the littoral fauna have been less studied. Quintero-Romero (2016) indicated that the eruption caused a clear effect on the abundance and diversity of mesoinvertebrates, with a decline of amphipods, decapods and echinoderms and an increase of polychaetes, especially of nereids. Further studies are needed in order to improve the knowledge on how the eruptions impacted the littoral ecosystem close to Tagoro.

11.8 Recommendations and Further Steps

The exceptionality of the volcanic eruption that gave rise to Tagoro, provides an excellent opportunity to improve the knowledge on newly formed volcanic habitats and their colonization, from the very first stage with pioneer organisms, up to the climax community formation with slow-growing coral species, as well as to analyze the ongoing ecological processes in between. This is one of the very few cases for the NE Atlantic and for other parts of the world.

Benthic and demersal colonization studies in newly formed volcanic habitats are extremely limited worldwide (Putts et al. 2019) and especially for the NE Atlantic. Sotomayor-García et al. (2019) reported high abundances for some pioneer macrofauna and megafauna species on Tagoro; these were mainly fast-growing, mobile and dominant species of the local fauna of El Hierro Island. Nevertheless, monitoring programs for studying this colonization must be carried out on a long-term scale due to the slow-growing nature of some organisms such as the colonial scleractinians and antipatharians. In this line, long-term monitoring programs will provide important insights into the benthic ecosystems responses to this catastrophic events (Lagger et al. 2017), which are not always detected in the marine realm. The new hard and mixed substrates formed after the Tagoro eruption are catalogued as Reefs (habitat code 1170) according to the EU Habitats Directive, which would allow to include this area into the annual monitoring program that is run as part of the Marine Strategy Framework Directive in Spanish waters. This monitoring program, which should be carried out by using non-invasive methods in order to preserve these newly formed faunal communities, would provide a very important temporal evolution information on the community succession and colonization on recently formed habitats, as well as on the impact that hydrothermal venting has on benthic and pelagic neighboring communities.

Furthermore, the new habitats with extreme characteristics formed around the principal and secondary cones, including hydrothermal vents and sulphurous-like fields (Danovaro et al. 2017; Sotomayor-Garcia et al. 2019, 2020), are singular habitats whose monitoring represents a unique opportunity to enhance the understanding of the bacterial and faunistic communities inhabiting them, their behavior, and the interactions with adjacent communities. No information is available yet for the meiofauna communities at Tagoro as well as for those organisms that could get some benefit from the bacterial communities in the most extreme habitats.

Previous studies in volcanic CO_2 vent sites show the resilience of many species to elevated CO_2 and low pH through different physiological and ecological mechanisms (Foo et al. 2018). However calcifying species, which are susceptible to skeletal dissolution in low pH conditions, are reduced in abundance and diversity compared to controls at other vent sites around the world (Fabricius et al. 2011, 2015; Goffredo et al. 2014). As the pH decreases, an increase in the dominance of generalists and small-bodied species results in a simplification of the food webs and trophic complexities (Vizzini et al. 2017). As indicated by Fraile-Nuez et al. (2018), five years after the eruption there were still present important physical–chemical anomalies surrounding the principal cone, including a pH decrease of − 1.25 units generating a pH gradient in the area. These special conditions make Tagoro a unique natural laboratory, of relatively easy access, where regular monitoring can be carried out to improve our understanding of how future ocean acidification may impact marine biota.

Another question that still remains unsolved is how all these eruptive events have affected the pelagic and demersal megafauna communities, including fisheries resources and the fishing fleet operating in the area at short and long-term. In this line, the location of Tagoro, just a nautical mile away from the Marine Reserve "Mar de Las Calmas" (ES7020057), could include this newly formed volcano within the Special area of conservation (SAC) with an "Integral-area" category in order to carry out experimental work on different aspects. On one hand, this would reduce the impact of professional fishing on these bottoms, allowing the development of the colonizing processes in a pristine environment; on the other hand, the large amount of nutrients and elemental resources released by the eruption would support the arrival of benthic, demersal and pelagic organisms, promoting an increase of biomass that will eventually export (i.e. spillover) from Tagoro to surrounding areas (Gell and Roberts 2003; Goñi et al. 2008; Sánchez Lizaso et al. 2000). In this sense, periodical monitoring of the marine resources and catches by the artisanal fleet would be desirable.

Besides the Marine Reserve "Mar de Las Calmas", other protection figures such as El Hierro Global Geopark, in the framework of the UNESCO International Geosciences and Geoparks Programme, will contribute to preserve this geological heritage and natural laboratory for studying marine recolonization in the Macaronesia.

Acknowledgements The publication of this work has been possible thanks to the funding of several institutions and organizations: Ministerio de Economía y Competitividad del Gobierno de España (MINECO) and FEDER through VULCANO-I (CTM2012-36317), VULCANO-II (CTM2014-51837-R) projects; the The Spanish Institute of Oceanography (IEO) through VULCANA-I (IEO-2015-2017) project; and to the special funding from IEO for the use of the ROV Liropus 2000 in several expeditions. The authors want to thank Tamia Brito Izquierdo (Marine Reserve Punta de la Restinga-Mar de Las Calmas) and Juan González Padrón for sharing images of dead fishes that appeared close to the shore of La Restinga as result of the eruptions. The authors are also grateful to Pablo J. González (book editor) and some reviewers for suggestions and corrections made to earlier drafts of this manuscript.

References

Acosta L, Casillas R, Folch A (2012) Análisis multidisciplinar del fenómeno sismo-volcánico de El Hierro (julio 2011) 91 p. https://ddd.uab.cat/record/102180

Aguilar de Soto N, Johnson M, Arranz P, Escánez A, Reyes C, Schiavi A, Madsen P, Brito A (2013) El Hierro el nacimiento de un volcán. Actas VIII Semana Científica Telesforo Bravo Instituto de Estudios Hispánicos de Canarias, pp 81–110

Alain K, Zbinden M, Le Bris N, Lesongeur F, Quérellou J, Gaill F, Cambon-Bonavita MA (2004) Early steps in microbial colonization processes at deep-sea hydrothermal vents. Env Microb 6(3):227–241. https://doi.org/10.1111/j.1462-2920.2003.00557.x

Ariza A, Kaartvedt S, Røstad A, Garijo JC, Arístegui J, Fraile-Nuez E, Hernández-León S (2014) The submarine volcano eruption off El Hierro Island: effects on the scattering migrant biota and the evolution of the pelagic communities. PLoS ONE 9(7):102354. https://doi.org/10.1371/journal.pone.0102354

Bacallado JJ, Báez M, Brito A, Cruz T, Dominguez F, Moreno E, Perez JM (1984) Fauna (marina y terrestre) del archipiélago Canario. Colección gran biblioteca Canaria. Editorial Edirca. Las Palmas de Gran Canaria. 356 p

Bortone S, van Tassell J, Brito A, Falcón J, Bundrick C (1991) A visual assessment of the inshore fishes and fishery resources off El Hierro, Canary Islands: a baseline survey. Sci Mar 53:529–541

Cardigos F, Colaço A, Dando PR, Ávila SP, Sarradin PM, Tempera F, Conceição P, Pascoal A, Santos RS (2005) Shallow water hydrothermal vent field fluids and communities of the D. João de Castro Seamount (Azores). Chem Geol 224:153–168. https://doi.org/10.1016/j.chemgeo.2005.07.019

Collie JS, Escanero GA, Valentine PC (1997) Effects of bottom fishing on the benthic megafauna of Georges Bank. Mar Ecol Prog Ser 155:159–172. https://doi.org/10.3354/meps155159

Couto RP, Rodrigues AS, Neto AI (2015) Shallow-water hydrothermal vents in the Azores (Portugal). J Int Coast Z Man 15:495–505

Cowen RK, Sponaugle S (2009) Larval dispersal and marine population connectivity. Annu Rev Mar Sci 1:443–466. https://doi.org/10.1146/annurev.marine.010908.163757

Cowen RK, Lwiza KMM, Sponaugle S, Paris CB, Olson DB (2000) Connectivity of marine populations: open or closed? Science 287:857–859. https://doi.org/10.1126/science.287.5454.857

Danovaro R, Canals M, Tangherlini M, Dell'Anno A, Gambi C, Lastras G, Amblas D, Sanchez-Vidal A, Frigola J, Calafat A, Pedrosa-Pàmies R, Rivera J, Rayo X, Corinaldesi C (2017) A submarine volcanic eruption leads to a novel microbial habitat. Nat Ecol Ev 1:144. https://doi.org/10.1038/s41559-017-0144

de Ronde CE, Hannington MD, Stoffers P, Wright IC, Ditchburn RG, Reyes AG, Baker ET, Massoth GJ, Lupton JE, Walker SL, Geene RR, Soong CW, Ishibashi J, Lebon GT, Bray CJ, Resing JA (2005) Evolution of a submarine magmatic-hydrothermal system: Brothers volcano, southern Kermadec arc, New Zealand. Econ Geol 100:1097–1133. https://doi.org/10.2113/gsecongeo.100.6.1097

Desbruyères D, Almeida A, Biscoito M, Comtet T, Khripounoff A, Le Bris N, Sarradin PM, Segonzac M (2000) A review of the distribution of hydrothermal vent communities along the northern Mid-Atlantic Ridge: dispersal vs. environmental controls. Hydrob 440:201–216. https://doi.org/10.1007/978-94-017-1982-7_19

Dolan MF, Grehan AJ, Guinan JC, Brown C (2008) Modelling the local distribution of cold-water corals in relation to bathymetric variables: adding spatial context to deepsea video data. Deep Sea Res I 55:1564–1579. https://doi.org/10.1016/j.dsr.2008.06.010

Dunstan PK, Johnson CR (1998) Spatio-temporal variation in coral recruitment at different scales on Heron Reef, southern Great Barrier Reef. Cor Reefs 17:71–81

Embley RW, Lupton JE, Massoth GJ, Urabe T, Tunnicliffe V, Butterfield DA, Shibata T, Okano O, Kinoshita M, Fujioka K (1998) Geologic, chemical, and biologic evidence for recent volcanism at 17.51S: East Pacific Rise. Ear Plan Sci Let 163:131–147

Fabricius KE, Langdon C, Uthicke S, Humphrey C, Noonan S, De'Ath G, Okazaki R, Muehllehner N, Glas MS, Lough JM (2011) Losers and winners in coral reefs acclimatized to elevated carbon dioxide concentrations. Nat Cl Ch 1(3):165–169. https://doi.org/10.1038/nclimate1122

Fabricius KE, Kluibenschedl A, Harrington L, Noonan S, De'Ath G (2015) In situ changes of tropical crustose coralline algae along carbon dioxide gradients. Sci Rep 5(1):1–7. https://doi.org/10.1038/srep09537

Fernández de Puelles ML, Gazá M, Cabanellas-Reboredo M, González-Vega A, Herrera I, Presas-Navarro C, Arrieta JM, Fraile-Nuez E (2021) Abundance and Structure of the zooplankton community during a post-eruptive process: the case of the submarine Volcano Tagoro (El Hierro; Canary Islands), 2013–2018. Front Mar Sci 8:692885. https://doi.org/10.3389/fmars.2021.692885

Foo SA, Byrne M, Ricevuto E, Gambi MC (2018) The carbon dioxide vents of Ischia, Italy, a natural system to assess impacts of ocean acidification on marine ecosystems: an overview of research and comparisons with other vent systems. Oceano Mar Bio 56:237–310. https://doi.org/10.1201/9780429454455-4

Fraile-Nuez E, González-Dávila M, Santana-Casiano JM, Arístegui J, Alonso-González IJ, Hernández-León S, Blanco ML, Rodríguez-Santana A, Hernandez-Guerra A, Gelado-Caballero MD, Eugenio F, Marcello J, de Armas D, Domínguez-Yanes JF, Montero MF, Laetch DR, Vélez-Belchí P, Ramos A, Ariza V, Comas-Rodríguez I, Benítez-Barrios VM (2012) The submarine volcano eruption at the island of El Hierro: physical-chemical perturbation and biological response. Sci Rep 2:486

Fraile-Nuez E, Santana-Casiano JM, González-Dávila M, Vázquez JT, Fernández-Salas LM, Sánchez-Guillamón O, Palomino D, Presas-Navarro C (2018) Cyclic behavior associated with the degassing process at the shallow submarine Volcano Tagoro, Canary Islands, Spain. Geosc (Switzerland) 8(12). https://doi.org/10.3390/geosciences8120457

Fricke H, Giere O, Stetter K, Alfredsson GA, Kristjansson JK, Stoffers P, Svavarsson J (1989) Hydrothermal vent communities at the shallow subpolar Mid-Atlantic ridge. Mar Biol 102:425–429. https://doi.org/10.1007/BF00428495

Gaines S, Roughgarden J (1985) Larval settlement rate: a leading determinant of structure in an ecological community of the marine intertidal zone. Proc Natl Acad Sci USA 82:3707–3711. https://doi.org/10.1073/pnas.82.11.3707

Gell FR, Roberts CM (2003) Benefits beyond boundaries: the fishery effects of marine reserves. Trends Ecol Evo 18(9):448–455. https://doi.org/10.1016/S0169-5347(03)00189-7

Gil-Díaz T, Haroun R, Tuya F, Betancor S, Viera-Rodríguez MA (2014) Effects of ocean acidification on the brown alga Padina Pavonica: decalcification due to acute and chronic events. PLoS ONE 9 (9):108630. https://doi.org/10.1371/journal.pone.0108630

Goffredo S, Prada F, Caroselli E, Capaccioni B, Zaccanti F, Pasquini L, Fantazzini P, Fermani S, Reggi M, Levy O, Fabricius KE, Dubinsky Z, Falini G, Pichat B, Centro Enrico Fermi I, del Viminale P (2014) Biomineralization control related to population density under ocean acidification Europe PMC Funders Group 4(7):593–597. https://doi.org/10.1038/nclimate224

Gómez-Letona M, Arístegui J, Ramos AG, Montero MF, Coca J (2018) Lack of impact of the El Hierro (Canary Islands) submarine volcanic eruption on the local phytoplankton community. Sci Rep 8:4667. https://doi.org/10.1038/s41598-018-22967-6

Goñi R, Adlerstein S, Alvarez-Berastegui D, Forcada A, Reñones O, Criquet G, Polti S, Cadiou G, Valle C, Lenfant P, Bonhomme P, Pérez-Ruzafa A, Sánchez-Lizaso JL, García-Charton JA, Bernard G, Stelzenmiller V, Planes S (2008) Spillover from six western Mediterranean marine protected areas: evidence from artisanal fisheries. Mar Eco Pro Ser 366:159–174. https://doi.org/10.3354/meps07532

González FJ, Rincón-Tomás B, Somoza L, Santofimia E, Medialdea T, Madureira P, López-Pamo E, Hein JR, Marino E, de Ignació C, Reyes J, Hopper M, Reitner J (2020) Low-temperature, shallow-water hydrothermal vent mineralization following the recent submarine eruption of Tagoro volcano (El Hierro, Canary Islands). Mar Geo 430:106333. https://doi.org/10.1016/j.margeo.2020.106333

González-Vega A, Fraile-Nuez E, Santana-Casiano JM, González-Dávila M, Escánez-Pérez J, Gómez-Ballesteros M, Tello O, Arrieta JM (2020) Significant release of dissolved inorganic nutrients from the shallow submarine Volcano Tagoro (Canary Islands) based on seven-year monitoring. Front Mar Sci 6:829. https://doi.org/10.3389/fmars.2019.00829

Gulmann LK, Beaulieu SE, Shank TM, Ding K, Seyfried WE Jr, Sievert SM (2015) Bacterial diversity and successional patterns during biofilm formation on freshly exposed basalt surfaces at diffuse-flow deep-sea vents. Front Microbiol 6:901. https://doi.org/10.3389/fmicb.2015.00901

Hernández CA, Sangil C, Hernández JC (2016) A new CO_2 vent for the study of ocean acidification in the Atlantic. Mar Pollut Bull 109(1):419–426. https://doi.org/10.1016/j.marpolbul.2016.05.040

Hernandez JM, Rolán E, Swinnen F, Gómez R, Pérez JM (2011) Moluscos y conchas marinas de Canarias: Solenogastres, Caudofoveata, Polyplacophora, Gastropoda, Bivalvia, Cephalopoda y Scaphopoda; ConchBooks: Hackenheim, Germay. ISBN 978-3-939767-36-7

Jones CG, Lawton JH, Shachak M (1994) Organisms as ecosystem engineers. In: Ecosystem management. Springer, New York. https://doi.org/10.1007/978-1-4612-4018-1_14

Kaiser MJ, Cheney K, Spence FE, Edwards DB, Radford K (1999) Fishing effects in northeast Atlantic shelf seas: Patterns in fishing effort, diversity and community structure VII. The effects of trawling disturbance on the fauna associated with the tubeheads of serpulid worms. Fish Res 40:195–205. https://doi.org/10.1016/S0165-7836(98)00212-4

Kelley DS, Baross JA, Delaney JR (2002) Volcanoes, fluids, and life at mid-ocean ridge spreading centers. Annu Rev Earth Planet Sci 30:385–491. https://doi.org/10.1146/annurev.earth.30.091201.141331

Lagger C, Servetto N, Torre L. Sahade R (2017) Benthic colonization in newly ice-free soft-bottom areas in an Antarctic fjord. PLoS ONE 12(11). https://doi.org/10.1371/journal.pone.0186756

Levin LA, Baco AR, Bowden DA, Colaco A, Cordes EE, Cunha MR, Demopoulos AWJ, Gobin J, Grupe BM, Le J, Metaxas A, Netburn AN, Rouse GW, Thurber AR, Tunnicliffe V, Van Dover CL, Vanreusel A, Watling L (2016) Hydrothermal vents and methane seeps: rethinking the sphere of influence. Fron Mar Sci 3:72. https://doi.org/10.3389/fmars.2016.00072

Long DJ, Baco AR (2014) Rapid change with depth in megabenthic structure-forming communities of the Makapu'u deep-sea coral bed. Deep Sea Res II 99:158–168. https://doi.org/10.1016/j.dsr2.2013.05.032

Longpré MA, Stix J, Klügel A, Shimizu N (2017) Mantle to surface degassing of carbon- and sulfur-rich alkaline magma at El Hierro, Canary Islands. Earth Planet Sci Let 460:268–280. https://doi.org/10.1016/j.epsl.2016.11.043

López C (2013) Crónica de una erupción Submarina. La crisis Sismo-Volcánica de El Hierro 2011–2012. Anuario Astronómico Del Observatorio de Madrid, pp 433–453

Marcus J, Tunnicliffe V, Butterfield DA (2009) Post-eruption succession of macrofaunal communities at diffuse flow hydrothermal vents on axial volcano. Juan de Fuca Ridge, Northeast Pacific. Deep-Sea Res II Top Stud Oceanogr 56:1586–1598. https://doi.org/10.1016/j.dsr2.2009.05.004

Martín-Sosa P, Cansado S (2006) Informe seguimiento de las pesquerías en el ámbito de las Reservas Marinas del Archipiélago Canario: Información pesquera en el entorno de la Reserva Marina de Punta de La Restinga-Mar de Las Calmas (El Hierro) Periodo 2003–2005. Instituto Español de Oceanografía

Meyer JL, Akerman NH, Proskurowski G, Huber JA (2013) Microbiological characterization of post-eruption "snowblower" vents at Axial Seamount, Juan De Fuca Ridge. Front Microbiol 4:153

Mienis F, de Stigter HC, White M, Duineveld G, de Haas H, van Weering TCE (2007) Hydrodynamic controls on coldwater coral growth and carbonate mound development at the SW and SE Rockall Trough Margin, NE Atlantic Ocean. Deep-Sea Res.

I 54:1655–1674. https://doi.org/10.1016/j.dsr.2007.05.013

Mullineaux LS, Peterson CH, Micheli F, Mills SW (2003) Successional mechanism varies along a gradient in hydrothermal fluid flux at deep-sea vents. Eco Monogr 73(4):523–542. https://doi.org/10.1890/02-0674

Mullineaux LS, Adams DK, Mills SW, Beaulieu SE (2010) Larvae from afar colonize deep-sea hydrothermal vents after a ctastrophic eruption. Proc Natl Acad Sci USA 107:7829–7834. https://doi.org/10.1073/pnas.0913187107

Mullineaux LS, Le Bris N, Mills SW, Henri P, Bayer SR, Secrist RG, Siu N (2012) Detecting the influence of initial pioneers on succession at deep-sea vents. PLoS ONE 7:50015. https://doi.org/10.1371/journal.pone.0050015

Mullineaux LS, Mills SW, Goldman E (1998) Recruitment variation during a pilot colonization study of hydrothermal vents (9° 50′ N, East Pacific Rise). Deep Sea Res Part II Top Stud Ocean 45(1–3):441–464. https://doi.org/10.1016/S0967-0645(97)00045-3

O'Brien CE, Giovannelli D, Govenar B, Luther GW, Lutz RA, Shank TM, Vetriani C (2015) 'Microbial biofilms associated with fluid chemistry and megafaunal colonization at post-eruptive deep-sea hydrothermal vents. Deep Sea Res Part II Top Stud Oceano 121:31–40. https://doi.org/10.1016/j.dsr2.2015.07.020

Oceana (2010) Propuesta de áreas marinas de importancia ecológica, Islas Canarias. Oceana y Fundación Biodiversidad, Madrid, 300 pp

Platt W, Connell JH (2003) Natural disturbances and directional replacement of species. Ecol Monogr 73:507–522. https://doi.org/10.1890/01-0552

Putts MR, Parrish FA, Trusdell FA, Kahng SE (2019) Structure and development of Hawaiian deep-water coral communities on Mauna Loa lava flows. Mar Ecol Prog Ser 630:69–82. https://doi.org/10.3354/meps13106

Quintero-Romero MV (2016) Estado de las poblaciones de mesoinvertebrados en sustratos rocosos despuésde la erupción ubmarina en la isla de El Hierro. Bachelor thesis, Universidad de la Laguna, 25 pp

Rivera J, Lastras G, Canals M, Acosta J, Arrese B, Hermida N, Micallef A, Tello O, Amblas D (2013) Construction of an oceanic island: insights from the El Hierro (Canary Islands) 2011–2012 submarine volcanic eruption. Geol 41(3):355–358. https://doi.org/10.1130/G33863.1

Roberts JM, Wheeler AJ, Freiwald A (2006) Reefs of the deep: the biology and geology of cold-water coral ecosystems. Science 312:543–547. https://doi.org/10.1126/science.1119861

Rossi S, Bramanti L, Gori A, Orejas C (2017) An overview of the animal forests of the world. In: Rossi S, Bramanti L, Gori A et al (eds) Marine animal forests. The ecology of benthic biodiversity hotspots. Springer, Cham, pp 1–26

Rouse GW, Pleijel F (2001) Polychaetes. Oxford University Press, Oxford, p 354

Sánchez-Lizaso JL, Goñi R, Reñones O, García-Charton JA, Galzin R, Bayle JT, Sánchez-Jerez P, Perez-Ruzafa A, Ramos AA (2000) Density dependence in marine protected populations: a review. Envir Cons 27(2):144–158. https://doi.org/10.1017/S0376892900000187

Sangil C (2013) Cambios en la biodiversidad vegetal submarina del Mar de Las Calmas tras la erupción volcánica de La Restinga: una oportunidad para profundizar en el conocimiento de los ecosistemas marinos de Canarias. In: Afonso Carrillo J (ed), El Hierro: el nacimiento de un volcán. Actas VIII Semana Científica Telesforo Bravo, Instituto de Estudios Hispánicos de Canarias, Puerto de La Cruz, pp 55–82

Santana-Casiano JM, González-Dávila M, Fraile-Nuez E, de Armas D, González AG, Domínguez-Yanes JF, Escánez J (2013) The natural ocean acidification and fertilization event caused by the submarine eruption of El Hierro. Sci Rep 3:1140. https://doi.org/10.1038/srep01140

Santana-Casiano JM, Fraile-Nuez E, González-Dávila M, Baker ET, Resing JA, Walker SL (2016) Significant discharge of CO_2 from hydrothermalism associated with the submarine volcano of El Hierro Island. Sci Rep 6:25686. https://doi.org/10.1038/srep25686

Shank TM, Fornari DJ, Von Damm KL, Lilley MD, Haymon RM, Lutz RA (1998) 'Temporal and spatial patterns of biological community development at nascent deep-sea hydrothermal vents (9°50′N, East Pacific Rise). Deep Sea Res Part II Top Stud Oceano 45(1–3):465–515. https://doi.org/10.1016/S0967-0645(97)00089-1

Shanks AL (2009) Pelagic larval duration and dispersal distance revisited. Biol Bull 216:373–385. https://doi.org/10.1086/BBLv216n3p373

Smith CR, De Leo FC, Bernardino AF, Sweetman AK, Martinez-Arbizu P (2008) Abyssal food limitation, ecosystem structure and climate change. Tree 23:518–528. https://doi.org/10.1016/j.tree.2008.05.002

Snelgrove PVR (1999) Getting to the bottom of marine biodiversity: sedimentary habitats: ocean bottoms are the most widespread habitat on Earth and support high biodiversity and key ecosystem services. Bio Sc 49:129–138. https://doi.org/10.2307/1313538

Somoza L, González FJ, Barker SJ, Madureira P, Medialdea T, De Ignacio C, Lourenzo N, León R, Vázquez JT, Palomino D (2017) Evolution of submarine eruptive activity during the 2011–2012 El Hierro event as documented by hydroacoustic images and remotely operated vehicle observations. Geochem Geophys Geosyst 18:3109–3137. https://doi.org/10.1002/2016GC006733

Sotomayor-García A, Rueda JL, Sánchez-Guillamón O, Urra J, Vázquez JT, Palomino D, Fernández-Salas LM, López-González N, González-Porto M, Santana-Casiano JM, González-Dávila M, Presas-Navarro C, Fraile-Nuez E (2019) First macro-colonizers and survivors around tagoro Submarine Volcano, Canary Islands, Spain. Geoscience (Switzerland) 9(1). https://doi.org/10.3390/geosciences9010052

Sotomayor-Garcia A, Rueda JL, Sánchez-Guillamón O, Vázquez JT, Palomino D, Fernández-Salas LM, López-González N, González-Porto M, Urra J, Santana-Casiano JM, González-Dávila M, Fraile-Nuez E (2020) Geomorphic features, main habitats and associated biota on and around the newly formed Tagoro submarine volcano, Canary Islands. Seafloor geomorphology as benthic habitat, pp 835–846. https://doi.org/10.1016/b978-0-12-814960-7.00051-8

Soule SA, Fornari DJ, Perfit MR, Rubin KH (2007) New insights into mid-ocean ridge volcanic processes from the 2005–06 eruption of the East Pacific Rise, 9° 46’-56’N. Geology 35:1079–1082

Sousa W (2001) Natural disturbance and the dynamics of marine benthic communities. In: Bertness MD, Gaines SD, Hay ME (eds) Marine community ecology. Sinauer Associates, Inc., Sunderland, pp 85–130

Speer KG (1989) A forced baroclinic vortex around a hydrothermal plume. Geophys Res Lett 16:461–464. https://doi.org/10.1029/GL016i005p00461

Tolstoy M, Cowen JP, Baker ET, Fornari DJ, Rubin KH, Shank TM, Waldhauser F, Bohnenstiehl DR, Forsyth DW, Holms RC, Love B, Perfit MR, Weekly RT, Soule SA, Glazer B (2006) A sea-floor spreading event captured by seismometers. Science 314:1920–1922. https://doi.org/10.1126/science.1133950

Tunnicliffe V, Embley RW, Holden JF, Butterfield DA, Massoth GJ, Juniper SK (1997) ‘Biological colonization of new hydrothermal vents following an eruption on Juan de Fuca Ridge. Deep Sea Res Part I Oceano Res Papers 44(9–10):1627–1644. https://doi.org/10.1016/S0967-0637(97)00041-1

Van Dover CL (2000) The ecology of hydrothermal vents. Princeton University Press, Princeton

Vázquez JT, Palomino D, Sánchez-Guillamón O, Fernández-Salas LM, Fraile-Nuez E, Gómez-Ballesteros M, López-González N, Tello-Antón MO, Lozano P, Santana-Cassiano JM, González-Dávila M (2016) Preliminary geomorphological analysis of the Tagoro Volcano underwater eruption (submarine slope of El Hierro Island), abstract book V symposium international of marine science. Universidad de Alicante, Alicante

Vizzini S, Martínez-Crego B, Andolina C, Massa-Gallucci A, Connell SD, Gambi MC (2017) Ocean acidification as a driver of community simplification via the collapse of higher-order and rise of lower-order consumers. Sci Rep 7(1). https://doi.org/10.1038/s41598-017-03802-w

Wagner D, Luck DG, Toonen RJ (2012) The biology and ecology of black corals (Cnidaria: Anthozoa: Hexacorallia: Antipatharia). Adv Marine Biol 63:67–132. Elsevier

Witt V, Ayris PM, Damby DE, Cimarelli C, Kueppers U, Dingwell DB, Wörheide G (2017) Volcanic ash supports a diverse bacterial community in a marine mesocosm. Geobiology 15(3):453–463. https://doi.org/10.1111/gbi.12231

Part IV

Volcanism and Society: Cascading Risks and Geoheritage

Identification and Management of Indirect Volcanic Risks: Citizens' Rockfall Observatory on the Island of El Hierro

12

I. Galindo, I. Montoya-Montes, J. C. García López-Davalillo, R. Sarro, M. Llorente, N. Sánchez, J. C. Santamarta, N. Cruz-Pérez, A. Ortega, and R. M. Mateos

Abstract

Secondary volcanic hazards (SVH) are not usually considered in volcanic hazard analysis, nor are they specifically included in volcanic risk management plans. However, SVH may cause more damage than primary volcanic hazards (PVH). The magmatic unrest on El Hierro Island in 2011–2012 is a perfect example of how SVH can be one of the leading causes of damage during magmatic unrest. Rockfalls are common on the island of El Hierro, mainly controlled by the heterogeneous lithology and the steep topography. Heavy rainfall and strong wind are usually the main triggering factors. However, during the 2011–2012 El Hierro Island magmatic unrest most of them were triggered due to earthquakes. Rock falls caused roadblocks and damage to road infrastructure. Two reports analysing rockfall hazards and associated risks during the emergency were based on expert knowledge and highlighted the need for, a comprehensive inventory of rockfalls, their

I. Galindo (✉) · I. Montoya-Montes · N. Sánchez
Instituto Geológico y Minero de España (IGME-CSIC), Unidad Territorial de Canarias, Alonso Alvarado, 43, 2A, Las Palmas de Gran Canaria, 35003 Las Palmas, Spain
e-mail: ines.galindo@csic.es

I. Montoya-Montes
e-mail: i.montoya@igme.es

N. Sánchez
e-mail: n.sanchez@igme.es

J. C. García López-Davalillo · R. Sarro
Instituto Geológico y Minero de España (IGME-CSIC), La Calera, 1, 28760 Tres Cantos, Spain
e-mail: jc.garcia@igme.es

R. Sarro
e-mail: r.sarro@igme.es

M. Llorente
Instituto Geológico y Minero de España (IGME-CSIC), Ríos Rosas 23, 28003 Madrid, Spain
e-mail: m.llorente@igme.es

J. C. Santamarta · N. Cruz-Pérez
Departamento de Ingeniería Agraria, Náutica, Civil y Marítima, Universidad de La Laguna (ULL), La Laguna, Tenerife, Spain
e-mail: jcsanta@ull.edu.es

N. Cruz-Pérez
e-mail: ncruzper@ull.edu.es

A. Ortega
La Palma Research Centre, Avenida Venezuela 19, 38760 Los Llanos de Aridane, Isla de La Palma, Tenerife, Spain
e-mail: ariadna.ortega@lapalmacentre.eu

R. M. Mateos
Instituto Geológico y Minero de España (IGME-CSIC), Unidad Territorial de Granada, Urb. Alcázar del Genil, 4-Edif. Zulema, Bajo, 18006 Granada, Spain
e-mail: rm.mateos@igme.es

P. J. González (ed.), *El Hierro Island*, Active Volcanoes of the World,
https://doi.org/10.1007/978-3-031-35135-8_12

processes characterisation, recurring event timing estimations and an analysis of their triggering factors. The need for better rockfall understanding on El Hierro island and the extensive experience of the local Civil Protection agents that were already working with citizens during the volcanic emergency, led to the development of a Rockfall Citizen's Observatory. The observatory aims to engage citizens in the study of this geological hazard, hence providing a substantial increase in the amount of high quality data for rockfall risk analysis.

Keywords

Secondary volcanic hazards · Rockfall · Landslide · Volcanic eruptions · Triggering · Assessment · Canary Islands

12.1 Introduction

The management of magmatic unrest and eruptions is complex since several hazards can occur simultaneously and affect different areas. Volcanic hazard analyses are mainly focused on primary volcanic hazards (PVH) such as lava flows, ash fallout or pyroclastic density currents. However, associated with magmatic and non-magmatic unrest, many other secondary volcanic hazards (SVH) can occur, such as earthquakes, tsunamis, floods, lahars, landslides or rockfalls (Rouwet et al. 2014). Unlike PVH, SVH can also occur before and after the eruptive phase of the unrest, even in areas far from the volcanic vent (s). These hazards may cause more damage than PVH (Tilling 1989). Therefore, the identification of secondary volcanic hazards and their study before eruptions is of extreme importance for safety reasons, as well as analysing the possible relationship they have with the primary volcanic hazards and their associated risks.

During the Tagoro submarine eruption just off the island of El Hierro, which began in October 2011 after several months of numerous seismic events, and ended in March 2012, some seismic events triggered rockfalls causing extensive damage (Llorente and Galindo 2013; Llorente et al. 2015). Previous to this magmatic unrest, the inhabitants of El Hierro were aware of this hazard, since they are common natural processes in the island mainly due to the existence of steep scarps and cliffs, and the volcano-structural complexity of the island (Llorente et al. 2015; Rossi et al. 2021). Steep scarps in El Hierro are mainly related to coastal cliffs and mega-landslide scars like those of El Julan, Las Playas and El Golfo (Fig. 12.1) (Leon et al. 2017 and references therein). These escarpments are very steep or even almost vertical, with heights of several hundred-metres.

El Hierro inhabitants have always coped with rockfalls, adapting their uses and customs to this natural hazard. New anthropogenic actions in the territory, such as the modernization of roads or the use of trails for sports and tourist activities increase the exposure to these types of hazards and therefore, increase the risk. However, it is very difficult to collect information to make a thorough analysis of rockfall source areas, as well as knowing their magnitude and frequency, in addition to keeping this information up to date. During the rockfall events between 2011 and 2013, the residents of El Hierro were equally impressed and impacted by this spectacular natural hazards, so much so that they recorded them and published photos of the rocks falling on social media, providing a useful way to collect information for scientists and risk managers. This gave rise to the idea of launching a citizen observatory project towards improving risk analysis associated to rockfalls hazard.

The concept of citizen science was first used in 1989 in the United States (Gharesifard et al. 2017). It has evolved to focus primarily on geohazard management, encouraging citizens to take an active part in risk prevention, particularly when they face recurring hazards (Grainger 2017). Various programmes have been successfully developed where citizen contributions have been a key factor towards developing adequate protection strategies. Some examples, among many others, include the following:

(a) Gas emission control from the transport sector, which has also developed a monitoring programme for future observatories on the same subject (Zaldei et al. 2017);
(b) Flood risk management (Wehn et al. 2015);
(c) Measurements of soil moisture (Mazumdar et al. 2018);
(d) Weather patterns in countries that suffer water scarcity (Gharesifard and Wehn 2016);
(e) Mapping fire effects (Kirchhoff et al. 2021);
(f) Biodiversity observation (Callaghan et al. 2021);
(g) Circular economy (Hielscher and Jaeger-Erben 2021).

One of the advantages of citizen observatories, beyond data collection, is that data can then be processed by scientists involving people in the processes affecting them and their territory and later have an effect in the public decision-making that results from these experiences (Sánchez-Renteria et al. 2016).

Citizens' engagement in a rockfall observatory on El Hierro is an opportunity to improve rockfall science and risk management in the island. This observatory is a pilot project towards the Canary Islands Rockfall Observatory. Here we present the first steps of this pilot project that might contribute to the well-being of the society as a whole, participating as a community in the management of rockfall risk, as well as its assessment and monitoring. The results of this pilot project might help to improve the implementation of other natural hazard citizen observatories.

12.2 Rockfalls and the Tagoro Magmatic Unrest

Tagoro eruption began in October 2011 offshore, south of El Hierro Island (Fig. 12.1), ending on 5 March 2012. However, from July 2011, high seismic activity was registered, and it continued after the eruption. From 2012 to 2014, several magmatic intrusions were detected intruding El Hierro crust (Díaz-Moreno et al. 2015; Sanchez-Pastor et al. 2018). During the eruption, earthquakes were initially located in the central part of the island, migrating to the south (Telesca et al. 2016). Dominguez Cerdeña et al. (2018) also detected migrations during the six post-eruptive episodes and suggested an anticlockwise rotation of the post-eruptive intrusions.

Before the eruption started, and due to the likelihood of an increase in seismic intensity and frequency, resulting in the triggering of rockfalls, the Canary Government decided to close one of the main roads on the island at Los Roquillos tunnel on 27 September 2011 (Europa Press 2011; Fig. 12.1). Thus, the Canary Government through the Emergency Plan for Volcanic Activity in the Canary Islands (PEVOLCA; Gobierno de Canarias 2018) requested a report to the Geological and Mining Institute of Spain (IGME-CSIC) regarding the instability of the South entrance to Los Roquillos tunnel. As mentioned before, rockfalls are a frequent geological event in this area because of steep slopes and the presence of weathered and fractured materials. Several events have occurred in this area since Los Roquillos tunnel opening in 2003. Heavy rain being the main trigger of instability events that often reached the road. No records of rockfalls were reported during the period of greatest seismic intensity. However, Ferrer et al. (2011) pointed out that seismicity, as a potential trigger for rockfall, should not be neglected in case of shallow (<10 km) earthquakes exceeding 4.5 magnitudes occurring near Los Roquillos tunnel. Therefore, due to the unstable conditions in the area, added to the seismic activity taking place, safety was not guaranteed and it was recommended to adopt urgent stabilization and control measurements to prevent rockfalls within this area (Ferrer et al. 2011).

On 8th October, a 4.3 magnitude earthquake located SW of El Pinar triggered rockfalls near another tunnel, affecting the tunnel entrance and the road from Pozo de la Salud to Lomo Negro (EFE 2011). Los Roquillos tunnel remained closed until October 13th when access was allowed in a controlled manner. Between October 17th and 20th, field observations carried out by IGME-CSIC in the El Golfo area enabled a preliminary evaluation of the cliff stability along

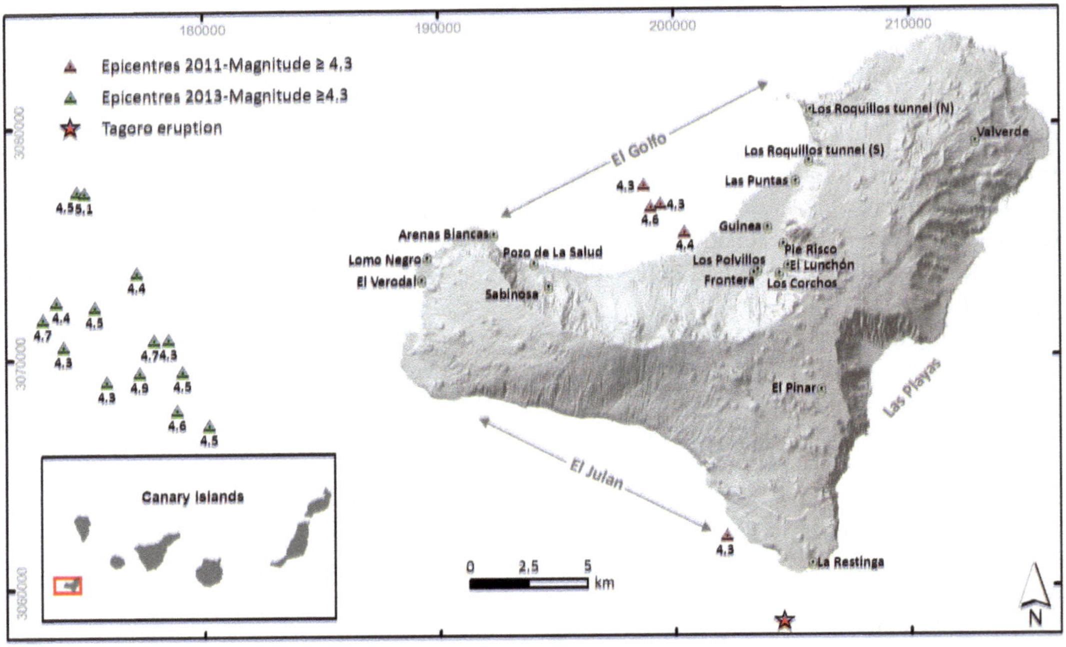

Fig. 12.1 Location map of El Hierro island and 2011 Tagoro eruption, including the position of epicentres with magnitude > 4.3 for 2011 and 2013 (Earthquake data and Digital Elevation Model from Spanish Geographical Institute)

the road HI-5, near Los Roquillos tunnel southern endpoint [Los Roquillos tunnel (S) in Fig. 12.1]. IGME-CSIC highlighted a complex scenario for the analysis of rockfall hazard, particularly in the identification of rockfall source areas. The report concluded that the studied hillside did not meet the conditions required to guarantee safety against rocks falling, hence making obvious the need to adopt rockfall control and stabilization measures both over the tunnel access as well as in the rest of the studied route. Additionally, IGME-CSIC warned of rockfall hazard induced by earthquakes if they exceeded 4.5 magnitudes, in the event of their epicentres being close to the tunnel and their foci located < 10 km depth. By the end of October, the epicentres of the earthquakes with a magnitude > 4 were located NW of Frontera (Fig. 12.1). Therefore, on November 5th, the tunnel was closed again, except for security and emergency personnel.

The closure of the tunnel was one of the most controversial measures taken during the emergency, since this road connects the two most populated regions of the island, Valverde and Frontera. The tunnel allows a reduced commuting time by half compared with other roads. Local administrations demanded the opening of the tunnel and the Frontera Town Hall declared economic and social emergency (La Provincia 2011). On November 14th, passage was allowed with time restrictions until November 25th, when the tunnel was reopened.

Another emergency action, aimed at preventing rockfall damage, was the evacuation of areas prone to rockfall. The first evacuation took place on September 27th overnight in Frontera municipality (Twenty minutos 2011). Initially, 300 people were expected to be evacuated, but only 53 were effectively evacuated from El Lunchón, Pie Risco, Los Corchos, Las Puntas and Guinea areas (Fig. 12.1). At the beginning of November, when the seismic activity increased again, a second preventive evacuation was ordered in Frontera. Thus, two further evacuations took place on November 4th involving 11 families in Las Puntas neighborhood and 51 people from Los Polvillos area the next day. On

14th November, the residents of the Los Polvillos area were allowed to return home and on 25th November, the rest of the residents from Las Puntas area were allowed to return home too.

In March 2013, during a new seismic volcanic event, rockfalls occurred in many places throughout the island, but they were mainly taking place on the west part of El Hierro (Fig. 12.2). On this occasion, the highest seismic energy released and the largest ground deformation associated with a post-eruptive intrusion was reported off the west coast of El Golfo (Domingez-Cerdeña et al. 2018; Fig. 12.1). Managers of the PEVOLCA requested a second technical report from IGME-CSIC (Llorente and Galindo 2013). Before this point, there was no verified evidence of a clear relationship between the seismic activity registered during the emergency and the rockfall occurrence. Thus, areas affected by rockfalls were examined in the field in order to locate detached boulders and characterize rock type, boulder size, impact marks on both anthropogenic structures and natural elements, as well as to identify source areas (Fig. 12.3). The vast majority of the detached boulders in 2013 were smaller than 1 m^3, however, some reached 2.5 m^3 (Llorente and Galindo 2013). Contributing factors were steep slopes, lithological heterogeneity and vegetation. Triggering factors were rainfalls, moderate wind, moisture changes, haloclasty weathering, and seismicity. Moreover, in order to find the relationship with seismic activity and rockfalls, the rockfalls registered by emergency services and seismic data from the Spanish Geographical Institute (IGN) were compared.

From this analysis, it was confirmed that some of the earthquakes with magnitudes higher than 4.4 mbLg triggered rockfalls (Llorente and Galindo 2013). Besides the normal triggering factors (rainfalls, strong winds, etc.), rockfalls seem to be also influenced by other factors such as seismic events frequency or deformation rates. Difficulties were found in correlation due to lack of a complete rockfall database, the variety of contributing and triggering factors, as well as the influence of epicentral distance, depth and location of earthquake with the surface effects. The report suggested that rockfall warnings should be activated in case of meteorological alerts (rainfall, high wind) and high probability of earthquake occurrence with magnitudes exceeding 4.3 mbLg. The report also strongly recommended actions towards increasing general awareness, and installing monitoring cameras in areas prone to rockfalls.

On 27th December 2013, the longest seismic swarm since October 2011 was recorded, 20 km away from the western coast of El Hierro. A 5.1 mbLg magnitude earthquake caused rockfalls, which damaged infrastructures, and caused the closure of roads: one between Frontera and Sabinosa, and the other between Pozo de la Salud and Lomo Negro. Due to the closing of the roads, 30 people were unable to leave Las Arenas Blancas and El Verodal beaches (Fig. 12.1; RTVE 2013).

12.3 Rockfalls Research in El Hierro After the Tagoro Eruption

Previous to the Tagoro eruption, the only complete island mapping to be conducted was for talus cones (Lain et al. 2011). After the magmatic unrest started, several studies focused on analysing rockfall susceptibility and hazard on the island were published (Ferrer et al. 2011; Fernandez-Hernández et al. 2012; Llorente and Galindo 2013; Llorente et al. 2015; Melillo et al. 2020; Rossi et al. 2021). However, during the eruption, an unpublished semi-quantitative heuristic rockfall susceptibility zonation of El Hierro was made available to the PEVOLCA, the Emergency Plan for the Canary Islands. The susceptibility zonation map obtained was validated by some historical events (yellow dots in Fig. 12.2) including the 2011 rockfall events (Fernandez-Hernández et al. 2012). This map revealed that 12% of the entire island showed medium to very high susceptibility. East and west parts of El Golfo valley, as well as Las Playas valley and the adjacent cliffs to the south, present high and very high susceptibility values, with slopes > 70°. The area of El Verodal also counts with very high susceptibility values. In

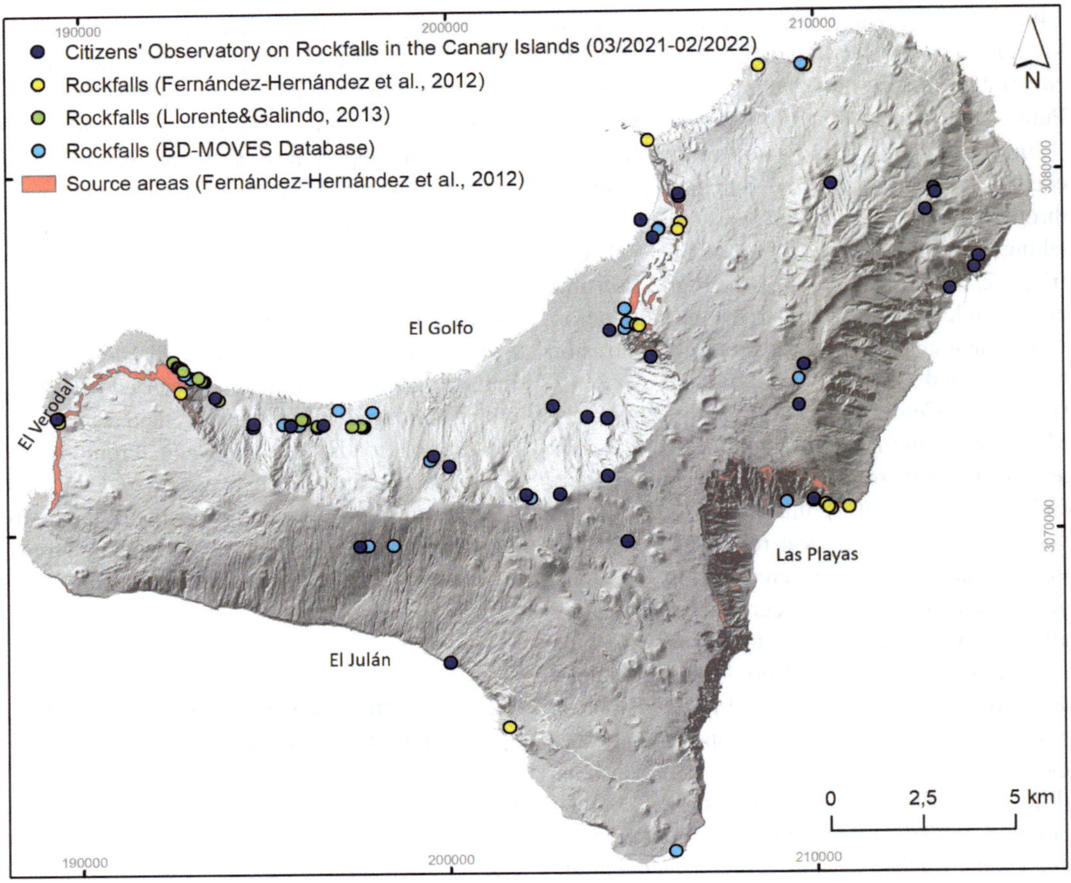

Fig. 12.2 Rockfall events compilation (*Sources*: Citizen Observatory on Rockfalls in the Canary Islands, Fernández-Hernández et al. 2012, Llorente and Galindo 2013, and BD-Moves database Topographical base: 2002 Digital Elevation Model from Spanish Geographical Institute)

Llorente et al. (2015) developed another rockfall susceptibility map with a slightly more conservative approach and included a higher number of index-based non-weighted models (Fernández-Hernández et al. 2012) and a higher number of areas with high susceptibility category.

The latest approach to rockfall susceptibility and hazard at El Hierro is based on mathematical models that allow the calculation of boulders travelling trajectories and path-lengths in different triggering scenarios (Rossi et al. 2021). To better identify the rockfall source areas, Rossi et al. (2021) proposed a probabilistic modelling framework applying a combination of multiple statistical models using the source area locations mapped in the field as the dependent variable and a set of thematic data as independent variables. The models use as input morphometric parameters derived from the Digital Elevation Model and lithological data as an expression of the mechanical behaviour of the rocks. The analysis of different training and validation scenarios allowed to test the model sensitivity to the input data, select the optimal model-training configuration, and evaluate the model applicability outside of the training areas. The final map obtained from the model for the entire island of El Hierro provided the probability of a given location being a potential source area and can be used as the input for rockfall runout simulation modelling. This last susceptibility analysis shows a high probability of source areas appearing along El

Golfo and Las Playas escarpments. It covers on the one hand, a larger surface if compared with the high susceptibility categorized areas in Fernández-Hernández et al. (2012); and on the other hand, it presents similar aspect to the heuristic model published by Llorente et al. (2015).

12.4 Citizens' Rockfall Observatory of El Hierro

Within the AGEO research project framework, three citizen observatories on rockfalls are being implemented in three areas of the Canary Islands: San Bartolomé de Tirajana in Gran Canaria, San Cristobal de La Laguna in Tenerife, and in the entire island of El Hierro (Santamarta Cerezal et al. 2021). The physiography and geology of these three areas means these are sites susceptible to rockfalls, and are therefore perfect locations to analyse how meteorological factors influence the occurrence of rockfalls. El Hierro rockfalls citizen observatory might also allow to analyse the relationship between rockfalls and seismicity.

El Hierro rockfalls citizen observatory raises special interest among the citizens of the island, and successful results are expected. To begin with, the low population density in the island contributes to a closer-knit community, which, in turn, will allow us to reach a high number of participants. Recruiting volunteers should be easier than in any other more populated islands in the Canaries. Moreover, the experience gained during the magmatic unrest makes the inhabitants of El Hierro be especially aware of geohazards as it is discussed in the following section. Thus, if citizens are provided with some basic training on rockfall hazard and some advice on how to participate in the observatory is given, it will encourage their involvement in every single task they may be asked to do and consequently, an improvement on the risk management of their island should be noticed. The rockfalls observatory in El Hierro intends to:

(i) Engage the local population in order to increase the awareness about rockfall hazard and reporting of rockfall events;
(ii) Compile rockfalls data in order to get an updated inventory of events, elaborate susceptibility maps and create 3D simulations of rock blocks (boulders) trajectories;
(iii) Contribute to the nation-wide IGME-CSIC mass movements database, BD-MOVES (https://info.igme.es/BD2DMoves/);
(iv) Help to improve risk management plans.

12.4.1 Stakeholders and Citizens' Engagement

The population of El Hierro is small with around 11,423 inhabitants (2022 data, Instituto Canario de Estadística), a factor that has favoured the cohesion of the community and its organisational capacity to manage emergencies and report rockfalls. Their previous experience dealing with the Tagoro magmatic unrest has created an opportunity to improve citizen´s awareness of geohazards, making this region especially interesting to collaborate with this citizen science initiative. The experience of the local Civil Protection dealing with citizens during the emergency is also an added value since they played the role of the "groupers", citizens accepted and respected by the community who helped to coordinate the evacuations during the Tagoro emergency. This organisational citizen figure was responsible for guiding people to the meeting points and keeping calm during emergencies. There were three groupers per meeting point, one of them being in charge of communications with the City Council coordinator of Civil Protection, informing about community behaviour, incidents, evacuation times, etc. All this information was passed on to the Civil Protection Coordination Centre. Therefore, groupers maintain direct contact with the communities and are very aware of any kind of hazard. The scheme has worked so well that it could be implemented in other situations of hazard or risk for the population, or in order for them to be involved in risk awareness and knowledge processes and the collection of information and events to catalogue occurring rockfalls.

Fig. 12.3 **a** Las Playas valley scarp. **b** Rockfalls and impact mark; **c** Fallen boulders along the road from Tigaday to Sabinosa; **d** Water pipe bent by a rockfall

The idea of the citizen observatory arose from this experience in El Hierro during the eruption. However, we expect three types of citizens to be involved: the general public, expert citizens, and groupers. The general public is neither non-specialists nor trained citizens; experts will include citizens trained during the project or with previous specific knowledge; groupers will be trained by popular leaders working in closed collaboration with the local Civil Protection. Groupers will also play an important role in engaging people to participate in the observatory. Basic training will be provided about the project, rockfall hazard (process, triggering factors, damages, etc.), and the platform for data collection. Thus, groupers might provide other people with the mentioned training, which, in the end, will result in obtaining more accurate data.

Information and training provided to each group will be different and adapted to each level. The aim is to obtain high-quality data that allow the characterization of rockfalls in El Hierro, as well as their triggering factors. Data provided by citizens will include the date, location of the event, pictures and videos, descriptions about damages, triggering factors, etc. Schools at all levels are also receiving training to participate in the observatory through activities in which they are being involved in monitoring the terrain (fracture identification, fracture opening measurements, notification of rock fragments falls, etc.). The Copernicus program (Dell'Acqua and De Vecchi 2017), one of the tools, provided by the European Union to collaborate with Earth observation projects, is also being introduced during training sessions.

The first step in the setup of the observatory has been to contact the relevant stakeholders with a view of managing strategies to reach the population and engage them to collaborate with the observatory. Currently, we are tightly working with the El Hierro Island Council on creating the AGEO's citizens network and organizing dissemination activities in order to involve the public in the observatory.

The upcoming activities will be focused on maintaining citizens' engagement, since one of the most significant challenges for the observatory is to keep motivation and retain long-term involvement, furthering this way the project goal.

El Hierro observatory will give the participants feedback on how data are included in the database and represented in a viewer, so the citizens will see on an event map the data they have sent. People will also receive the list of events registered, some news related to rockfall, statistics, and alerts (meteorological, seismic and rockfall events, roads conditions and closures), etc. Stakeholders will also benefit of other products such as updated inventory of events, susceptibility maps or 3D simulations.

Activities planned in the schools of El Hierro take the form of challenges or games, intended to engage students in the project, ensuring they enjoy the activities by providing extra motivation stemming from the competition. Initiatives involving games are highly popular in these projects because they have proven to be especially effective when gathering data being the main objective of the citizen scientists as they are useful for strengthening collaboration links between society and science. Therefore, opportunities to gain and retain contributors are provided through the use of gamification techniques (Ponti et al. 2018; Veeckman et al. 2019).

12.4.2 Rockfall Analysis

Records of rockfalls are being included in the BDMOVES mass movements' Spanish national database (Fig. 12.2). In order to have events maps previous to the launching of the observatory, we have compiled and validated information on rockfalls in the Canary Islands from previous databases and news. Preliminary results for the island of El Hierro show that a total of 76 historical rockfalls have been reported from 2005 to February 2021, mainly concentrated on the western side of the El Golfo escarpment (Fig. 12.2). Furthermore, from March 2021 (starting point of the observatory data collection) up to date, 34 rockfalls events have been reported in El Hierro. The most frequent damage is indirect and usually resulting in road closures. Nonetheless, direct damage (houses, cars, roads, paths, etc.) is also registered. Fortunately, casualties or loss of life are the least common consequences.

Nearly 70% of the historical events (previous to the Observatory set up) correspond to rockfalls that occurred between 2011 and 2013 during the seismic volcanic crisis, but only a few of them are clearly correlated with earthquakes. However, during this period, citizens were more sensitive to natural hazards due to the unrest period. Although many more rockfalls have occurred during this period they are usually linked to storms and news only report the events in a general light, e.g. "floods and rockfalls affected El Hierro" (ABC 2020; La Provincia 2021). Without additional information, it is impossible to locate and characterize the events. However, relating the occurrence of rockfalls to rain thresholds is of the utmost importance to establish rockfall alerts to reduce rockfall risk (Melillo et al. 2020).

12.5 Summary

Volcanic eruptions in active volcanic zones may produce a series of associated secondary hazards. Damages derived by secondary hazardous events such as road closures due to rockfalls and landslides invading the roadways many times can turn the affected population into collateral victims of the eruption, who are often more numerous than those directly affected by an eruption (lava flows, pyroclasts fall, etc.). Costs of economic activities interrupted by such closures should not be neglected (Llorente 2014). Considering these aspects, some classifications of risk level found in the Civil Protection Territorial Plans in the Canaries should be reviewed, in particular the exposure, as they are just considering people directly injured by the volcanic event and therefore, underestimating the damage indexes.

The Tagoro eruption showed that for the effective management of any seismic or meteorological emergency on El Hierro Island, rockfall hazard knowledge and readiness are essential to reduce risks. Planning instruments as susceptibility maps, in which the rockfall source areas, possible trajectories and deposit areas of the detached blocks are shown, hazard maps, in which the probability of the hazard occurrence is assessed for different triggering scenarios, and risk maps, in which the exposed elements and their valuation are included, are basic tools that must be used to achieve the knowledge and preparedness needed. But modelling rockfall phenomena is very complex and requires several inputs, including precise location of source areas, quality topographic and geological data, and a complete record of past events allowing reliable validation of the models for predictive use.

Despite rockfalls being recognized as a frequent geological hazard process, with high-moderate associated risk in the Civil Protection plans in the Canary Islands (Cabildo de Gran Canaria 2005). These plans do not consider it as a primary hazard, labelling it as a secondary hazard associated. Therefore, the islands do not have a specific plan in case of rockfalls or other mass movement emergencies; only if other hazards occur and their plans are activated rockfalls mitigation measures will start to be implemented. Regarding the PEVOLCA, rockfalls are considered as an indirect hazard derived from volcanic activity. Besides this mention of rockfalls in a classification of direct and indirect hazards, no other mention appears in the entire document. Thus, there is a PEVOLCA review to integrate rockfalls and other SVH in the volcanic risk analysis and mapping.

Due to the high velocity of the rockfall process plus the small size of the boulders sometimes falling in remote locations, it becomes extremely difficult to obtain a full report of every event occurred in a region. Citizen scientists enable us to significantly increase the number of updated samples through events reporting if compared to traditional research (Gardiner et al. 2012). Likewise, larger temporal and spatial scales are also allowed by citizen science data (Callaghan et al. 2019). Initiatives like El Hierro observatory improve and update event inventories, an essential tool to carry out any precise analysis or calculation of hazards and risks. Just one year of citizens' data collection has gathered 50% of historical data registered in 16 years. The more updated the data the more accurate the results. Updated event maps are not only indispensable tools to help in prevention and mitigation stages, but also to build derived maps that help in the analysis of the process.

Great benefits arise from citizen science initiatives: low cost or free reports, huge quantity of data from varied locations, increased awareness of the environment, empowered societies with a sense of community and learning encouragement (Lee et al. 2020). However, participant recruitment and long-term engagement is a challenge. Mechanisms such as advertisements, press releases, broad online presence, online citizen science platforms, feedback on project success, rewards and recognition and acknowledging volunteer contributions play a vital role in participants' retention (Bonney et al. 2009; Crimmins et al. 2014; De Moor et al. 2019; Veeckman et al. 2019). Stakeholders involved in El Hierro observatory are carefully managing all those initiatives in order to keep active communication channels open between people and observatory managers. Dynamic systems of incentives are necessary to get high recruitment rates and thus, replace users who continuously leave the project (Crall et al. 2017; Callaghan et al. 2019).

A not negligible issue that will also be tackled in El Hierro observatory is the measurement of citizens' engagement as a means of tracking participation over time. De Moor et al. (2019) or Crall et al. (2017) presented it on their research, enabling to assess whether fluctuations in participation are related to recruiting and outreach strategies. Not only are small societies like El Hierro population usually more involved in every event related to their surrounding environment, but also present a strong commitment with the social group they belong to, making collaboration among citizens more frequent and a part of their daily life.

The experience during the Tagoro eruption demonstrates that management of volcanic unrest should be improved through the understanding of all geological hazards related to magma intrusion, storage and eruption; both primary and secondary. To improve management during future volcanic emergencies Civil Protection plans might consider including in its emergency plan the study and mapping of rockfalls, as well as other secondary hazards at the same level as primary hazards. Thus, a holistic approach should be used when dealing with volcanic risk analysis in order to improve volcanic multi-hazards monitoring and management which will consequently, mitigate volcanic risks.

Acknowledgements This research received funding from the INTERREG Atlantic Area Programme through the AGEO project (EAPA_884/2018). We also thank the editors of the book Series and the book (Corrado Cimarelli and Pablo J. González) for their extensive proof-reading of early manuscripts to help increase the clarity of this chapter.

References

ABC (2020) El temporal de viento y lluvia se ceba con El Hierro. https://www.abc.es/opinion/abci-temporal-viento-y-lluvia-ceba-hierro-200912180300-1132610483405_noticia.html?ref=https:%2F%2Fwww.google.com%2F. Accessed 21 Apr 2021

Barredo JI, Benavides A, Hervás J, Van Westen CJ (2000) Comparing heuristic landslide hazard assessment techniques using GIS in the Tirajana Basin, Gran Canaria Island, Spain. Int J Appl Earth Obser Geoinform 2:9–23. https://doi.org/10.1016/s0303-2434(00)85022-9

Becerril L, Galve JP, Morales JM, Romero C, Sánchez N, Martí J, Galindo I (2016) Volcano-structure of El Hierro (Canary Islands). J Maps 12(1):43–52. https://doi.org/10.1080/17445647.2016.1157767

Blahůt J, Balek J, Klimeš J, Rowberry M, Kusák M, Kalina J (2019) A comprehensive global database of giant landslides on volcanic islands. Landslides 16:2045–2052. https://doi.org/10.1007/s10346-019-01275-8

Blahůt J, Olejár F, Rott J, Petružálek M (2020) Current stability modelling of an incipient San Andrés giant landslide on El Hierro Island, Canaries, Spain: first attempt using limited input data. Acta Geodyn Geomater 17(197):89–99. https://doi.org/10.13168/AGG.2020.0006

Bonney R, Cooper CB, Dickinson J, Kelling S, Phillips T, Rosenberg KV, Shirk J (2009) Citizen science: a developing tool for expanding science knowledge and scientific literacy. Bioscience 59(11):977–984. https://doi.org/10.1525/bio.2009.59.11.9

Cabildo de Gran Canaria (2005) Plan Territoral Insular de Emergencias de Protección Civil de Gran Canaria. Cabildo de Gran Canaria, Las Palmas de Gran Canaria, p 512

Callaghan CT, Rowley JJL, Cornwell WK, Poore AGB, Major RE (2019) Improving big citizen science data: moving beyond haphazard sampling. PLoS Biol 17(6): e3000357. https://doi.org/10.1371/journal.pbio.3000357

Callaghan CT, Watson JE, Lyons MB, Cornwell WK, Fuller RA (2021) Conservation birding: a quantitative conceptual framework for prioritizing citizen science observations. Biol Conser 253:108912. https://doi.org/10.1016/j.biocon.2020.108912

Crall A, Kosmala M, Cheng R, Brier J, Cavalier D, Henderson S, Richardson AD (2017) Volunteer recruitment and retention in online citizen science projects using marketing strategies: lessons from Season Spotter'. JCOM 16(01):A01. https://doi.org/10.22323/2.16010201

Crimmins TM, Weltzin JF, Rosemartin AH, Surina EM, Marsh L, Denny EG (2014) Focused campaign increases activity among participants in nature's notebook, a citizen science project. Nat Sci Educ 43(1):64–72

De Moor T, Rijpma A, Prats López M (2019) Dynamics of engagement in citizen science: results from the "Yes, I do!"-project. Cit Sci Theory Pract 4(1):38:1–17. https://doi.org/10.5334/cstp.212

Dell'Acqua F, De Vecchi D (2017) Potentials of active and passive geospatial crowdsourcing in complementing sentinel data and supporting copernicus service portfolio. Proc IEEE 105(10):1913–1925. https://doi.org/10.1109/JPROC.2017.2727284

Díaz-Moreno A, Ibáñez JA, De Angelis S, García-Yeguas A, Prudencio J, Morales J, Tuvè T, García L (2015) Seismic hydraulic fracture migration originated by successive deep magma pulses: the 2011–2013 seismic series associated to the volcanic activity of El Hierro Island. J Geophys Res Solid Earth 120:7749–7770. https://doi.org/10.1002/2015JB012249

Domínguez Cerdeña I, García-Cañada L, Benito-Saz MA, Del Fresno C, Lamolda H, Pereda de Pablo J, Sánchez Sanz C (2018) On the relation between ground surface deformation and seismicity during the 2012–2014 successive magmatic intrusions at El Hierro Island. Tectonophysics 744:422–437. https://doi.org/10.1016/j.tecto.2018.07.019

EFE Agency (2011) El terremoto de 4,3 grados en El Hierro fue sentido en casi toda la isla. https://www.economiahoy.mx/flash/noticias/3436885/02/21/El-terremoto-de-43-grados-en-El-Hierro-fue-sentido-en-casi-toda-la-isla.html. Accessed 21 Apr 2021

Europa Press (2011) El Hierro reclama un informe estructural del túnel de Los Roquillos tras la crisis sísmica. https://www.europapress.es/islas-canarias/noticia-hierro-reclama-informe-estructural-tunel-roquillos-crisis-sismica-20111006181314.html. Accessed 21 Apr 2021

Fernandez-Hernández M, Paredes C, Castedo R, Llorente M, Vega-Panizo R (2012) Rockfall detachment susceptibility map in El Hierro Island, Canary Islands, Spain. Nat Hazards 64:1247–1271. https://doi.org/10.1007/s11069-012-0295-1

Ferrer M, Mateos RM, García López-Davalillo JC (2011) Condiciones de estabilidad de la ladera situada sobre la carretera HI5 (Valverde-Frontera) en la zona de acceso a la Boquilla sur del túnel de Los Roquillos. In: Technical Report. Instituto Geológico y Minero de España, Madrid

Fritz S, Fonte CC, See L (2017) The role of citizen science in earth observation. Remote Sens 9(4):1–14. https://doi.org/10.3390/rs9040357

Gardiner MM, Allee LL, Brown PMJ, Losey JE, Roy HE, Smyth RR (2012) Lessons from lady beetles: accuracy of monitoring data from US and UK citizen-science programs. Front Ecol Environ 10(9):471–476. https://doi.org/10.1890/110185

Gee MJR, Watts AB, Masson DG, Mitchell NG (2001) Landslides and the evolution of El Hierro in the Canary Islands. Mar Geol 177(3–4):271–293. https://doi.org/10.1016/S0025-3227(01)00153-0

Gharesifard M, Wehn U (2016) To share or not to share: drivers and barriers for sharing data via online amateur weather networks. J Hydrol 535:181–190. https://doi.org/10.1016/j.jhydrol.2016.01.036

Gharesifard M, Wehn U, Van der Zaag P (2017) Towards benchmarking citizen observatories: features and functioning of online amateur weather networks. J Environ Manag 193:381–393. https://doi.org/10.1016/j.jenvman.2017.02.003

Gobierno de Canarias (2018) Plan Especial de Protección Civil y Atención de Emergencias por riesgo volcánico en la Comunidad Autónoma de Canarias (PEVOLCA). Boletín Oficial De Canarias 158:26312–26542

Grainger A (2017) Citizen observatories and the new earth observation science. Remote Sens 9(2):1–30. https://doi.org/10.3390/rs9020153

Hielscher S, Jaeger-Erben M (2021) From quick fixes to repair projects: Insights from a citizen science project. J Clean Prod 278:123875. https://doi.org/10.1016/j.jclepro.2020.123875

Kirchhoff C, Callaghan CT, Keith DA, Indiarto D, Taseski G, Ooi MKJ, Le Breton TD, Mesaglio T, Kingsford RT, Cornwell WK (2021) Rapidly mapping fire effects on biodiversity at a large-scale using citizen science. Sci Total Environ 755:142348. https://doi.org/10.1016/j.scitotenv.2020.142348

Klimeš J, Yepes J, Becerril L, Kusák M, Galindo I, Blahut J (2016) Development and recent activity of the San Andrés landslide on El Hierro, Canary Islands, Spain. Geomorphology 261:119–131. https://doi.org/10.1016/j.geomorph.2016.02.018

La Provincia (2011) Concentración para abrir el túnel de Los Roquillos. https://www.laprovincia.es/fiestas/2011/10/13/concentracion-abrir-tunel-roquillos-10652105.html. Accessed 21 Apr 2021

La Provincia (2021) La lluvia y el viento producen desprendimientos, caída de árboles, cortes eléctricos y de telefonía. https://www.laprovincia.es/gran-canaria/2021/01/06/lluvia-viento-producen-desprendimientos-caida-27110683.html. Accessed 21 Apr 2021

Lain L, Galindo I, Rodriguez C, Sarro R, Becerril L, Llorente M, Romero MC (2011) Convenio específico de colaboración entre el Consejo Insular de Aguas de El Hierro y el Instituto Geológico y Minero de España para la realización de Proyectos y trabajos de investigación en materia de peligrosidad geológica. In: Technical Report. Instituto Geológico y Minero de España, Madrid

Lee KA, Lee JR, Bell P (2020) A review of citizen science within the earth sciences: potential benefits and obstacles. Proc Geol Assoc 131(6):605–617. https://doi.org/10.1016/j.pgeola.2020.07.010

León R, Somoza L, Urgeles R, Medialdea T, Ferrer M, Biain A, García-Crespo J, Mediato JF, Galindo I, Yepes J, González FJ, Gimenez-Moreno J (2017) Multi-event oceanic island landslides: new onshore-offshore insights from El Hierro Island, Canary Archipelago. Mar Geol 393:156–175

Llorente M, Galindo I (2013) Informe sobre los desprendimientos de roca ocurridos en la isla de El Hierro a finales del mes de marzo de 2013 y su relación con la sismicidad. In: Technical Report. Instituto Geológico y Minero de España, Madrid

Llorente M, Galindo I, Paredes C (2015) Desprendimientos de rocas en la isla de El Hierro. In: Hernández-Gutiérrez LE, Santamarta-Cerezal JC (eds) Ingeniería Geológica en Terrenos Volcánicos Métodos, Técnicas y Experiencias en las Islas Canarias. Ilustre Colegio Oficial de Geólogos, pp 316–333

Llorente M (2014). Evaluación cuantitativa de pérdidas por peligros geológicos. Caso del archipiélago de Canarias: inundaciones, sismicidad y vulcanismo. PhD Thesis. Universidad Politécnica de Madrid, Madrid

López C, Blanco MJ, Abella R, Brenes B, Cabrera Rodríguez VM, Casas B, Domínguez Cerdeña I, Felpeto A, Fernández de Villalta M, del Fresno C, García O, García-Arias MJ, García-Cañada L, Gomis Moreno A, González-Alonso E, Guzmán Pérez J, Iribarren I, López-Díaz R, Luengo-Oroz N, Meletlidis S, Moreno M, Moure D, Pereda de Pablo J, Rodero C, Romero E, Sainz-Maza S, Sentre Domingo MA, Torres PA, Trigo P, Villasante-Marcos V (2012) Monitoring the volcanic unrest of El Hierro (Canary Islands) before the onset of the 2011–2012 submarine eruption. Geophys Res Lett 39:1–7. https://doi.org/10.1029/2012GL051846

Mazumdar S, Wrigley SN, Neil I, Ciravegna F (2018) Harnessing location-based services for effective. Int J Spat Data Infrastruct Res 13:101–108. https://doi.org/10.2902/1725-0463.2018.13.art10

Melillo M, Gariano SL, Peruccacci S, Sarro R, Mateos RM, Brunetti MT (2020) Rainfall and rockfalls in the Canary Islands: assessing a seasonal link. Nat Hazards Earth Syst Sci 20:2307–2317. https://doi.org/10.5194/nhess-20-2307-2020

Ponti M, Hillman T, Kullenberg C, Kasperowski D (2018) Getting it right or being top rank: games in citizen science. Cit Sci Theory Pract 3(1):1–12. https://doi.org/10.5334/cstp.101

Rossi M, Sarro R, Reichenbach P, Mateos RM (2021) Probabilistic identification of rockfall source areas at regional scale in El Hierro (Canary Islands, Spain). Geomorphology 381:107661. https://doi.org/10.1016/j.geomorph.2021.107661

Rouwet D, Sandri L, Marzocchi W, Gottsmann J, Selva J, Tonini R, Papale P (2014) Recognizing and tracking volcanic hazards related to non-magmatic unrest: a review. J Appl Volcanol 3(1):1–17

RTVE (2013) El Hierro registra un terremoto de magnitud 5.1, el mayor registrado en la isla desde 2011. https://www.rtve.es/noticias/20131227/hierro-registra-terremoto-magnitud-51-se-sentido-cuatro-islas-occidentales/834582.shtml. Accessed 21 Apr 2021

Sánchez-Pastor P, Obermann A, Schimmel M (2018) Detecting and locating precursory signals during the 2011 El Hierro, Canary Islands, submarine eruption. Geophys Res Lett 45:10288–10297. https://doi.org/10.1029/2018GL079550

Sánchez-Rentería G, Bonilla-Escobar FJ, Fandiño-Losada A, Gutiérrez-Martinez MI (2016) Citizen security observatories: tools for decision making and governability. Revista Peruana de Medicina Experimental y Salud Publica, 33(2):362–367. https://doi.org/10.17843/rpmesp.2016.332.2203

Santamarta Cerezal JC, Ortega Rodríguez A, Rodríguez-Martín J, Galindo Jiménez I, Sánchez Jiménez N, Montoya Montes I, Leyva Campos S, Cruz-Pérez N (2021) La gestión de riesgos geológicos en el Área Atlántica dentro del programa europeo Interreg AGEO. Ingeopress 287:24–27. https://doi.org/10.21125/inted.2021.0250

Sarro R, Mateos RM, Reichenbach P, Aguilera H, Riquelme A, Hernández-Gutiérrez LE, Martín A, Barra A, Solari L, Monserrat O, Alvioli M, Fernández-Merodo JA, López-Vinielles JA, Herrera G (2020) Geotechnics for rockfall assessment in the volcanic island of Gran Canaria (Canary Islands, Spain). J Maps 16:605–613. https://doi.org/10.1080/17445647.2020.1806125

Telesca L, Lovallo M, López C, Molist JM (2016) Multiparametric statistical investigation of seismicity occurred at El Hierro (Canary Islands) from 2011 to 2014. Tectonophysics 672:121–128. https://doi.org/10.1016/j.tecto.2016.01.045

Tilling RI (1989) Volcanic hazards and their mitigation: progress and problems. Rev Geophys 27(2):237–269

Twenty minutos (2011) Evacúan a 53 personas en El Hierro ante el riesgo de desprendimientos por seísmos Noticia del 27 de septiembre de 2011 https://www.20minutos.es/noticia/1171178/0/hierro/seismos/volcan/. Accessed 21 Apr 2021

Urgeles R, Canals M, Baraza J, Alonso B, Masson D (1997) The most recent megalandslides of the Canary Islands: El Golfo debris avalanche and Canary debris flow, west El Hierro Island. J Geophys Res 102 (B9):20305–20323. https://doi.org/10.1029/97JB00649

Veeckman C, Talboom S, Gijsel L, Devoghel H, Duerinckx A (2019) Communication in Citizen Science. A practical guide to communication and engagement in citizen science. SCIVIL, Leuven

Wehn U, Rusca M, Evers J, Lanfranchi V (2015) Participation in flood risk management and the potential of citizen observatories: a governance analysis. Environ Sci Policy 48:225–236. https://doi.org/10.1016/j.envsci.2014.12.017

Zaldei A, Camilli F, De Filippis T, Di Gennaro F, Di Lonardo S, Dini F, Gioli B, Gualtieri G, Matese A, Nunziati W, Rocchi L, Toscano P, Vagnoli C (2017) An integrated low-cost road traffic and air pollution monitoring platform for next citizen observatories. Transp Res Proced 57:531–538

13 Tagoro, the Youngest Submarine Volcano in the Spanish Geoheritage Inventory: Scientific Value, Geoconservation and Opportunities for Geotourism

Juana Vegas, Inés Galindo, Juan-Tomás Vázquez, Ricardo León, Nieves Sánchez, Esther Martín-González, and Carmen Romero

Abstract

The submarine eruption of La Restinga, now known under the name of Tagoro volcano, began in early October 2011 offshore the southern coast of El Hierro island and ended in early March 2012. This eruption produced a volcanic cone, hornitos, a thick pyroclastic apron that prograded towards the base of the volcanic edifice and volcanic products that emerged from the sea floor over sea level, such as lava balloons and low-density vesicular pyroclasts (xeno-pumices/restingolitas) that popped up in the sea surface. All these submarine volcanic elements and processes, which have been monitored and studied by direct and indirect scientific techniques, have made it possible to gain scientific knowledge of the whole eruption, taking into consideration that no other submarine eruption had been studied before in Spain. Tagoro is the youngest Spanish submarine volcano included in the national geoheritage inventory (Spanish Inventory of Geological Sites of Interest), being representative and the best example of an underwater eruption, as well as one that has been studied and monitored by various research teams since the beginning of the magmatic unrest. It is essential that geoheritage inventories include submarine geosites as an essential part of geological knowledge of the planet. For this reason, a new methodology has been designed for assessment because most of the criteria used in terrestrial geosites are not applicable. This submarine volcano, despite its inaccessibility, is a geotouristic resource for the El Hierro UNESCO Global Geopark (UGGp), which uses new technologies for its interpretation, through virtual reality presentation so that the visitors can

J. Vegas (✉) · R. León
Instituto Geológico y Minero de España (IGME-CSIC), Ríos Rosas 23, 28003 Madrid, Spain
e-mail: j.vegas@igme.es

R. León
e-mail: r.leon@igme.es

I. Galindo · N. Sánchez
Instituto Geológico y Minero de España (IGME-CSIC), Unidad Territorial de Canarias, Alonso Alvarado, 43, 2A, Las Palmas de Gran Canaria, 35003 Las Palmas, Spain
e-mail: i.galindo@igme.es

N. Sánchez
e-mail: n.sanchez@igme.es

J.-T. Vázquez
Instituto Español de Oceanografía (IEO-CSIC), Centro Oceanográfico de Málaga, Puerto Pesquero S/N, 29640 Fuengirola, Spain
e-mail: juantomas.vazquez@ieo.es

E. Martín-González
Tenerife Natural History Museum, Fuente Morales 1, 38003 Santa Cruz de Tenerife, Canary Islands, Spain
e-mail: mmartin@museosdetenerife.org

C. Romero
Geography Department, La Laguna University, Campus de Guajara S/N, La Laguna, Tenerife, Spain
e-mail: mcromero@ull.edu.es

P. J. González (ed.), *El Hierro Island*, Active Volcanoes of the World,
https://doi.org/10.1007/978-3-031-35135-8_13

access it at the Interpretation Centre. It is necessary to promote educational and tourist use by means of new technologies such as augmented reality, an e-library and a specialized section on its website that can be easily accessed with QR codes located in La Restinga port and in the UGGp network of hotels and rural houses to connect with the audience. Its protection is guaranteed as it lies within the 'Punta de La Restinga-Mar de Las Calmas' Marine Reserve, but the geological values of Tagoro should be specifically included in the regulations as one of the natural resources that contribute to this protected marine area.

Keywords

Submarine volcanic geosites · Underwater geoheritage · Assessment · Geoturism · Canary Islands

13.1 Introduction

In Spain, the Canary Islands are the geological domain where volcanic-type geosites predominate, with a wide representation of volcano-derived types, environments, morphologies, structures, processes, rocks and minerals in this insular geological framework, ranging from the Oligocene to the present (Robertson and Stillman 1979; Vera et al. 2004). Volcanoes, along with dinosaur fossils, are very probably among the most attractive geological features for people. This is the case of the Tagoro volcano, on the island of El Hierro, which erupted between 10th October 2011 and 5th March 2012 (López et al. 2012; Rivera et al. 2013; Somoza et al. 2017). Add to this that the Tagoro is an underwater volcano and the component of mystery and attraction is even stronger for all those who followed this eruption, from scientists, local population, politicians and civil protection bodies to society in general.

It is common for geoheritage inventories to be carried out at any scale from the local to the global, for exclusive inclusion in land geosites. Currently, the inclusion of geosites from the underwater or submarine environment is not considered in many inventories, especially those carried out at a national scale (Rovere et al. 2011; Galindo et al. 2019a, 2019b, 2019c). The inventories that most frequently include submarine geosites tend to be UNESCO Global Geoparks that offer these geosites as an opportunity for geotourism, mainly through scuba diving (e.g. Lima et al. 2013; Galindo et al. 2017, 2018). This scarce representation of submarine geosites is due to the fact that (1) there is no standardized methodology to identify and classify underwater geosites, (2) many of the assessment criteria are not adapted to underwater geosites, (3) it is more difficult to carry out traditional field work in situ, and (4) in general, these geosites are not given the same attention as land-based geosites with a longer tradition for nature tourism.

In the development of the Canary Islands geoheritage inventory, submarine geosites have been included due to their strong connection with the geodiversity and geological processes of these volcanic islands (LIGCANARIAS project, Galindo et al. 2020). This regional inventory is being included in the Spanish Inventory of Geological Sites of Interest (IELIG, acronym in Spanish).

Tagoro volcano is a unique geosite of very high heritage value at national scale and included in the Spanish National Inventory (IELIG) because it is the best example of the evolution of a submarine volcano. Since its birth, Tagoro volcano has been extensively and continuously documented by scientific monitoring and research campaigns (Carracedo et al. 2015), and it now constitutes a reference model for submarine geoheritage sites. It has a high value as a model because it is the most recent Spanish volcano that has erupted in the twenty-first century and, for this reason, the best scientific documentation of volcanic processes is available in real time, allowing one to determine the evolution and development of this underwater volcano (Somoza et al. 2017 and references herein). Since these active volcanic processes are from being regarded as a social problem in the Canary

Islands, the Tagoro can be one of the best educational and geotourist resource of the El Hierro UNESCO Global Geopark (UGGp). This objective will be achieved thanks to a promotion plan developed through good planning and economic investment, which combines science, new technologies and marketing, thus favouring the management of this geosite in the future. It is necessary to protect and, above all, to disseminate this geosite as a great natural resource for society, a model for science and formal education at all educational levels. Despite this, the Tagoro volcano can become one of the main attractions of this UGGp and one of the pillars for the sustainable development of this territory.

13.2 Geological Setting

El Hierro Island is the youngest island of the Canary Archipelago (Fig. 13.1). It is located in the south-westernmost emerged part of a ridge of volcanic islands and seamounts, measuring more than one thousand kilometres in length (from Lars at the NE to Tropic at the SW seamounts), named the Canary Islands Volcanic Province (Geldmacher et al. 2005). The base of the volcanic edifice is located between 3600 m in water depth in the eastern part and 4000 m in water depth in the western part. The summit is located at 1503 m. The volcanic edifice lies on an approximately 156-Ma-old (Jurassic) oceanic crust (Roeser 1982). The oldest rock has been recovered in the southeast part of the submarine volcanic edifice (El Hierro ridge) at 133 Ma (van den Bogaard 2013). In the subaerial environment, the oldest has been dated at 1.12 Ma (Guillou et al. 1996; Bellido-Mulas et al. 2012). These differences indicate the presence of two superimposed volcanic buildings.

The origin of the Canary Islands is still controversial, but it is commonly accepted that it is the result of a mantle plume upwelling close to the northwest African continental margin. This mantle plume dynamic produced lithospheric melting periods with the subsequent generation of volcanic edifices from the Jurassic until the present (the Tagoro volcano on El Hierro and 2021 volcano on La Palma being the most recent examples) (van den Bogaard 2013). El Hierro Island shows a radial distribution of dykes and eruptive fissures (Becerril et al. 2015). Three main eruptive cycles on land can be described: (i) The Tiñor Edifice (1.2–0.88 Ma), the first cycle and outcropped along the Las Playas escarpment; (ii) The El Golfo edifice (0.54–0.176 Ma), covering the Tiñor edifice; and (iii) The Rift Volcanism (0.158 Ma-present), filling the El Golfo and Las Playas flank collapses (Guillou et al. 1996; Bellido-Mulas et al. 2012). Finally, the most recent volcanic events are related to the probable eruption in 1973 at Lomo Negro (Hernández Pacheco 1982) and the Tagoro eruption at La Restinga offshore in 2011 (López et al. 2012; Troll et al. 2012; Martí et al. 2013; Rivera et al. 2013; Somoza et al. 2017).

The Tagoro fissure submarine edifice consisted in two main construction phases (from explosive to effusive) intercalated by collapse events (López et al. 2012; Rivera et al. 2013; Somoza et al. 2017). The shallow part of Tagoro shows chaotic heterometric boulders. Its base presents large accumulations of lava balloons (description in: Somoza et al. 2017) and extensive deep volcaniclastic aprons of ash and vesicular lapilli generated during the explosive phases. In the latter phases of effusive activity, secondary cones of lava ponds and flows were deposited over the lava balloons. Hornitos constitute the final volcanic activity characterized by the presence of hydrothermal alterations and bacterial mats at the top of the volcanic edifice (Somoza et al. 2017; Gonzalez et al. 2020).

13.3 Geoheritage Assessment Criteria

This geosite has been inventoried as 'Tagoro 2011–2012 submarine volcanic eruption' (code C7008) within the LIGCANARIAS project (IGME 2021a). This project carries out the geosite inventory in the geological domain of the Canary Islands, which is one of the domains that make up the Spanish Inventory of Geological Sites of Interest (IELIG) (IGME 2021b). The

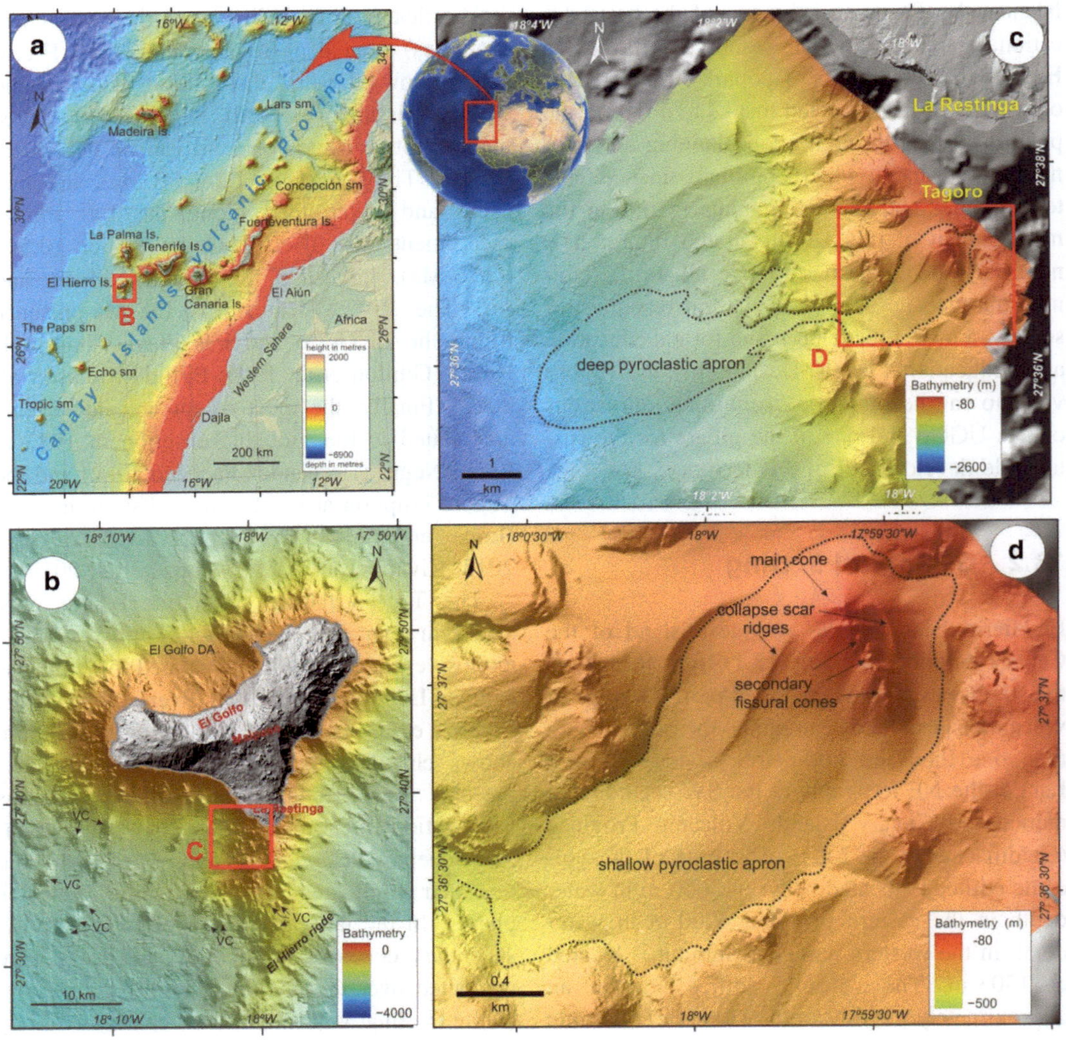

Fig. 13.1 **a** Situation map of Canary Islands Island (Spain). The black square indicates the position of El Hierro island. **b** Study area where the 2011–2012 submarine eruption occurred south of El Hierro, offshore from La Restinga (red square). **c** Bathymetric map of 'The Tagoro 2011–2012 submarine volcanic eruption' Geosite. Black dotted line comprises the geosite boundary with a volcanic cone and deep pyroclastic apron that is included in the geoheritage inventory for the Canary Islands. **d** Detail of the Tagoro main volcanic cone and shallow pyroclastic apron. *Source* of bathymetry and topography data: **a** GEBCO Digital Atlas (https://www.gebco.net/data_and_products/gebco_digital_atlas/); **b** and background of **c**, the Instituto Hidrográfico de la Marina (IHM) for the bathymetry and MDT05-Centro Nacional de Información Geografica (CNIG-IGN) (https://centrodedescargas.cnig.es/CentroDescargas/) for the topography; **c**, **d**, from the Instituto Español de Oceanografía (IEO), acquired along the VULCANA-0319 cruise

starting methodology of García-Cortés et al. (2019) for the selection and assessment of geosites was adapted to this geological domain due to the specific geological characteristics of oceanic volcanic islands (Vegas et al. 2019).

The first phase of this methodology consisted in identifying the twelve geological frameworks for the Canary Islands (Galindo et al. 2017), which are representative of the geodiversity of this archipelago. One of them is the framework

"Geological elements submerged below sea level" and it was defined because the largest volume of an oceanic volcanic island is submarine and it is necessary to make visible geosites that are representative of the ocean floor. It includes all the volcanic, erosive and sedimentary deposits (including fossils) and morphologies with heritage value found today below sea level, both those that were formed in an underwater or submarine environment (lava deltas, pillow lavas, turbidite systems, dykes, submarine volcanoes, underwater debris flow deposits, etc.), as well as those that originally formed on the surface of the islands and which are currently submerged due to the rise of the sea level during the Holocene.

However, it is not possible to apply the standard assessment methodology in this geological framework as many of the conventional criteria, such as 'accessibility by roads', 'density of population' or 'facilities for public use', were not applicable, gave very ambiguous results, or could not be used directly. Above all, it was not easy to apply the usual criteria to calculate educational and tourist value because their results are not discriminatory. A specific methodology was carried out to assess this type of geosites (Table 13.1).

13.4 Assessment of Scientific, Educative and Touristic Values for the Tagoro Submarine Volcanic Eruption

Following the methodology proposed by the Spanish Inventory of Geological Sites of Interest, the scientific, educational and touristic value of each geosite is assessed independently (Garcia-Cortés et al. 2019). Therefore, for this geosite these three values are also calculated separately and the criteria designed for submarine geosites have been assessed for the Tagoro (Table 13.2), where very high results are shown for scientific and educative value. However, the touristic value is low due to the underwater geosites located below the scuba diving range being more difficult to access for most visitors.

13.4.1 Scientific Value

Six main criteria are evaluated to calculate the scientific value: representativeness (R), scientific knowledge of the geosite (K), state of conservation (C), study techniques used in geosite research (TE), rarity (A) and diversity (D); with K being the criterion that gives it the highest value. This assessment of scientific value for submarine geosites is quite similar compared to the method for terrestrial geosites. The criterion that is unique to submarine geosites is TE, since scientific research of these geosites relies mostly on indirect techniques (Galindo et al. 2017).

The scientific value of the Tagoro submarine eruption is 8.75 points out of 10, which is a very high value within the 314 geosites of the Canary Islands inventory. This geosite is unique and representative as a model for the inventory of the geological domain of the Canary Islands and is also unique in the Spanish geoheritage inventory. However, when considering the possibility that the Tagoro submarine eruption could be of international relevance, there are few similar cases in the scientific literature. Similar submarine eruptions have also occurred at Pantelleria, Italy (Foerstner 1891; Washington 1909); Myojinsho, Japan (Fiske et al. 1998); Capelinhos, Azores (Machado et al. 1962; Gaspar et al. 2003); Taal, Philippines (Moore et al. 1966); Socorro island, Mexico (Siebe et al. 1995); Kavachi volcano, Solomon Islands (Baker et al. 2002); Ambae, Vanuatu (Németh et al. 2006); Hunga Haapai, Tonga (Vaughan and Webley 2010); Surtsey, Iceland (Romagnoli and Jakobsson 2015); or Nishino-Shima, Izu Volcanic Arc, Japan (Kaneko et al. 2019). The magmatic unrest that started before the onset of the eruption and continued later on, provided a firm basis to better understand magmatic intrusions and shallow magma storage processes associated with this type of eruptions (e.g. Becerril et al. 2013; Martí et al. 2013; Barnett and Gudmundsson 2014).

Given that the 2011–2012 El Hierro event that originated the submarine Tagoro volcano is extremely significant, being among the first submarine eruptions worldwide to be fully monitored in real time since the beginning of

Table 13.1 Specific criteria developed for the assessment of scientific, educative and touristic value of submarine geosites applied in the Canary Islands geoheritage inventory (LIGCANARIAS Project, IGME 2021a)

Assessment criteria	Points	Scientific value (weight)	Educational value (weight)	Touristic value (weight)
Representativeness (R)				
Not useful as a model to represent, even partially, a geological feature or process	0	× 20	× 20	× 0
Useful as a model to represent part of a geological feature or process	1	× 20	× 20	× 0
Useful as a model to represent an entire geological feature or process	2	× 20	× 20	× 0
Best known example of an entire feature or process in the geological domain inventoried (Canary Islands)	4	× 20	× 20	× 0
Scientific knowledge of the geosite (K)				
There are no scientific papers, geological maps, reports, or PhD theses on the geosite	0	× 30	× 10	× 0
There are technical geological reports, geological maps or papers citing the existence of the geosite, but within other more general scientific publications	1	× 30	× 10	× 0
Technical geological reports or PhD theses or papers in national scientific journals or abstracts at national conferences are available for the geosite	2	× 30	× 10	× 0
Geosite investigated by several research teams, the subject of PhD theses and/or papers published in SCI international scientific journals	4	× 30	× 10	× 0
State of conservation (C)				
Strongly degraded: the geosite is practically destroyed or eroded	0	× 10	× 10	× 0
Degraded: the geosite shows significant deterioration and erosion	0	× 10	× 10	× 0
Altered: the geosite has damage that prevents an appreciation of some characteristics of geological interest	1	× 10	× 10	× 0
Favourable with alterations: some damage that does not decisively affect the value or geological interest of the geosite	2	× 10	× 10	× 0
Favourable: the geosite is practically intact and is well preserved	4	× 10	× 10	× 0
Study techniques used in geosite research (TE)				
Only an indirect geophysical technique has been used	0	× 15	× 10	× 0
Two types of indirect geophysical techniques have been used	1	× 15	× 10	× 0
Three or more types of indirect geophysical techniques have been used	2	× 15	× 10	× 0
Indirect geophysical techniques and direct sampling techniques have been used (e.g. direct sampling by scuba diving in coastal and neritic areas; or with cores, robots or dredges in bathyal and abyssal areas)	4	× 15	× 10	× 0

(continued)

Table 13.1 (continued)

Assessment criteria	Points	Scientific value (weight)	Educational value (weight)	Touristic value (weight)
Rarity (A)				
There are several similar geosites in the geological domain of the Canary Islands	0	× 15	× 10	× 0
There are less than 5 similar geosites in the geological domain of the Canary Islands	1	× 15	× 10	× 0
The best known example in the geological domain of the Canary Islands	2	× 15	× 10	× 0
The only known example within the geological domain of the Canary Islands	4	× 15	× 10	× 0
Diversity (D)				
The geosite only presents the primary type of geological interest	0	× 10	× 10	× 0
The geosite presents another type of geological interest in addition to the primary type, but it is not relevant	1	× 10	× 10	× 0
The geosite presents 2 other types of geological interest in addition to the primary type, or only 1, but they are relevant	2	× 10	× 10	× 0
The geosite presents 3 other types of geological interest in addition to the primary type, or only 2, but they are relevant	4	× 10	× 10	× 0
Spectacularity or scenic (B)				
It does not meet, by default, the following three criteria	0	× 0	× 10	× 20
(1) Very high relief amplitude (>1000 m), or (2) complete geological elements that allow their perfect identification in official bathymetric maps (year 2018, for Spain's scale) or through other indirect techniques, or (3) presence of varied geological elements (high geodiversity) in the geosite	1	× 0	× 10	× 20
Presence of two of the three previous characteristics	2	× 0	× 10	× 20
Presence of the first three characteristics	4	× 0	× 10	× 20
Accessibility (Ac)				
Without any possibility of access	0	× 0	× 0	× 30
No direct access, but underwater geosites can be explained from boats that make tourist routes	1	× 0	× 0	× 30
Geosites that can be visited with a snorkel: between 6 and 8 m deep	2	× 0	× 0	× 30
Underwater geosites located at a depth of between 40 and 20 m with direct access through recreational diving	4	× 0	× 0	× 30
Size of the singular geological elements of the geosite (E)				
The geosite has numerous geological elements of centimetre and metric size (fragile and vulnerable for public use)	0	× 0	× 0	× 10
The geosite has geological elements of decametric size (not vulnerable to controlled visits, but sensitive to other more aggressive public uses)	1	× 0	× 0	× 10
The geosite has geological elements of hectometre size (they could be deteriorated by massive visits)	2	× 0	× 0	× 10

(continued)

Table 13.1 (continued)

Assessment criteria	Points	Scientific value (weight)	Educational value (weight)	Touristic value (weight)
The geosite has geological elements of kilometre size without any impact on its conservation due to its public use	4	× 0	× 0	× 10
Possibilities for educational use (PE)				
No possibilities of educational activities. The geosite is a geological process/element that is not included in educational curricula	0	× 0	× 20	× 0
Possibility of carrying out specific educational activities for university studies	1	× 0	× 20	× 0
Possibility of carrying out educational activities for secondary schools and universities	2	× 0	× 20	× 0
Educational activities are regularly carried out for all educational levels	4	× 0	× 20	× 0
Potential for touristic activities (PT)				
No potential for touristic activities	0	× 0	× 0	× 40
Potential geotourism activities through new technologies (App, augmented reality on visitor centres, etc.) and by diving and/or snorkelling	1	× 0	× 0	× 40
Existing geotourism activities through new technologies (App, augmented reality on visitors´ centres, etc.)	2	× 0	× 0	× 40
There are geotourism activities from tourist companies on the geosite through diving and/or snorkelling	4	× 0	× 0	× 40
Results		Σ_S	Σ_E	Σ_T
Value (0–10)		$V_S = \Sigma_S/40$	$V_E = \Sigma_C/40$	$V_T = \Sigma_T/40$

unrest by several research teams (Fraile-Nuez et al. 2012; López et al. 2012; Rivera et al. 2013; Somoza et al. 2017), it could be considered an outstanding universal value because Tagoro´s natural significance is so exceptional as to transcend national boundaries and to be of common importance for present and future generations of all humanity.

In addition to the eruptive process itself, volcanic products that emerged and been recovered for the El Hierro UGGp museum are the so-called 'lava balloons', measuring more than 1 m in diameter and erupted in November 2011; and the xeno-pumice balls ejected and floating on the sea surface during the first days of the eruption (late October 2011). The xeno-pumice balls appeared as black volcanic "bombs" that exhibit cores of white and porous pumice-like material (Troll et al. 2012) and have appeared in other underwater eruptions (Gaspar et al. 2003). These kinds of volcanic rock contribute to the geosite having a secondary petrological interest.

13.4.2 Educational Value

The score obtained for the educational value is 7.5 points out of 10, a high value for this use. The criteria that have been taken into account for its assessment are 8: Representativeness (R), Scientific knowledge of the geosite (K), Study techniques used in geosite research (TE), Rarity (A), Diversity (D), Spectacularity or Scenery (B), State of conservation (C) and Possibilities for educational use (PE). The PE criterion indicates a

Table 13.2 Assessment of scientific, educative and touristic values for the 'Tagoro submarine volcanic eruption' geosite, El Hierro island

Geosite: *'Tagoro submarine volcanic eruption'* assessment criteria	Points	Scientific value (weight)	Educational value (weight)	Touristic value (weight)
Representativeness (R)	3	60	60	0
Scientific knowledge of the geosite (K)	4	120	40	0
State of conservation (C)	4	40	40	0
Study techniques used in geosite research (TE)	4	60	40	0
Rarity (A)	2	30	20	0
Diversity (D)	2	20	20	0
Spectacularity or Scenery (B)	2	0	20	40
Accessibility (Ac)	1	0	0	30
Size of the singular geological elements (E)	4	0	0	40
Possibilities for educational use (PE)	2	0	40	0
Potential for touristic activities (PT)	2	0	0	80
Results		340	300	190
Value (0–10)		8.75	7.5	4.75

possibility of carrying out educational activities for secondary schools and universities.

This geosite is a clear example of geological heritage with great potential for educational activities in the field of formal education, which have not yet been implemented. This may be due to the fact that little time has elapsed since the publication of the scientific papers and knowledge transfer is needed to prepare educational materials and activities adapted to educational curricula, especially in primary and secondary education. Fieldwork or laboratory practices in the formal education of university degrees in Earth Sciences is much easier since the teaching staff is specialized in this topic.

Tagoro also meets all the requirements to be used in environmental education promoted by local administrations and UGGp managers, which means that the implementation of activities to promote communication and awareness of the impact of submarine eruptions, especially among the local population, can be carried out. In 2011 and 2012, during the volcanic emergency, many educational and awareness-raising activities were carried out on the island before and during the eruption. However, since the emergency officially ended, these activities have not continued. This is a considerable mistake because showing this geosite could help the community to be more resilient against natural hazards as the IGCP-692 'Geoheritage for Geohazard Resilience' is being demonstrated in other volcanic regions of the world (http://www.geopoderes.com/).

13.4.3 Tourism Value

Only four main criteria are evaluated to calculate the touristic value for submarine geosites: spectacularity or scenery (B), accessibility (Ac), size of the singular geological elements (E), Potential for touristic activities (PT). The score obtained for this value is 4.75 points out of 10, a medium–low value for this touristic use at this moment.

The PT criterion is the one that weighs the most in the total value for tourist use, but it must be taken into account that submarine geosites, a priori, have a lower tourist potential than those geosites located in the interior of the islands or on their coastline, which visitors can access by themselves and at any time of the year. For this reason, the tourist value of underwater geosites is always lower and, in turn, this value tends to be in inverse proportion to the depth of the geosites.

During the whole duration of the eruption and after its end, scientific tourism increased (Dóniz-Páez 2014). The tourist phenomenon increases when there are volcanic eruptions that do not pose a serious danger to people, such as the Hawaiian type (Bird et al. 2010; Erfurt-Cooper 2011) or the Icelandic type (Bird and Gísladóttir 2014). The most important touristic activity around Tagoro is in the El Hierro UGGp Interpretation Centre, where the lava balloons and xeno-pumices recovered on the sea surface during the eruption are displayed. These crustal xenoliths from the El Hierro 2011–2012 eruption underline the role of partial melting and assimilation of pre-island sedimentary layers in the early shield-building phase of ocean islands (Berg 2016). Xeno-pumices are known by the popular name of 'restingolitas', after La Restinga, the nearest town to the eruption (Troll et al. 2012), and they are also exhibited in several stores, cafés and restaurants in the municipality of Frontera (Fig. 13.2). In the second building of this Centre, visitors can enjoy an audiovisual that reproduces the eruption of 2011–2012 in La Restinga. In this way, through this infrastructure an approach is created for visitors to the underwater geosites, with audiovisuals, videos and real rocks serving as the best experience for one learn about these unique volcanic processes.

The Tagoro also serves as a claim and promotion of the Hero Fest (www.hero-fest.com), an alternative music festival held at the UGGp, and for the second edition of August 2018 a poster was designed showing the explosive underwater volcano (Fig. 13.2).

13.5 Actions Proposed to Promote the Public Use of the Tagoro Submarine Geosite

Although the Tagoro volcano is 88 m below sea surface, this is no impediment for developing an educational and tourist strategy around this submarine eruption, since nowadays new technologies allow this geosite to be brought closer to visitors and students. The development of new tourism products and experiences will increase its value for public use.

13.5.1 Environmental Education

Environmental education and geoconservation have developed a mutually beneficial relationship in the last decade (Vegas and Díez-Herrero 2018): environmental education advances by sensitizing and raising awareness of actions to protect the geoheritage; it supports the process of defining standards and regulations through communication and participation; and finally, it enables the implementation and effectiveness of the initiatives adopted, through dissemination and training.

Many city councils in Spain have environmental education programs that include geoheritage as a resource (Vegas and Díez-Herrero 2021; Díez-Herrero et al. 2021). For this reason, Tagoro is an excellent educational resource for the nearest municipalities, such as La Restinga and El Pinar. For the UGGp El Hierro itself, it is also a great opportunity for educational activities. In its Interpretation Centre, it is possible to launch environmental education activities whose focal point is the Tagoro eruption, its consequences, the social impact and the potential for socioeconomic development. Education increases the resilience of the population to volcanic phenomena regarding the impact and hazards of submarine eruptions as promoted by the IGCP-692 Project 'Geoheritage for Geohazard Resilience' (IGCP-692 2021).

13.5.2 Formal Primary, Secondary and High School Education

Volcanism is usually included in the Earth Science curricula in secondary and high school cycles in Spain (mainly included in the subjects of natural environment; biology and geology; geography for secondary-level schools and

Fig. 13.2 **a, b** Restingolites from the Tagoro submarine eruption that are exposed in a pharmacy in Frontera (El Hierro UGGp). **c, d** Poster display in El Hierro airport advertising the alternative music festival Hero Fest of 2018, whose icon is the Tagoro submarine volcano that erupted in 2011–2012 in La Restinga (Herofest 2021)

geology in the 2nd year of high schools, corresponding to student ages of 17 years old). This type of volcanic eruption in the submarine environment associated with islands is not usually included because such eruptions are less frequent compared to the volcanism of the oceanic crust and mid-ocean ridges. This geosite is a magnificent example of submarine volcanic eruptions originating at shallow depth, which are characterized by higher explosiveness with the development of lava-balloons and a large amount of pyroclasts of different grain sizes, compared to deep volcanic eruptions whose likelihood of effusive behaviour increases with water depth. In the revision and updating of textbooks for these subjects in Spain, Tagoro should be included as an example of submarine eruptions for all the graphic material available, which perfectly illustrates the entire eruptive process. Moreover, similar eruptions have taken place in other submarine environments for comparison with El Tagoro: Surtsey 1963 is likewise the result of an eruption that started below water surface (Thorarinsson and Vonnegut 1964). The Ambae 2005 eruption formed a pyroclastic mound in the caldera lake emerged from the lake, but a few months later it shoaled (Global Volcanism Program 2005); and there is the case of Ferdinandia in the Straight of Sicily in the nineteenth century (Garrison et al. 2021).

Since the Tagoro eruption, in the Canary Islands this eruption has been used as an educational resource, even in the time of the pandemic (2020), this phenomenon being explained in primary education and linked to the popular Canarian legend of the island of 'San Borondón' (an island in the sea that appears and disappears).

13.5.3 Information and Communication Technologies

Information and Communication Technologies (ICT) are a new way and undoubtedly the best tool for the dissemination and outreach of the submarine geoheritage. Through technologies such as software applications (apps), virtual reality, social media and web pages, the barrier between people and underwater geosites can be eliminated, allowing new forms of educational and geotouristic uses. In many cases, ICT may be the only way to get to know these inaccessible geosites and favour the transfer of scientific knowledge to society.

Augmented reality (AR) is an enhanced version of the real physical world that is achieved through the use of digital visual elements, sound, or other sensory stimuli delivered via technology (Hayes 2021). In the case of Tagoro, AR has a great potential, allowing us to see how this eruption occurred with a real experience thanks to the large amount of audiovisual materials that were recovered by scientists with very diverse methods. Having an AR experience with Tagoro can be a magnificent resource to implement in the Interpretation Centre of the El Hierro UGGp.

The apps for mobile phones and tablets are a tool for the outreach of geological processes and geoheritage. The City Council of El Pinar (El Hierro) designed an App in 2018 with the aim of spreading the word about the eruption of the Tagoro submarine volcano in 2011. This initiative, developed through the Department of New Technologies, Communication, Heritage and Tourism, aims to give evidence of what happened in the volcanic eruption from a positive perspective and bring all the public closer to the island's seabed. All this information, collected in the app, is told through a tale that includes an interactive design with real photos of the seabeds, in addition to providing knowledge about the conquest of El Hierro and other historical data. With this initiative of the local government, the aim is to make the natural heritage and history known to the local population and visitors in a fun and entertaining way.

Another of the most widely used ICT in geotourism consists in web pages and their combination with QR codes for access from mobiles and tablets for transmitting content, recreations and maps. The installation of QR plates in the Port of La Restinga to access the website of the UGGp of El Hierro and thus be able to view the Tagoro audiovisual in 2011–2012 would be highly recommended to see what the eruption was like with real images. They have a low cost and a low visual impact, it being very easy to replace and update the information on the web, significantly reducing the high cost of replacing conventional panels (Vegas and Díez-Herrero 2018). An e-library could be created on the UGGp website, where access to publications and online resources on the Tagoro geosite could be compiled.

Social media has changed the landscape of human communication and interaction where scientists, researchers, and scientific organizations are now gradually engaging in these online platforms (Jan Landicho 2020). For this reason, the use of the UGGp′s social media can keep news about the activities that develop around the Tagoro geosite.

13.5.4 Guided Boat Tours and Scuba Diving in the Tagoro Geosite

Scuba diving makes this type of geotourism a unique and exclusive product compared to other UGGps. In Spain, the UGGps that have specialized diving routes and activities are Cabo de Gata, Lanzarote and the Chinijo archipelago. This is a unique tourist product that can be made accessible to all audiences through routes made by boat in the Tagoro geosite, with the possibility of implementing virtual reality on these routes in the interests of an exclusive experience. The association between diving companies and the UGGp will favour new products and routes where dive monitors and guides can show the underwater eruption based on scientific data. The UGGp restaurants are other companies that can benefit from a geosite like Tagoro, offering gastronomic experiences inspired by the submarine eruption with the creation of new dishes such as a 'restingolite soup',' lava croquettes balloons', etc.; which can be combined with the boat tours.

13.6 Tagoro Geosite Challenges and Future Perspectives

Despite the fact that 30% of existing UGGps have a maritime/submarine part, only some of them consider the marine environment to be a full part of their geoheritage and responsibilities

(Martini et al. 2021). Therefore, it is important that the UGGps begin to consider their marine environment and integrate promotion and actions in this respect. In this sense, the El Hierro UGGp has, in the Tagoro eruption, one of the best resources of the submarine geoheritage of Spain that can be displayed worldwide.

What is most valuable is the extensive scientific knowledge available about the eruption, which was monitored from its inception and followed at all times until the volcanic activity ended. All these scientific data constitute the basis for generating a development strategy for tourism, educational and conservation activities for this geosite. The achievement of these objectives by the UGGp managers in the short and medium term depends on four major challenges.

Challenge #1: *Protection*. The volcanic cone of this geosite is included within the 'Punta de La Restinga-Mar de Las Calmas' Marine Nature Reserve, which was established in 1996 by a Ministerial Order of 24th January 1996 and Decree 30/1996. This Reserve covers an area of 1180 hectares. The Tagoro is not included in the reserve's maximum protection zone and is included in 'Waters beyond baselines', where only professional hook fishing, local traditional fishing and scuba diving under authorization from the Regional Government are allowed. The protection and declaration of this marine area was due exclusively to biodiversity. Since 2016, the new volcano appears on the official hydrographic charts as 'Tagoro' ('a meeting place made of a stone circle' in the indigenous Canarian language). We propose extending the nature reserve's protection of geodiversity and geoheritage to the marine environment. It is crucial for the preservation of singular submarine geological structures, such as submarine hornitos or lava balloons, from submarine facilities (e.g. cables, foundations, wind farms) or mining exploitations that may have a severe impact on their conservation. The geosite of the Tagoro submarine eruption of 2011–2012 should be formally included in the Marine Reserve.

Challenge #2: *Visibility* (i.e. popularization). The first step to take is for the El Hierro UGGp to also include the Tagoro geosite within its inventory and make it visible on its website, with a special section where it is easily explained and with diagrams showing how the eruption occurred. In addition, it can include an e-library section with scientific articles and videos recorded during the eruption. It is necessary to reactivate a communication strategy around the Tagoro geosite, using marketing techniques and ICT. It has been proven that through the use of social media, science is brought closer to more people and in real time. Making this type of volcanism in oceanic environments visible will make it possible to publicize these marine eruptions that are more difficult to observe and know by the local population and the general public.

Challenge #3: *Active and positive resource for sustainable tourism*. El Hierro offers nature tourism, as opposed to the mass tourism promoted by other Canary Islands. For this reason, it has a network of exclusive small hotels and rural houses where visitors can connect with nature and which are respectful with the environment. This small business infrastructure can also be used as an ally to promote the Tagoro geosite. Therefore, information points of the activities around this geosite could be created by installing plates with the QR code that allows access to the UGGp website with all the information of the companies that carry out routes and dives and of the Interpretation Centre, where people can learn about the eruption in depth. Hotel establishments have a unique tourism brand in the Tagoro geosite, since there are very few places in the world where it is possible to do tourism around a volcanic eruption under the sea that occurred recently in the twenty-first century.

Challenge #4: *A key geosite in the El Hierro UNESCO Global Geopark focused on geological hazards*. The UGGps have to promote activities around the 'Top 10 Focus Areas': Natural Resources, Geological Hazards, Climate Change, Education, Science, Women and social responsibility, Sustainable Development, Local and

Indigenous Knowledge, and Geoconservation (UNESCO 2016). The Tagoro is the geosite that constitutes the best example of geological hazards for this UGGp. This volcanic eruption initially triggered a series of small earthquakes that alerted the authorities, scientists and the local population during the first months. These earthquakes were caused by the volcanic tremor prior to the eruption and are very well identified in the seismic network records as being very different from earthquakes caused by the movement of faults in other geotectonic areas. At that time, the area where the eruption would occur was not yet known, as neither was whether it would happen on the island itself or under the sea, as finally happened. The management carried out during the eruption by the authorities is also a model to show, especially in the educational field, as it is a magnificent example for future eruptions. It is necessary to promote positive activities in the local population that increase resilience to future submarine eruptions in El Hierro and can also be shown in the Canary Islands because similar submarine eruptions can occur on other islands. Resilience is fully embedded in the institutional and management structure of the UGGp (Dierickx et al. 2016; Fassoulas et al. 2018).

13.7 Summary

This chapter presents for the first time the results of the underwater/submarine geoheritage assessment methodology applied for the case of the Tagoro volcano geosite, which is included in the Spanish Inventory of Geological Heritage (IELIG). This submarine geosite stands out for its high scientific value, as it is the best model of a submarine eruption in Spain that has been monitored since the unrest phase. Due to its high scientific value, it is necessary to modify the regulation of this protected marine area to include the new geological elements derived from the eruption, thus fully guaranteeing its conservation.

The great challenges facing the managers of the UNESCO Global Geopark so that El Tagoro can be used for tourism and education include the use of ICT that will allow an approach to a resource under the sea. It is necessary to invest in an exclusive promotion and outreach so that it becomes one of the key tourist attractions on El Hierro island.

Acknowledgements This research received funding from the Canaries Agency for Research, Innovation, and Society of Information (ACIISI) under the Government of the Canary Islands through the LIGCANARIAS project (ProID2017010159); EMODnet Bathymetry HRSM3 (EASME/EMFF/2019/1.3.1.9/Lot1/SI2.836043) and from VULCANA-II (IEO-2018-2020) project.

References

Barnett ZA, Gudmundsson A (2014) Numerical modeling of dykes deflected into sills to form a magma chamber. J Volcanol Geotherm Res 281:1–11

Becerril L, Galindo I, Gudmundsson A, Morales JM (2013) Depth of origin of magma in eruptions. Sci Rep 3(1):1–6

Becerril L, Galindo I, Martí J, Gudmundsson A (2015) Three-armed rifts or masked radial pattern of eruptive fissures? The intriguing case of El Hierro volcano (Canary Islands). Tectonophysics 647–648:33–47. https://doi.org/10.1016/j.tecto.2015.02.006

Bellido-Mulas F, Gómez Sainz de Aja JA, Barrera JL (2012) Mapa geológico digital continuo E. 1: 50.000, Zona Canarias - El Hierro (Zona-2916). In: GEODE Mapa Geológico Digital continuo de España. Sistema de Información Geológica Continua: SIGECO. IGME. Ed: J Navas. http://cuarzo.igme.es/sigeco/default.htm. Accessed 20 Oct 2012

Berg SE (2016) Silicic magma genesis in basalt dominated oceanic settings. Examples from Iceland and the Canary Islands. In: Digital comprehensive summaries of Uppsala dissertations from the faculty of science and technology. Acta Universitatis Upsaliensis, Uppsala, pp 1338–1354

Bird DK, Gísladóttir G (2014) Southern iceland: volcanoes, tourism and volcanic risk reduction. In: Erfurt-Cooper P (ed) Volcanic tourist destinations geoheritage, geoparks and geotourism. Springer, Berlin, pp 35–46

Bird D, Gisladottir G, Dominey D (2010) Volcanic risk and tourism in southern Iceland: implications for hazard risk and emergency response education and training. J Volcanol Geotherm Res 189:33–48

Carracedo JC, Troll VR, Zaczek K, Rodriguez-Gonzalez A, Soler V, Deegan FM (2015) The 2011–2012 submarine eruption off El Hierro, Canary Islands: new lessons in oceanic island growth and volcanic crisis management. Earth Sci Rev 150:168–200

Dierickx F, Pavlova I, Gaines S (2016) Natural hazards in UNESCO global geoparks. Eur Geoparks Mag 13:16–17

Díez Herrero A, Sacristan N, González S, Meixoeiro J, Revuelta J, Martínez A, Fernandez Luna D, Arroyo MI, Iglesias J (2021) Educación ambiental con patrimonio geológico: algunas experiencias prácticas. De Re Metallica 36:87–98

Dóniz-Páez FJ (2014) Reflexiones en torno al turismo volcánico. El caso de Islas Canarias, España. PASOS Revista De Turismo y Patrimonio Cultural 12(2):467–478. https://doi.org/10.25145/j.pasos.2014.12.034

Erurt-Cooper P (2011) Geotourism in volcanic and geothermal environments: playing with fire? Geoheritage 3:187–193

Fassoulas C, Watanabe M, Pavlova I, Amorfini A, Dellarole E, Dierickx F (2018) UNESCO Global Geoparks: living laboratories to mitigate natural induced disasters and strengthen communities' resilience. In: Antronico L, Marincioni F (eds) Natural hazards and disaster risk reduction policies, geographies of the anthropocene series books, Il Sileno Edizioni, Rende, pp 175−197

Fiske RS, Cashman KV, Shibata A, Watanabe K (1998) Tephra dispersal from Myojinsho, Japan, during its shallow submarine eruption of 1952–1953. Bull Volcanol 59(4):262–275

Foerstner H (1891) Das Gestein der 1891 bei Pantelleria entstandenen Vulcaninsel und seine Beziehungen zu den jüngsten Eruptivgesteinen der Nachbarschaft. Tschermaks Mineralogische Und Petrographische Mitteilungen 12(6):510–521

Fraile-Nuez E, González-Dávila M, Santana-Casiano JM, Arístegui J, Alonso-González IJ, Hernández-León S, Blanco MJ, Rodríguez-Santana A, Hernández-Guerra A, Gelado-Caballero MD, Eugenio F, Marcello J, De Armas D, Domínguez-Yanes JF, Montero MF, Laetsch DR, Vélez-Belchí P, Ramos A, Ariza AV, Comas-Rodríguez I, Benítez-Barrios VM (2012) The submarine volcano eruption at the island of El Hierro: physical-chemical perturbation and biological response. Sci Rep 2:486. https://doi.org/10.1038/srep00486

Galindo I, Romero C, Llorente M, Rubio JC, Vegas J, Sánchez N, Díaz A (2017) Resultados preliminares del inventario de lugares de interés geológico submarinos del Geoparque Mundial UNESCO de Lanzarote y archipiélago Chinijo. In: Investigando el mar: viaje al planeta agua. Actas XII Semana Científica Telesforo Bravo. Instituto de Estudios Hispánicos de Canarias, pp 15–40

Galindo I, Romero C, Coello Bravo JJ, Sánchez N, Martín-González E, Vegas J (2019a) Propuesta de contextos geológicos regionales para el inventario de patrimonio geológico de las Islas Canarias. Geogaceta 65:39–42

Galindo I, Romero C, Llorente M, Rubio JC, Díaz GA, Sánchez N, Martín-González E, Mangas J, Vegas J (2019b) Geoheritage in the shallow submarine slopes of an oceanic volcanic edifice: a new option for diving geotourism. In: Mateo E, Martínez-Frías J, Vegas J (eds) Lanzarote and Chinijo Islands geopark: from earth to space. Springer International Publishing, Cham, pp 85–98

Galindo I, Vegas J, Romero C, Llorente M, Martín-González E, Rubio JC, Díaz A, Mangas J, Mateo E, Sánchez N (2019c) Geoheritage Inventory of the Lanzarote and Chinijo Islands UNESCO Global Geopark. In: Mateo E, Martínez Frías J, Vegas J (eds) Lanzarote and Chinijo Islands Geopark: from earth to space, geoheritage, geoparks and geotourism. Springer Nature, Cham, pp 31–45

Garrison C, Kilburn C, Smart D, Edwards S (2021) The blue suns of 1831: was the eruption of Ferdinandea, near Sicily, one of the largest volcanic climate forcing events of the nineteenth century? Clim past Disc 57:1–56. https://doi.org/10.5194/cp-2021-78

Gaspar JL, Queiroz G, Pacheco JM, Ferreira T, Wallenstein N, Almeida MH, Coutinho R (2003) Basaltic lava balloons produced during the 1998–2001 serreta submarine ridge eruption (Azores). Geophys Monogr Am Geophys Union 140:205–212

Geldmacher J, Hoernle K, Pvd B, Duggen S, Werner R (2005) New 40Ar/39Ar age and geochemical data from seamounts in the Canary and Madeira volcanic provinces: support for the mantle plume hypothesis. Earth Planet Sci Lett 237:85–101. https://doi.org/10.1016/j.epsl.2005.04.037

Global Volcanism Program (2005) Bulletin of the global volcanism network. In: Wunderman R (ed) Report on Ambae (Vanuatu), vol 30. Smithsonian Institution, London, p 11. https://doi.org/10.5479/si.GVP.BGVN200511-257030

Gonzalez FJ, Rincón-Tomás B, Somoza L, Santofimia E, Medialdea T, Madureira P, López-Pamo E, Hein JR, Marino E, de Ignacio C, Reyes J, Hoppert M, Reitner J (2020) Low-temperature, shallow-water hydrothermal vent mineralization following the recent submarine eruption of Tagoro volcano (El Hierro, Canary Islands). Mar Geol 430:106333. https://doi.org/10.1016/j.margeo.2020.106333

Guillou H, Carracedo JC, Torrado FP, Badiola ER (1996) K-Ar ages and magnetic stratigraphy of a hotspot-induced, fast grown oceanic island: El Hierro, Canary Islands. J Volcanol Geotherm Res 73:141–155. https://doi.org/10.1016/0377-0273(96)00021-2

Hernández Pacheco A (1982) Sobre una posible erupción en 1793 en la isla de El Hierro (Canarias). Estud Geol 38:15–25

Herofest (2021) http://www.hero-fest.com

IGCP-692 (2021) Geoheritage for resilience project. http://www.geopoderes.com/

IGME (2021a) Lugares de Interés Geológico de Canarias: Estudio, Inventario y Divulgación. LIG Canarias. http://www.igme.es/LIGcanarias/

IGME (2021b) Spanish inventory of geological sites of interest (IELIG). http://info.igme.es/ielig/

Kaneko T, Maeno F, Yasuda A, Takeo M, Takasaki K (2019) The 2017 Nishinoshima eruption: combined analysis using Himawari-8 and multiple high-resolution satellite images. Earth Planets Space 71:140. https://doi.org/10.1186/s40623-019-1121-8

López C, Blanco MJ, Abella R, Brenes B, Cabrera Rodríguez VM, Casas B, Domínguez Cerdeña I, Felpeto A, de Villalta MF, del Fresno C, García O, García-Arias MJ, García-Cañada L, Gomis Moreno A, González-Alonso E, Guzmán Pérez J, Iribarren I, López-Díaz R, Luengo-Oroz N, Meletlidis S, Moreno M, Moure D, de Pablo JP, Rodero C, Romero E, Sainz-Maza S, Sentre Domingo MA, Torres PA, Trigo P, Villasante-Marcos V (2012) Monitoring the volcanic unrest of El Hierro (Canary Islands) before the onset of the 2011–2012 submarine eruption: Monitoring El Hierro Prior 2001 Eruption. Geophys Res Lett 39:1846. https://doi.org/10.1029/2012GL051846

Martí J, Pinel V, López C, Geyer A, Abella R, Tárraga M, Rodríguez C (2013) Causes and mechanisms of the 2011–2012 El Hierro (Canary Islands) submarine eruption. J Geophys Res Solid Earth 118(3):823–839

Martini G, Zouros N, Zhang J, Xiaochi J, Komoo I, Border M, Watanabe M, Frey ML, Rangnes K, Tan Van T, Melo JPP, Patzak M, Hilario A, Nakada S, Sá AA (2021) UNESCO global geoparks in the "World after": a multiplegoals roadmap proposal for future discussion. Episodes 15:1002. https://doi.org/10.18814/epiiugs/2021/021002

Rivera J, Lastras G, Canals M, Acosta J, Arrese B, Hermida N, Micallef A, Tello O, Amblas D (2013) Construction of an oceanic island: Insights from the El Hierro (Canary Islands) 2011–2012 submarine volcanic eruption. Geology 41:355–358. https://doi.org/10.1130/G33863.1

Robertson AHF, Stillman CJ (1979) Submarine volcanic and associated sedimentary rocks of the Fuerteventura basal complex, Canary Islands. Geol Mag 116:203–214

Roeser HA (1982) Magnetic anomalies in the magnetic quiet zone off Morocco, In: von Rad U, Hinz K, Sarnthein M, Seibold E (eds) Geology of the northwest African continental margin. Springer, Berlin, pp 61–68. https://doi.org/10.1007/978-3-642-68409-8_4

Romagnoli C, Jakobsson SP (2015) Post-eruptive morphological evolution of island volcanoes: surtsey as a modern case study. Geomorphology 250:384–396

Rovere A, Vacchi M, Morri C, Firpo M (2011) Bringing geoheritage underwater: definitions, methods, and application in two Mediterranean marine areas. Environ Earth Sci 64:133–142

Siebe C, Komorowski JC, Navarro C, McHone J, Delgado H, Cortés A (1995) Submarine eruption near Socorro Island, Mexico: geochemistry and scanning electron microscopy studies of floating scoria and reticulite. J Volcanol Geotherm Res 68(4):239–271

Somoza L, González FJ, Barker SJ, Madureira P, Medialdea T, de Ignacio C, Lourenço N, León R, Vázquez JT, Palomino D (2017) Evolution of submarine eruptive activity during the 2011–2012 E l H ierro event as documented by hydroacoustic images and remotely operated vehicle observations. Geochem Geophys Geosyst 18:3109–3137. https://doi.org/10.1002/2016GC006733

Thorarinsson S, Vonnegut B (1964) Whirlwinds produced by the eruption of Surtsey volcano. Bull Am Meteorol Soc 45:440–444

Troll VR, Klügel A, Longpré MA, Burchardt S, Deegan FM, Carracedo JC, Wiesmaier S, Kueppers U, Dahren B, Blythe LS, Hansteen TH, Freda C, Budd DA, Jolis EM, Jonsson E, Meade FC, Harris C, Berg SE, Mancini L, Polacci M, Pedroza K (2012) Floating stones off El Hierro, Canary Islands: xenoliths of pre-island sedimentary origin in the early products of the October 2011 eruption. Solid Earth 3:97–110. https://doi.org/10.5194/se-3-97-2012

UNESCO (2016) UNESCO global geoparks, p 20. http://www.unesco.org/open-access/terms-use-ccbysa-en

van den Bogaard P (2013) The origin of the Canary Island Seamount Province: new ages of old seamounts. Sci Rep 3:1–7. https://doi.org/10.1038/srep02107

Vegas J, Díez-Herrero A (2018) Best practice guidelines for the use of the geoheritage in the city of Segovia/Manual de buenas prácticas para el uso del Patrimonio geológico en la ciudad de Segovia. Ayuntamiento de Segovia, p 120. http://www.segovia.es/educaPatriGeo/ELIGES/index.html

Vegas J, Díez-Herrero A (2021) An assessment method for urban geoheritage as a model for environmental awareness and geotourism (Segovia, Spain). Geoheritage 13:27

Vegas J, Sánchez N, Romero C, Galindo I (2019) Manual para la elaboración del Inventario de Lugares de Interés Geológico de Canarias. Instituto Geológico y Minero de España, p 52

Vera JA, Ancochea A., Barnolas Cortinas A, Bea Carredo F, Calvo Sorando JP (2004) Geología de España. Ed. Instituto Geológico y Minero de España, p 884

Washington HS (1909) ART. VIII. The submarine eruptions of 1831 and 1891 near Pantelleria. Am J Sci 27(158):131

MIX
Papier aus verantwortungsvollen Quellen
Paper from responsible sources
FSC® C105338

If you have any concerns about our products,
you can contact us on
ProductSafety@springernature.com

In case Publisher is established outside the EU,
the EU authorized representative is:
Springer Nature Customer Service Center GmbH
Europaplatz 3, 69115 Heidelberg, Germany

Printed by Libri Plureos GmbH
in Hamburg, Germany